Formal Languages and Automata Theory

Formal Languages and Automata Theory

Dr K Anuradha M Tech PhD
Professor and Head
Department of Computer Science and Engineering
Gokaraju Rangaraju Institute of Engineering and Technology, Bachupally
Hyderabad (AP)

Dr Y Vijayalata M Tech PhD
Professor of Computer Science and Engineering
Gokaraju Rangaraju Institute of Engineering and Technology, Bachupally
Hyderabad (AP)

CBS

CBS Publishers & Distributors Pvt Ltd

New Delhi • Bengaluru • Chennai • Kochi • Kolkata • Mumbai • Pune
Hyderabad • Nagpur • Patna • Vijayawada

Formal Languages
and
Automata Theory

ISBN: 978-81-239-2284-3

First Edition: 2013
Reprint: 2016

Published by Satish Kumar Jain and produced by Varun Jain for
CBS Publishers & Distributors Pvt Ltd
4819/XI Prahlad Street, 24 Ansari Road, Daryaganj, New Delhi 110 002, India.
Ph: 23289259, 23266861, 23266867   Fax: 011-23243014   Website: www.cbspd.com
e-mail: delhi@cbspd.com; cbspubs@airtelmail.in.
Corporate Office: 204 FIE, Industrial Area, Patparganj, Delhi 110 092
Ph: 4934 4934   Fax: 4934 4935   e-mail: publishing@cbspd.com; publicity@cbspd.com

Branches

- **Bengaluru:** Seema House 2975, 17th Cross, K.R. Road,
  Banasankari 2nd Stage, Bengaluru 560 070, Karnataka
  Ph: +91-80-26771678/79   Fax: +91-80-26771680   e-mail: bangalore@cbspd.com
- **Chennai:** 7, Subbaraya Street, Shenoy Nagar, Chennai 600 030, Tamil Nadu
  Ph: +91-44-26260666, 26208620   Fax: +91-44-42032115   e-mail: chennai@cbspd.com
- **Kochi:** Ashana House, No. 39/1904, A.M. Thomas Road, Valanjambalam,
  Ernakulam 682016, Kochi, Kerala
  Ph: +91-484-4059061-65, 67   Fax: +91-484-4059065   e-mail: kochi@cbspd.com
- **Kolkata:** 6/B, Ground Floor, Rameswar Shaw Road, Kolkata-700 014, West Bengal
  Ph: +91-33-22891126, +91-33-22891127, +91-33-22891128   e-mail: kolkata@cbspd.com
- **Mumbai:** 83-C, Dr E Moses Road, Worli, Mumbai-400018, Maharashtra
  Ph: +91-22-24902340/41   Fax: +91-22-24902342   e-mail: mumbai@cbspd.com
- **Pune:** Bhuruk Prestige, Sr. No. 52/12/2+1+3/2 Narhe, Haveli
  (Near Katraj-Dehu Road Bypass), Pune 411 041, Maharashtra
  Ph: +91-20-64704058/59, 32392277   Fax: +91-20-24300160   e-mail: pune@cbspd.com

Representatives

- **Hyderabad**   0-9885175004   • **Nagpur**   0-9021734563
- **Patna**   0-9334159340   • **Vijayawada**   0-9000660880

Printed at India Binding House, Noida, UP

PREFACE

This book deals with all three theories of computer science – Formal Languages, Automata and computation. The major contribution in this book is that it contains numerous solved problems. Most of the topics are explained with graphical notations.

Key features:

♦ Illustrates theoretical concepts with numerous examples

♦ Different types of machines are explained in detail

♦ The concepts explained in this book helps in understanding compiler construction.

♦ Can be used as text for both UG and PG courses of Computer science and Information technology

♦ End of the each chapter contains solved problems and exercise problems for self assessment

TABLE OF CONTENTS

viii

Chapter 1

Fundamentals

1.1 INTRODUCTION

What does the term "compute" mean?
What does the term "computer" mean?
Is there anything that a computer cannot do?

Such highly theoretical questions can be easily answered if one can understand the siginificance of need to study formal languages and automata theory and also the practical applications of this theory. Every working process in universe can be represented by means of models. The model can be theoretical or mathematical. These mathematical models can be termed as a solvable problem.

In theoretical computer science, automata theory is the study of abstract mathematical model and problems which it is able to solve. Finite automata, pushdown automata, linear bounded automata and turing machines are the models for study.

A formal language is an abstraction of the general programming languages. A formal language is necessary to interpret an abstract mathematical model. Formal languages and automata theory is based on mathematical computations.

1.2 BASIC CONCEPTS

1.2.1 SYMBOLS

A symbol is an **abstract entity** that cannot be defined formally just like point and line in geometry.

Examples of frequently used symbols are letters and digits, i.e., a, b, c......z and 0, 1, 2...9 respectively.

1.2.2 ALPHABETS

An **alphabet** is a finite, non-empty set of symbols. Conventionally, we use the symbol Σ for an alphabet.

Note: The symbols of an alphabet are indivisible.

Common alphabets include:

1) $\Sigma = \{0, 1\}$, the binary alphabets
2) $\Sigma = \{a, b, c.......z\}$, the set of all lower case letters
3)   The set of all ASCII characters or the set of all printable ASCII characters

1.2.3 STRINGS

A **string** (or word) is a finite sequence of symbols chosen from some alphabet. A string is denoted by **w**.

Example 1.1

1)   "01101" is a string from the binary alphabet $\Sigma = \{0, 1\}$
      "111" is another string from this alphabet
2)   If a, b, c, d are symbols then "abcd" is a string

Empty string:

The empty string is the string with zero occurrences of symbols .That is, the empty string consisting of zero symbols. Empty string is denoted by ε.

Length of a string:

The length of a string is the number of symbols in the string. The length of a string **w** is denoted by **|w|**.

Example 1.2

If "abc" is a string then |abc|=3
If "01101" is a string then |01101|=5
The length of empty string is 0 i.e.,$|\varepsilon|=0$

Powers of an alphabet:

If Σ is an alphabet, we can express the set of all strings of certain length from that alphabet by using an exponential notation.

We define Σ^k to be set of strings of length **k,** each of whose symbols is in Σ.

The set of all strings over an alphabet Σ is denoted by Σ^*, where Σ^* is called **kleene closure**.

i.e., $\Sigma^* = \Sigma^0 \cup \Sigma^1 \cup \Sigma^2 \cup \Sigma^3 \cup ....$

The set of all non-empty strings over an alphabet Σ is denoted by Σ^+, where Σ^+ is called as **positive closure**.

i.e., $\Sigma^+ = \Sigma^1 \cup \Sigma^2 \cup \Sigma^3 \cup \ldots\ldots\ldots$

These can also be represented as,

$\Sigma^+ = \Sigma^* - \{\varepsilon\}$ **or** $\Sigma^* = \Sigma^+ \cup \{\varepsilon\}$

Note: $\Sigma^0 = \{\varepsilon\}$ regardless of what alphabet Σ is, i.e., ε is the only string whose length is **0**

Example 1.3

If $\Sigma = \{0, 1\}$

then $\Sigma^1 = \{0,1\}$ is the set of all strings of length **1**.

$\Sigma^2 = \{00, 01, 10, 11\}$ is the set of all strings of length 2.

$\Sigma^3 = \{000,001,010,011,100,101,110,111\}$ is the set of all strings of length 3

and so on.

$\Sigma^+ = \Sigma^1 \cup \Sigma^2 \cup \Sigma^3 \cup \ldots\ldots\ldots$

Example 1.4

If $\Sigma = \{0, 1\}$

then $\Sigma^* = \{\varepsilon,0,1,00,01,10,11,000,011,1100,1001\ldots\ldots\}$

Operations on strings

Concatenation Operation: It is a binary operation.

The concatenation of two strings is the string formed by writing the first string followed by second string, with no intervene space.

Let **x** and **y** be two strings. Then **xy** denotes the concatenation of **x** and **y**, i.e., the string formed by making a copy of **x** and following it by a copy of **y**.

Let **x** be a string composed of **i** symbols $x = a_1 a_2 \ldots .. a_i$ and **y** be a string composed of **j** symbols $y = b_1 b_2 \ldots .. b_j$. Then $xy = a_1 a_2 \ldots .. a_i b_1 b_2 \ldots .. b_j$ is a string of length **i+j**.

Example 1.5

Let   x=01101 and y=110

then   xy =01101110

yx =11001101

For any string **w**, the equations $\varepsilon w = w \varepsilon = w$ holds, i.e., ε is the identity for concatenation.

Properties of Concatenation:

Property 1: Concatenation on a set Σ^* is associative, since for each x, y, z in Σ^*, x(yz)=(xy)z.

Property 2: ε is an identity element for concatenation, $x\varepsilon = \varepsilon x = x$, $\forall x \in \Sigma^*$.

Property 3: Concatenation has left and right cancellation laws.

$$\text{For x, y, z in } \Sigma^*, zx = zy \Rightarrow x = y \text{ (left cancellation)}$$
$$xz = yz \Rightarrow x = y \text{ (right cancellation)}$$

Property 4: For x, y in Σ^* we have $| xy | = | x | + | y |$

where |x|, |y|, |xy| denote the lengths of the strings x,y and xy respectively.

Transpose operation:

We extend the concatenation operation to define the transpose operation as follows. For any **x** in Σ^* and **a** in Σ, $(xa)^T = a(x)^T$

Example 1.6 $(aaabab)^T = b(aaaba)^T = babaaa$

Palindrome:

A palindrome is a string which is same whether written forward or backward.

Example 1.7   malayalam, 10101.

A palindrome of even length can be obtained by concatenation of a string and its transpose.

Prefix and suffix of a string:

A **prefix** of a string is a substring of leading symbols of that string.
If **w** is a prefix of **y**, then there exists y^l in Σ^* such that $y=wy^l$.

Similarly, **suffix** of a string is a substring of trailing symbols of that string.
If **w** is a suffix of **y**, then there exists y^l in Σ^* such that $y=y^l w$.

A prefix or suffix of a string, other than the string itself is called a proper prefix or suffix.

Property: $x \varepsilon = \varepsilon x = x$ (ε is identity)

Example 1.8

The string "123" has four prefixes.
The four prefixes for the string "123" are $\varepsilon,1,12,123$.

Example 1.9

The string "123" has four suffixes.
The four suffixes for the string "123" are ε,3, 23, and 123.

1.2.4 LANGUAGES

A set of strings all of which are chosen from some Σ^*, where Σ is a particular alphabet is called a **language**.

If Σ is an alphabet and $L \subseteq \Sigma^*$, then L is a language over Σ i.e., a language is a subset of Σ^*.

Note 1: A language over Σ need not include strings with all the symbols of Σ.
     2: The language may be finite or infinite, but the alphabet should be finite.

Example 1.10

1) The empty set $\emptyset$ and the set consisting of empty string $\{\varepsilon\}$ are languages.

$\emptyset$ and $\{\varepsilon\}$ are distinct because the latter has a member while the former does not have any member.

2) The set of palindromes over the alphabet $\Sigma = \{0,1\}$ is an infinite language. Some members of this language are ε, 0, 1, 00, 11, 010, 1101011, 0110....

3) The language of all strings consisting of **n** 0's followed by **n** 1's for some $n \geq 0$, is given as
$$L = \{\varepsilon, 01, 0011, 000111,.....\}$$

4) The language for the set of all binary numbers whose value is a prime is given as
$$L = \{10, 11, 101, 111, 1011.....\}$$

5) The language for the set of all strings of 0's and 1's, with an equal number of each is given as
$$L = \{\varepsilon, 01, 10, 0101, 1001, 0011....\}$$

Set formers as way to define language:

Definition 1: We can define a language using a "set-former"
$$L = \{w \mid \text{something about } w\}$$

Examples 1.11

     a) $L = \{w \mid w$ consists of an equal number of 0's and 1's$\}$
     b) $L = \{w \mid w$ is a binary number whose decimal value is prime$\}$
     c) $L = \{w \mid w$ is a syntactically correct C-program$\}$

Definition 2: We can replace **w** by some expression with parameters and describe the string in the language by stating conditions on the parameters.

Examples 1.12

a) $L = \{0^n 1^n | n \geq 1\}$, n is the parameter.

This language consists of the strings L= {01, 0011, 000111, ...}

b) $L = \{0^i 1^j | 0 \leq i \leq j\}$. This language consists of strings with some 0's followed by atleast as many 1's.

c) $L = \{0^m 1^n | m \geq 0 \text{ and } n \geq 1\}$

1.3 FINITE STATE MACHINE

1.3.1 INTRODUCTION

The finite automaton is a mathematical model of a system, with discrete inputs and outputs. It is a useful design tool for finite state systems. It has a finite number of internal states and the system can be in any one of these finite states.

If its output depends only on the input then it is called *automaton without memory*. If its output depends not only on input but also on the state then it is called *automaton with a finite memory*.

Example 1.13

1) A primary example is a switching circuit, such as the control unit of a computer. A switching circuit is composed of finite number of gates, each of which can be in one of the two conditions usually denoted by 0 and 1.

2) Consider the shift registers as finite state machine

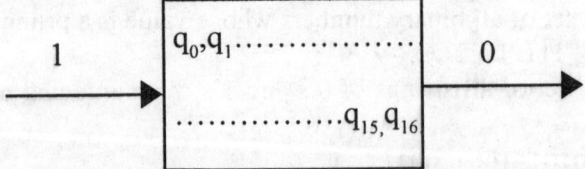

Problems with Automata Theory:

In automata theory, a *problem* is the question of deciding whether a given string is a member of some particular language or not.

Given a string **w** in Σ^*, decide whether or not **w** is in L.

Uses of automata theory:

1) Automata theory is used to know what is computable and what is not.

2) Automata theory is applied in compiler writing.

3) To know whether an efficient algorithm exists for some practical problems, we make use of automata theory.

1.3.2 Finite Automaton

Mathematical Definition: A finite automata can be represented by a 5-tuple $(Q, \Sigma, \delta, q_0, F)$ where

    i) $\mathbf{Q}$ is a finite non-empty set of states.

    ii) Σ is a finite non-empty set of inputs called input alphabet.

    iii) δ is a function which maps $Q \times \Sigma$ to Q and is called transition function i.e., $\delta(\mathbf{q, a})$ is a state for each state $\mathbf{q}$ and input symbol $\mathbf{a}$. The function describes the change of states during transition. This mapping is usually represented by transition table or transition diagram.

    iv) $\mathbf{q_0} \in \mathbf{Q}$ is the initial state.

    v) $\mathbf{F} \subseteq \mathbf{Q}$ is the set of final states. It is assumed that there may be more than one final state.

1.3.3 Finite Automaton Model

Diagrammatical representation of finite automaton

We picture a FA as a finite control, which is in some state from $\mathbf{Q}$, reading a sequence of symbols from Σ written on a tape as shown in the following figure.

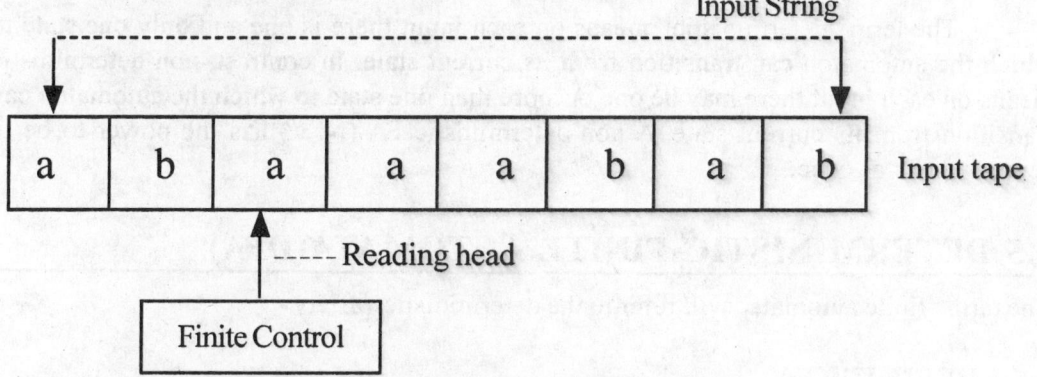

Model of Finite Automaton

In one move, the FA in state $\mathbf{q}$ and scanning symbol $\mathbf{a}$ enters state $\delta(\mathbf{q, a})$ and moves its head one symbol to the right. If $\delta(\mathbf{q, a})$ is an accepting state, then the FA is deemed to have accepted the string written in its input tape upto, but not including the position to which the head has just moved. If the head has moved off the right end of the tape, then it accepts the entire string.

The various components are explained as follows:

Input tape: The input tape is divided into squares, each square containing a single symbol from the input alphabet Σ.

Reading head: The head examines only one square at a time and moves one square either to the left or to the right. For further analysis, we restrict the movement of R-head only to the right side.

Finite control: The input to the finite control will be usually the symbol under the R-head, say **a** or the present state of the machine, and say **q**, to give the following outputs:

    a)   A motion of R-head along the tape to the next square (In a null move, i.e., R-head remains in the same square).

    b)   The next state of the finite state machine given by $\delta(\mathbf{q, a})$.

1.3.4 ACCEPTABILITY OF STRINGS BY FINITE AUTOMATA

A string **w** is accepted by a finite automaton $\mathbf{M = (Q, \Sigma, \delta, q_0, F)}$,

        if $\delta(\mathbf{q_0, w}) = \mathbf{q_f}$ for some $\mathbf{q_f}$ in **F**

This concludes that the string is accepted by the finite automata.

1.4 TYPES OF FINITE AUTOMATA

There are two types of finite automata:

      1) *Deterministic Finite Automata (DFA)*

      2) *Non- Deterministic Finite Automata (NFA)*

    The term "deterministic" means on each input there is one and only one state to which the automaton can transition from its current state. In contrast, non-deterministic means on each input there may be one or more than one state to which the automaton can transition from its current state. A non-deterministic FA (NFA) has the power to be in several states at once.

1.5 DETERMINISTIC FINITE AUTOMATA(DFA)

The term "finite automata" will refer to the deterministic variety.

1.5.1 DEFINITION

    Analytically, a finite automata (FA) can be represented by a 5-tuple $M = (Q, \Sigma, \delta, q_0, F)$ where

    i)   **Q** is a finite non-empty set of states.

    ii) Σ is a finite non-empty set of inputs called input alphabet.

    iii) δ is a function which maps $\mathbf{Q \times \Sigma}$ into **Q** and is called transition function i.e., $\delta(\mathbf{q,a})$ is a state for each state **q** and input symbol **a**. This is the function which describes the change of states during the transition. This mapping is usually represented by a transition table or transition diagram. The transition function δ takes two arguments i.e., state and input symbol and returns a state.

iv) $q_0 \in Q$ is the initial state.

v) $F \subseteq Q$ is the set of final states. It is assumed that there may be more than one final state.

Important properties of DFA:

- There is no transition on input symbol ε i.e., the empty input symbol.
- There is atmost one transition for each state **q** and input symbol **a.**

Example for DFA

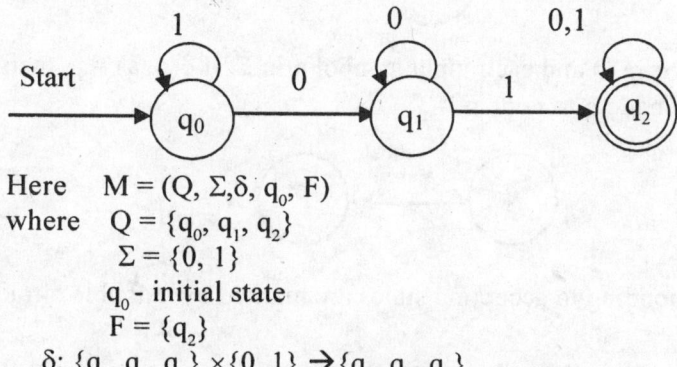

Here $M = (Q, \Sigma, \delta, q_0, F)$
where $Q = \{q_0, q_1, q_2\}$
        $\Sigma = \{0, 1\}$
        q_0 - initial state
        $F = \{q_2\}$
$\delta: \{q_0, q_1, q_2\} \times \{0, 1\} \rightarrow \{q_0, q_1, q_2\}$

In DFA, from each state there should be one arc labeled **0** and one arc labeled **1** From the above figure, we have $\delta(q_0, 0) = q_1$, $\delta(q_0, 1) = q_0$ and so on

1.5.2 TRANSITION DIAGRAM

Transition diagram is a directed graph associated with a FA. The vertices of the graph correspond to the states of the FA. If there is a transition from state **q** to state **p** on input **a**, then there is an arc labeled **a** from state **q** to state **p** in the transition diagram. The FA accepts a string **x** if the sequence of transitions correspond to the symbols of **x** leads from the start state to an accepting state(or final state).

Example 1.14 Consider the following transition diagram

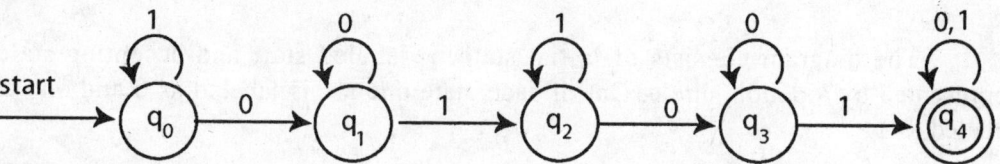

Solution :

From the above diagram:

a) For each state **q** in **Q** there is a node in transition diagram.

b) An arrow into the starting state **q₀**, labeled **start**

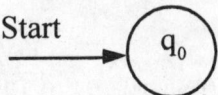

c) For each state **q** in **Q** and each input symbol **a** in Σ, if δ(q, a) = p, then there is an arc labeled **a** from node **q** to node **p**

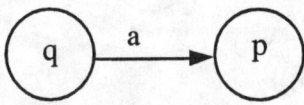

d) Nodes corresponding to accepting states are marked by a double circle

Example 1.15   The transition diagram for DFA accepting the string with substring 01.
Solution

$L = \{w|$ w is of the form x01y where x and y are strings consisting of 0's and 1's$\}$
$L = \{10101, 010, 101, 01\ldots\ldots\}$

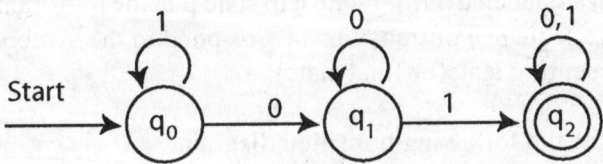

The diagram consists of initial state **q₀** labeled start and accepting state **q₂**, represented by a double circle. Out of each state one arc is labeled as **0** and one arc is labeled as **1**.

1.5.3 TRANSITION TABLE

A transition table is a conventional tabular representation of transition function and it takes two arguments and returns a value.

The rows of the table correspond to states and columns correspond to the inputs. The entry for the rows correspond to the state **q** and columns correspond to input **a** is the state $\delta(q, a)$. The initial state is marked with arrow and the final state is encircled.

Example 1.16 Construct the transition table for the following transition diagram

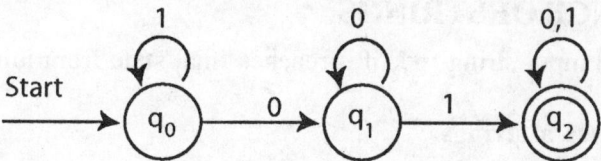

Transition table:

Q \ Σ	0	1
$\rightarrow q_0$	q_1	q_0
q_1	q_1	q_2
$\textcircled{q_2}$	q_2	q_2

In the above table, the initial state q_0 is marked with arrow and the final state q_2 is encircled.

The transition functions are defined as,
$$\delta(q_0, 0) = q_1, \quad \delta(q_0, 1) = q_0$$
$$\delta(q_1, 0) = q_1, \quad \delta(q_1, 1) = q_2$$
$$\delta(q_2, 0) = q_2, \quad \delta(q_2, 1) = q_2$$

Extending the transition function to strings:

The extended transition function is a function that takes state **q** and string **w** as arguments and returns state **p**. It is denoted by $\hat{\delta}$

$\hat{\delta}$ is a function mapping from $Q \times \Sigma^*$ to **Q**.

The extended transition function $\hat{\delta}$ is defined as follows:

1) $\hat{\delta}(q, \varepsilon) = q$. That is, if we are in state **q** and read no inputs, then we are in still state **q**.

2) $\hat{\delta}(q, w) = \hat{\delta}(q, xa) = \delta(\hat{\delta}(q, x), a)$, where **w** is a string of the form **xa**; that is, **a** is the last symbol of **w**, and **x** is the string consisting of all but the last symbol.

1.5.4 ACCEPTANCE OF STRINGS

For the given input string '**w**', if it reaches final state from initial state then it is acceptable.

How a DFA processes a strings

Model I: Using transition function δ

DFA starts with initial state q_0. Let $a_1, a_2 \ldots \ldots a_n$ be a sequence of input symbols, δ is transition function.

Step 1: $\delta(q_0, a_1) = q_1$, DFA in state q_0 on input a_1 enters state q_1.

Step 2: $\delta(q_1, a_2) = q_2$, DFA in state q_1 on input a_2 enters state q_2.

Similarly, $\delta(q_{i-1}, a_i) = q_i$ for each i $(1 \le i \le n)$

If q_n is a member of **F**, then the input string $a_1 a_2 \ldots \ldots a_n$ is accepted, otherwise it is rejected.

Model II: Using extended transition function $\hat{\delta}$.

Step 1: $\hat{\delta}(q_0, \varepsilon) = q_0$, $\hat{\delta} q_0$ is the initial state.

Step 2: $\hat{\delta}(q_0, w) = (q_0, xa) = \delta(\hat{\delta}(q_0, x), a) = q_f$, q_f is final state.

i.e., $\hat{\delta}(q_0, w) =$ **final state** , hence **w** is accepted by the DFA

Example 1.17

Consider the following transition table for DFA and check whether the string w=110101 is accepted by DFA using transition function δ and extended transition function $\hat{\delta}$

Transition table

Q \ Σ	0	1
→ q_0	q_2	q_1
q_1	q_3	q_0
q_2	q_0	q_3
q_3	q_1	q_2

In the above transition table, we can observe that initial and final states are same. Acceptance of the string w=110101 by DFA, using transition function δ and extended transition function $\hat{\delta}$

Model I: Using transition function δ

$$\delta(q_0, 110101) = \delta(q_1, 10101)$$

$$= \delta(q_0, 0101)$$

$$= \delta(q_2, 101)$$

$$= \delta(q_3, 01)$$

$$= \delta(q_1, 1)$$

$$= \delta(q_0, \varepsilon) = q_0 \quad \text{— accepting state}$$

$$\therefore \quad \delta(q_0, \underline{110101}) = q_0 \quad (\text{since } \delta(q_0, x)=q \text{ for some } q \in F)$$

$\therefore$ The string "110101" is accepted by DFA.

Model II: Using extended transition function $\hat{\delta}$

$$\hat{\delta}(q_0, \varepsilon) = q_0$$
$$\hat{\delta}(q_0, 1) = \hat{\delta}(q_0, \varepsilon 1) = \delta(\hat{\delta}(q_0, \varepsilon), 1) = \delta(q_0, 1) = q_1$$
$$\hat{\delta}(q_0, 11) = \delta(\hat{\delta}(q_0, 1), 1) = \delta(q_1, 1) = q_0$$
$$\hat{\delta}(q_0, 110) = \delta(\hat{\delta}(q_0, 11), 0) = \delta(q_0, 0) = q_2$$
$$\hat{\delta}(q_0, 1101) = \delta(\hat{\delta}(q_0, 110), 1) = \delta(q_2, 1) = q_3$$
$$\hat{\delta}(q_0, 11010) = \delta(\hat{\delta}(q_0, 1101), 0) = \delta(q_3, 0) = q_1$$
$$\hat{\delta}(q_0, 110101) = \delta(\hat{\delta}(q_0, 11010), 1) = \delta(q_1, 1) = q_0$$
$$\therefore \quad \hat{\delta}(q_0, 110101) = q_0$$

$\therefore$ The string "110101" is accepted by the DFA.

1.5.5 LANGUAGE RECOGNIZERS

The language of DFA, $M = (Q, \Sigma, \delta, q_0, F)$ is denoted by $L(M)$ and is defined as
$$L(M) = \{w \mid \hat{\delta}(q_0, w) \in F\} \text{ or } L(M) = \{w \in \Sigma^* \mid \delta(q_0, w) \in F\}$$

i.e., the language of **M** is the set of strings **w** that takes initial state q_0 to one of the accepting states.

Note: If **L** is **L(M)** for some DFA **M** then **L** is called a regular language.

Example 1.18

Design a DFA for the following language

$$L = \{ 0^m 1^n \mid m \geq 0 \text{ and } n \geq 1\}$$

Solution: This language consists of the strings that should end with **1**.

$$\therefore L(M) = \{1, 01, 011, 0011, 001, 1111, ...\}$$
$$\{0, 00, 100, 110, ...\} \notin L(M)$$

The DFA for the above language is as follows

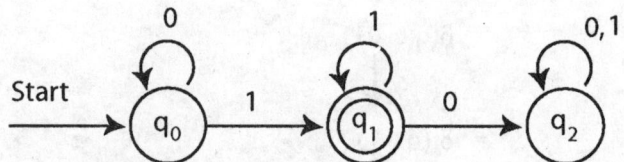

The transition table for above DFA is

Q \ Σ	0	1
→ q_0	q_0	q_1
(q_1)	q_2	q_1
q_2	q_2	q_2

Consider the string "011" and check whether it is accepted by the above DFA

$$\hat{\delta}(q_0, \varepsilon) = q_0$$
$$\hat{\delta}(q_0, 0) = \hat{\delta}(q_0, \varepsilon 0) = \delta(\hat{\delta}(q_0, \varepsilon), 0) = \delta(q_0, 0) = q_0$$
$$\hat{\delta}(q_0, 01) = \delta(\hat{\delta}(q_0, 0), 1) = \delta(q_0, 1) = q_1$$
$$\hat{\delta}(q_0, 011) = \delta(\hat{\delta}(q_0, 01), 1) = \delta(q_1, 1) = q_1 \text{ - final state}$$

$$\therefore \quad \hat{\delta}\ (q_0, 011) = q_1.$$

$\therefore$ "011" is accepted by the DFA.

1.6 NON–DETERMINISTIC FINITE AUTOMATA (NDFA / NFA)

1.6.1 DEFINITION

A non-deterministic finite automata is a 5-tuple **M= (Q, Σ, δ, q_0, F)** where

i) **Q** is a finite non-empty set of states.

ii) **Σ** is a finite non-empty set of inputs called input symbols.

iii) **δ** is a function which maps **Q$\times\Sigma$** into **2^Q** which is the power set of **Q** i.e., the set of all subsets of **Q**. It takes two arguments a state and input symbol, returns a set of states.

iv) **q_0** $\in$ **Q** is the initial state.

v) **F** $\subseteq$ Q is the set of final states.

Note: The difference between the deterministic and non –deterministic automata is only in δ.

For DFA, the outcome is a state, i.e., an element of **Q**.

For NFA, the outcome is a set of states i.e., a subset of **Q.**

Example 1.19   Consider the following transition diagram for NFA

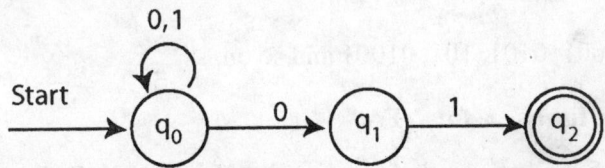

The five tuple form for the above NFA is

$Q = \{q_0, q_1, q_2\}$, $\Sigma = \{0, 1\}$

$q_0 =$ initial state

$F = \{q_2\}$

$\delta: Q \times \Sigma \rightarrow 2^Q$

$2^Q = \{\varepsilon, \{q_0\}, \{q_1\}, \{q_2\}, \{q_0, q_1\}, \{q_1, q_2\}, \{q_0, q_2\}, \{q_0, q_1, q_2\}\}$

$\delta (q_0, 0) = \{q_0, q_1\}$

$\{q_0, q_1\} \in 2^Q$

$\{q_0, q_1\} \subseteq Q$ and so on.....

Transition table for the above NFA

Q \ Σ	0	1
→ q_0	$\{q_0, q_1\}$	$\{q_0\}$
q_1	$\varnothing$	$\{q_2\}$
(q_2)	$\varnothing$	$\varnothing$

This NFA will accept the strings that end with "01"

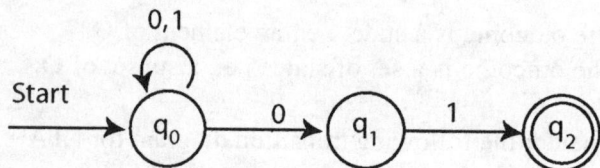

It accepts strings 01, 001, 0101, 101, 01001 and so on.

Extended transition function for NFA:

The function δ can be extended to a function $\hat{\delta}$ mapping $\mathbf{Q} \times \Sigma^*$ to 2^Q and reflecting sequence of inputs as follows:

1. $\hat{\delta}(q, \varepsilon) = \{q\}$
2. $\hat{\delta}(q, wa) = \{p \mid$ for some state r in $\hat{\delta}(q, w)$, p is in $\delta(r,a)\}$
3. $\delta(P, w) = \cup_{q \text{ in } P} \delta(q, w)$ for each set of states $P \subseteq Q$.

The important properties of NFA are as follows:

1. There may be a transition on input symbol 'ε' i.e. empty input symbol.
2. A state can have more than one transition on the input symbol 'a'.

1.6.2 ACCEPTANCE OF A STRING USING NFA

1. For the given input string **'w'**, if it reaches final state from initial state then it is acceptable. It happens sometimes and sometimes does not happen for the same input string.

2. The sequences of states for acceptability of a string are different.

How a NFA processes a string:

Consider the above transition diagram and check the acceptance of the input string "00101" using NFA

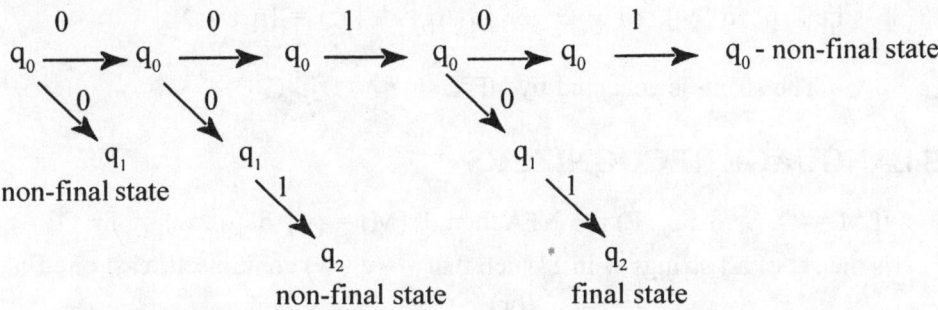

The states of NFA during the processing of input sequence "00101" is

$$q_0 \xrightarrow{0} q_0 \xrightarrow{0} q_0 \xrightarrow{1} q_0 \xrightarrow{0} q_1 \xrightarrow{1} q_2 \text{ (transition path)}$$

Acceptance of input string "00101" using NFA

i) Using transition function δ

$$\delta(q_0, 00101) = \delta(q_0, 0101)$$
$$= \delta(q_0, 101)$$
$$= \delta(q_0, 01)$$
$$= \delta(q_1, 1)$$
$$= \delta(q_2, \varepsilon)$$

$\therefore$ $\delta(q_0, 00101) = q_2$ - final state

$\therefore$ The string "00101" is accepted by NFA.

ii) Using extended transition function $\hat{\delta}$

$\hat{\delta}(q_0, \varepsilon) = \{q_0\}$,

$\hat{\delta}(q_0, 0) = \hat{\delta}(q_0, \varepsilon 0) = S\delta(\hat{\delta}(q_0, \varepsilon), 0) = S\delta(\{q_0\}, 0) = S\delta(q_0, 0) = \{q_0, q_1\}$

$\hat{\delta}(q_0,00) = \delta(\hat{\delta}(q_0,0),0) = \delta(\{q_0 q_1\}, 0) = \delta(q_0,0) \cup \delta(q_1,0) = \{q_0, q_1\} \cup \emptyset = \{q_0, q_1\}$

$\hat{\delta}(q_0,001) = \delta(\hat{\delta}(q_0,00),1) = \delta(\{q_0 q_1\}, 1) = \delta(q_0,1) \cup \delta(q_1,1) = \{q_0\} \cup \{q_1\} = \{q_0, q_2\}$

$\hat{\delta}(q_0,0010) = \delta(\hat{\delta}(q_0,001),0) = \delta(\{q_0 q_2\}, 0) = \delta(q_0,0) \cup \delta(q_2,0) = \{q_0, q_1\} \cup \emptyset = \{q_0, q_1\}$

$\hat{\delta}(q_0,00101) = \delta(\hat{\delta}(q_0,0010),1) = \delta(\{q_0 q_1\}, 1) = \delta(q_0,1) \cup \delta(q_1,1) = \{q_0\} \cup \{q_2\} = \{q_0, q_2\}$

It contains a final state q_2.

i.e. $\hat{\delta}(q_0, w) \cap F \neq \emptyset$ or $\{q_0, q_2\} \cap \{q_2\} = \{q_2\} \neq \emptyset$

∴ The string is accepted by NFA.

1.6.3 LANGUAGE RECOGNIZERS

If $M = (Q, \Sigma, \hat{\delta}, q_0, F)$ is a NFA then, $L(M) = \{w| \hat{\delta}(q_0, w) \cap F \neq \emptyset\}$ i.e. $L(M)$ is the set of all strings **w** in Σ^* such that $\hat{\delta}(q_0, w)$ contains at least one final state.

(or)

$L(M) = \{w \in \Sigma^* | \hat{\delta}(q_0, w)$ contains a state in $F\}$

Example 1.20

Design a NFA to accept the strings with 0's and 1's such that string contains either two consecutive 0's or two consecutive 1's.
Solution:

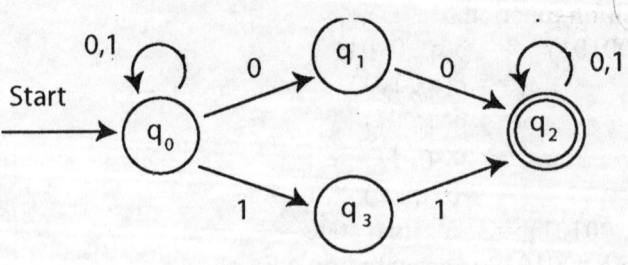

Consider a string "10100", the transition path is given as:

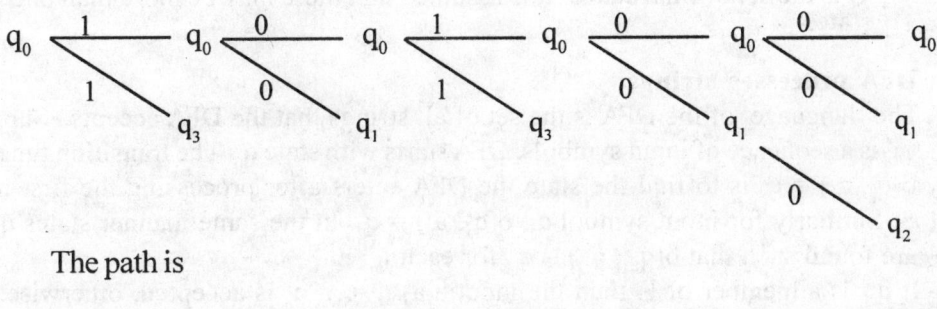

The path is

$$q_0 \xrightarrow{\ 1\ } q_0 \xrightarrow{\ 0\ } q_0 \xrightarrow{\ 1\ } q_0 \xrightarrow{\ 0\ } q_1 \xrightarrow{\ 0\ } q_2$$

Example 1.21

Design NFA which accepts set of strings over $\{0,1\}$, where 2nd symbol from right side is 1.

Solution

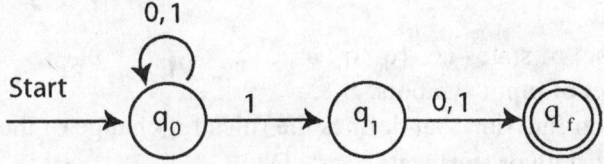

Solved Problems

Problem 1. Define finite automata and explain when it accepts strings as valid.

Solution:

Deterministic Finite Automata (DFA):

A finite automata can be represented by a 5-tuple $(Q, \Sigma, \delta, q_0, F)$ where

i) 'Q' is a finite non-empty set of states.

ii) 'Σ' is a finite non-empty set of inputs called input alphabet.

iii) 'δ' is a function which maps $Q \times \Sigma$ to Q and is called transition function i.e. $\delta(q, a)$ is a state for each state 'q' and input symbol 'a'. i.e. the function describes the change of states during transition. This mapping is usually represented by transition table or transition diagram.

iv) $q_0 \in Q$ is the initial state.

v) $F \subseteq Q$ is the set of final states. It is assumed that there may be more than one final state.

How a DFA processes strings:

The 'language' of the DFA is the set of all strings that the DFA accepts. Suppose $a_1 a_2 \ldots \ldots a_n$ is a sequence of input symbols, DFA starts with state q_0. The transition function 'δ', say $\delta(q_0, a_1) = q_1$ is to find the state the DFA enters after processing the first input symbol a_1. Similarly for input symbol a_2, $\delta(q_1, a_2) = q_2$. In the same manner states q_3, q_4 $\ldots \ldots q_n$ are found such that $\delta(q_{i-1}, a_i) = q_i$ for each i.

If q_n is a member of F, then the input $a_1 a_2 \ldots \ldots \ldots a_n$ is accepted, otherwise it is rejected.

Problem 2. Explain the terms deterministic finite automaton (DFA) and Non-deterministic finite automaton (NFA).

Solution:

Deterministic finite automaton (DFA)

A Deterministic Finite Automaton (DFA) is a 5-tuple machine M, represented as $M = (Q, \Sigma, \delta, q_0, F)$

where   Q = a finite set of states i.e. $\{q_0, q_1, q_2, \ldots\}$

Σ = a finite set of input symbols.

δ = a transition function that defines the rules for change of the states.

q_0 = an initial state or start state, $q_0 \in Q$

F = the set of final states or accepting states, $F \subseteq Q$.

A transition function δ takes two arguments, state and the input symbol and returns the output state i.e., $\delta: Q \times \Sigma \rightarrow Q$

If q_1 is the current state and '1' is the input symbol, then transition function is written as

$\delta(q_1, 1) = q_2$,here q_2 is the output state.

A DFA is represented by a transition graph in which the nodes represent states and labeled edges represent transition function.

If the transition function is $\delta(q_1, 1) = q_2$ then the transition graph is

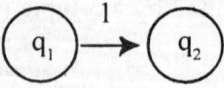

The important properties of DFA are:

1. There is no transition on input symbol 'ϵ' i.e. the empty input symbol.

2. There is atmost one transition for each state 'q' and input symbol 'a'.

Non-deterministic finite automaton:

A Non-deterministic finite automaton (NFA) is a 5-tuple machine, M represented as

$$M = (Q, \Sigma, \delta, q_0, F)$$

where Q = a finite set of states i.e. $\{q_0, q_1, q_2, ...\}$

Σ = a finite set of input symbols including the input symbol 'ε'

δ = a transition function that defines the rules for change of the states

q_0 = an initial state or start state, $q_0 \in Q$

F = the set of final states or accepting states, $F \subseteq Q$

A transition function δ takes two arguments, state and the input symbol and returns a set of output states which is a subset of 2^Q i.e., more than one output state.

$$\delta: Q \times \Sigma \rightarrow 2^Q$$

If q_1 is the current state and '1' is the input symbol, then transition function is written as

$$\delta(q_1, 1) = \{q_1, q_2, q_3\}$$

If the transition function is $\delta(q_1, 1) = \{q_1, q_2, q_3\}$ then the transition graph is

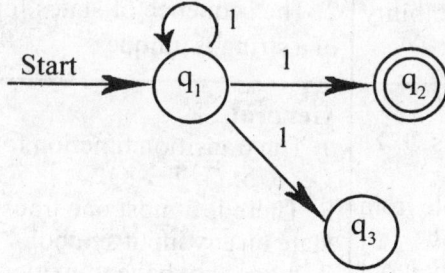

The important properties of NFA

1. There may be a transition on input symbol 'ε' i.e. the empty input symbol.
2. A state can have more than one transition on the input symbol 'a'.

Problem 3. Differentiate NFA and DFA w.r.t transition states and acceptance.

Solution:

Differences between NFA and DFA

NFA	DFA
Transition states: 1. For the given state and input there may be more than one transition state. 2. For an input, if there is no transition state, then it is dead state. 3. It cannot decide which state it has to go when an input is given to state, because a state can have more than one transition state. 4. For the same input it may reach final state and sometimes it may not reach.	**Transition states:** 1. For the given state and input there is only one transition state. 2. For any input, there is no dead state. 3. It will decide which state it has to go when an input is given with no ambiguity. 4. For the string (or input) of the acceptable language it always reaches the final state.
Acceptance: 1. For the given input string 'w', if it reaches final state from initial state then it is acceptable. It happens sometimes and sometimes does not happen for the same input string. 2. The sequences of states for acceptability of a string are different.	**Acceptance:** 1. DFA accepts a string if the automaton is in one of its final states, when the end of the string is reached. 2. The sequence of states for acceptability of a string is unique
General : 1. The transition function for NFA is $Q \times \Sigma$ to 2^Q i.e. $\delta: Q \times \Sigma \rightarrow 2^Q$ 2. More than one transition is possible from each state on the same input symbol. 3. It can have transition on the empty input symbol i.e., ε 4. It accepts an input string if there exist at least one path from start state to final state. 5. It takes more time to recognize an input string. 6. An NFA can be smaller than DFA. 7. The input strings which are accepted by DFA's may or may not be accepted by NFA's. 8. NFA can also be represented by transition table and transition graph.	**General :** 1. The transition function for DFA is $Q \times \Sigma$ to Q i.e. $\delta: Q \times \Sigma \rightarrow Q$. 2. There is atmost one transition from each state on any input symbol. 3. It does not have transition on empty input symbol i.e., ε. 4. It accepts the input string, if there is a path from start to final state. 5. It is faster than NFA. 6. A DFA can be much bigger than NFA. 7. The input strings which are accepted by NFA's are also accepted by DFA's. 8. DFA can be represented by transition table and transition graph.

Problem 4. Construct a DFA accepting the set of strings with an even number of 0's and an even number of 1's over an alphabet {0, 1}.

Solution:

The FA accepts a string '**x**', if the sequence of transitions corresponding to the symbols of **x** leads from the start state to an accepting state.

In the following figure,   Start state → q_0

Final state → q_0

The following figure shows that the finite automata (FA) accepts the set of strings which contain an even number 0's and 1's.

The strings that are accepted by FA are 001100, 10101010, 01010110 . . .

$\delta(q_0, x) = q_0$, q_0 - start state, q_0 - final state.

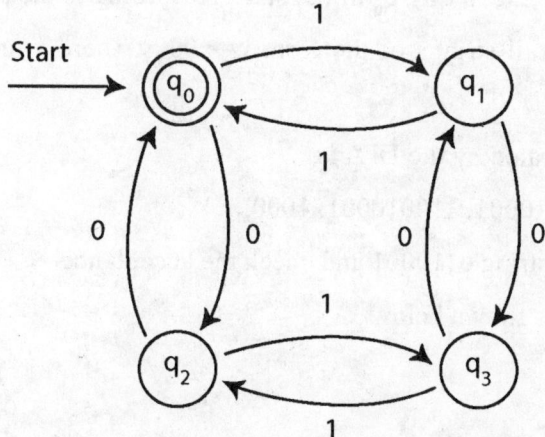

Here q_0 is a start state as well as final state. Note carefully that symmetry of 0's and 1's is maintained. We can associate meanings to each state as:

q_0 : state of even number of 0's and even number of 1's.

q_1 : state of odd number of 1's and even number of 0's.

q_2 : state of odd number of 0's and even number of 1's.

q_3 : state of odd number of 0's and odd number of 1's.

$$\downarrow \qquad\qquad \downarrow \qquad\qquad \downarrow$$
$$\delta(q_0, 001100) \;=\; \delta(q_2, 01100) = \delta(q_0, 1100)$$
$$\qquad\qquad\qquad\qquad \downarrow \qquad\qquad \downarrow$$
$$=\; \delta(q_1, 100) \;=\; \delta(q_0, 00)$$
$$\downarrow$$
$$=\; \delta(q_2, 0)$$
$$=\; q_0 - \text{final state.}$$

Problem 5. *Give a deterministic automata (DFA) accepting the set of all strings over the alphabet {0, 1} having three consecutive 0's.*

Solution:

Let DFA be M= $(Q, \Sigma, \delta, q_0, F)$, q_0-initial state. According to the problem we must design a DFA that accepts all strings of the form w=x000y,where x and y can be any combination of 0's and 1's

The language generated by the DFA is

L(M)={ 0110001, 110010001, 1000,...}

Consider the input string 0110001 and check the acceptance

The required DFA is shown below:

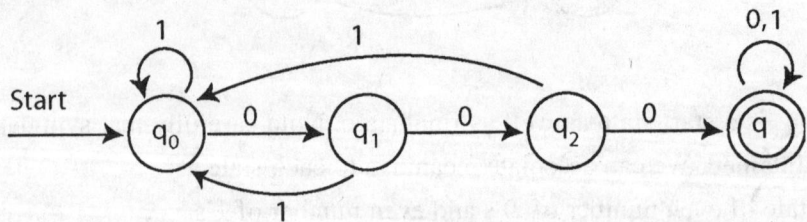

Consider the input string 0110001 and check the acceptance

$$\delta(q_0, 0110001) \; = \; \underset{\downarrow}{\delta}(q_1, 110001)$$

$$= \; \delta(q_0, 10001)$$

$$= \; \delta(q_0, 0001)$$

$$= \; \delta(q_1, 001)$$

$$= \; \delta(q_2, 01)$$

$$= \; \delta(q_3, 1)$$

$$= \; q_3 \qquad \text{- final state}$$

$\therefore$ The string 0110001 is accepted by the above finite automata.

Problem 6. Draw DFA which accepts even number of a's over an alphabet {a, b}

Solution:

The language L(M)={baa,aabb,babaaa,bbbaab,...}

  DFA to accept even number of a's:

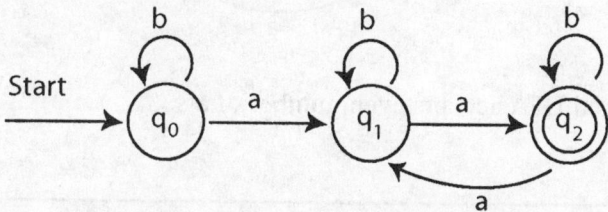

Various stages involved in the construction are shown below:

  1. To accept first two a's, we require minimum three states q_0, q_1, q_2.

Take q_2 as final state.

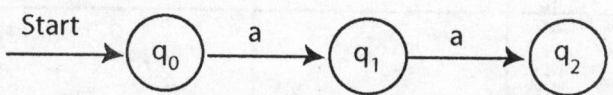

  2. Any number of b's are followed by first 'a' and in turn first 'a' is followed by any number of b's.

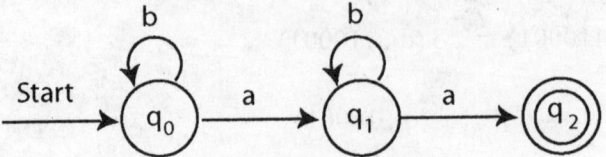

3. Second 'a' is followed by any number of b's

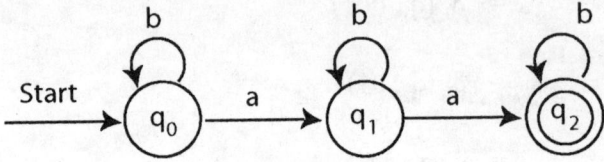

4. If any third 'a' is followed, then 'q_2' goes to 'q_1'

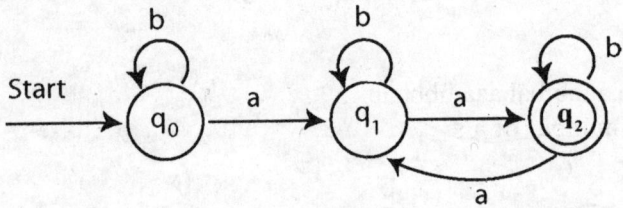

∴ It is the required DFA accepts even number of a's.

Transition table :

Q \ Σ	a	b
→q_0	q_1	q_0
q_1	q_2	q_1
ⓠq_2	q_1	q_2

Consider the string **'abab'** and check the acceptance, using extended transition

$$\hat{\delta}(q_0, \varepsilon) = q_0$$

$$\hat{\delta}(q_0, a) = \delta(\hat{\delta}(q_0, \varepsilon), a) = \delta(q_0, a) = q_1$$

$$\hat{\delta}(q_0, ab) = \delta(\hat{\delta}(q_0, a), b) = \delta(q_1, b) = q_1$$

$$\hat{\delta}(q_0, aba) = \delta(\hat{\delta}(q_0, ab), a) = \delta(q_1, a) = q_2$$

$$\hat{\delta}(q_0, abab) = \delta(\hat{\delta}(q_0, aba), b) = \delta(q_2, b) = q_2. \text{ accepted}$$

$\therefore$ The string 'abab' is accepted by DFA.

DFA which accepts odd number of a's over {a, b}

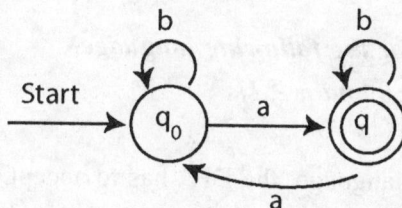

The above DFA accepts aaa, ababab, aaaaabb, bababa, babbabbabb......

Problem 7. *Given DFA which reads string from {a, b} and ends with 'aaa'*

Solution:

Let DFA be M = (Q, Σ, δ, q_0, F)

where q_0- initial state and it is given that $\Sigma = \{a, b\}$

According to the problem, we must design a DFA that accepts all the strings of the form
w = w^1aaa , where w^1 can be any combination of input symbols a's and b's.

$\therefore$ L(M) = {baaa, bbaaa, abaaaa, bbbaaaa, ...}

First we design the FA that fulfils the condition of the problem i.e., string ending with aaa. This is shown in figure below

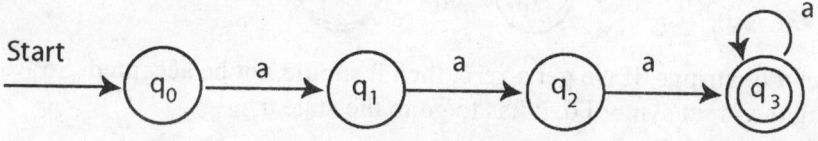

In DFA, there is a transition on every input symbol from every state. Here, we need to decide the transition on input symbol '**b**' from the states q_0, q_1, q_2, q_3. This completes the FA.

The required DFA is shown below:

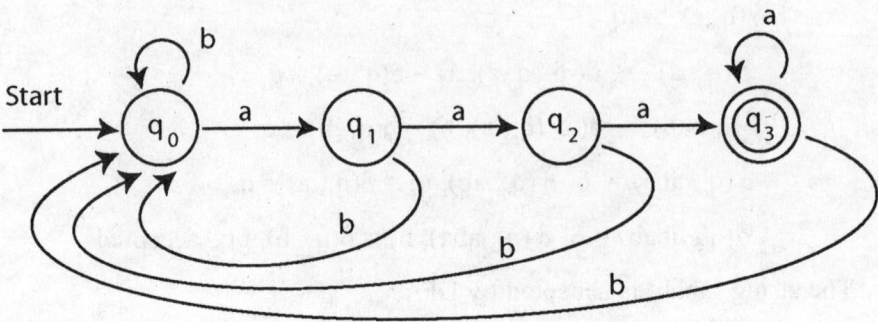

Problem 8. *Design a DFA for the following language*

$$L = \{0^m \, 1^n \mid m \geq 0 \text{ and } n \geq 1\}$$

Solution:

According to the given language, the DFA has to accept strings i.e. any number of zero's (including null zero's) followed by any number of 1's (at least one 1).

The language generated is $L(M) = \{1, 11, 01, 0011, 000111, \dots\}$

$$\{10, 0110, 01100, \dots\} \notin L(M)$$

First we consider accepting any number of 1's

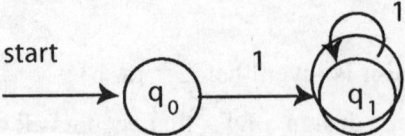

Then include any number of 0's before any number of 1's

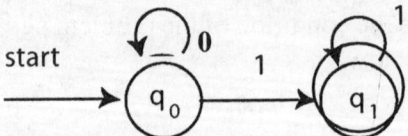

After this format of strings, if we get a zero, then it should not be accepted. So from 'q_1' state, on seeing the input symbol 0, it has to go to the state q_2

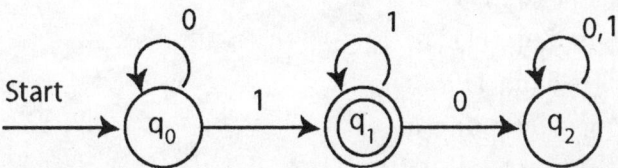

After seeing the input symbol '0' in q_1 and going to q_2 for all input symbols, it has to be in the same state. That means we never enter into final state from 'q_2' state.

Transition table for the above DFA:

Q \ Σ	0	1
→ q_0	q_0	q_1
ⓠq_1	q_2	q_1
q_2	q_2	q_2

From the transition table, the transition function 'δ' is defined as follows

$$\delta(q_0, 0) = q_0, \quad \delta(q_0, 1) = q_1$$
$$\delta(q_1, 0) = q_2, \quad \delta(q_1, 1) = q_1$$
$$\delta(q_2, 0) = q_2, \quad \delta(q_2, 1) = q_2$$

1. For the input string '011' the transition path is shown below

$$q_0 \xrightarrow{0} q_0 \xrightarrow{1} q_1 \xrightarrow{1} q_1 \text{- final state}$$

2. Consider a string 00111, check whether the string is accepted by above drawn DFA.

$$\delta(q_0, 00111) = \delta(q_0, 0111)$$

$$= \delta (q_0, 111)$$
$$= \delta (q_1, 11)$$
$$= \delta (q_1, 1)$$
$$= \delta (q_1, \varepsilon) = q_1$$

The string '00111' is accepted by the DFA.

Problem 9. Design FA which accepts only those strings which start with 1 and ends with 0.

Solution:

The FA will have a start state A from which only the edge with input '1' will go to next state. and the edge with '0' goes to the dead end.

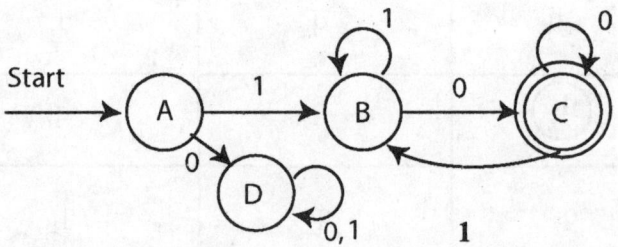

In state B if we read 1, we will be in B state, but if we read 0 at state B, we will reach to state C which is a final state. In state C if we read either 0 or 1 we will go to state C or B respectively. Note that the special care is taken for 0, if the input ends with 0 it will be in final state.

Problem 10. Design FA which checks the given unary number is divisible by 3.

Solution:

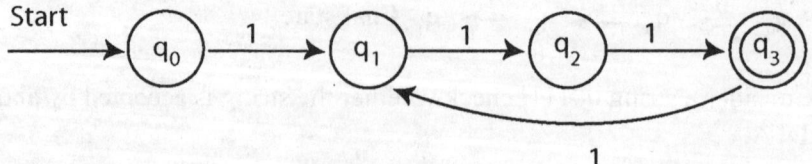

The unary number is made up of ones. The number 3 can be written in unary form as 111, number 5 can be written as 11111 and so on. The unary number which is divisible by 3 can be 111 or 111111111 and so on. The transition table is as follows

States ╲ Input	1
→ q_0	q_1
q_1	q_2
q_2	q_3
q_3	q_1

Consider a number 111111 which is equal to 6 i.e. divisible by 3. So after complete scan of this we reach to final state q_3.

start q_0 111111
         1q_1 11111
         11q_2 1111
         111q_3 111
         1111q_1 11
         11111q_2 1
         111111q_3 → now we are in final state.

Problem 11. Design a DFA to accept strings of a's and b's ending with 'abb' over
Σ= {a, b}.

Solution:

Initially, we will design DFA for the string **abb** as

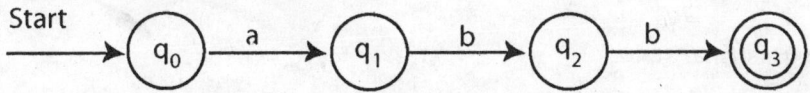

In this transition, q_0 is a start state and q_3 is a final state. Thus only after recognizing 'abb' the DFA will enter in final state.

We can design the complete DFA as follows

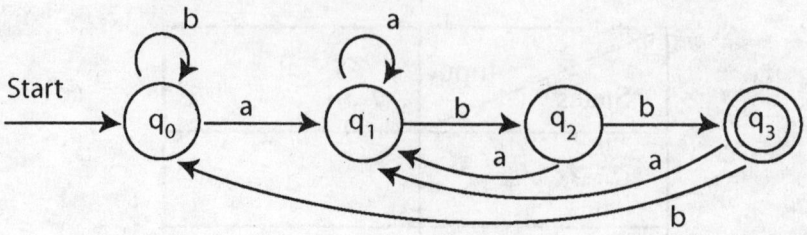

***Problem* 12. Design DFA over Σ = {a, b} for**

> ***i) (ab)ⁿ with n ≥ 0.***
>
> ***ii) (ab)ⁿ with n≥ 1.***

Solution:

This is a language of 'ab' in a pair, in (i), ε is accepted and in (ii), ε is not accepted.
The DFA for (i)

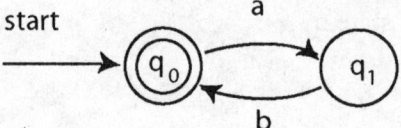

The DFA for (ii)

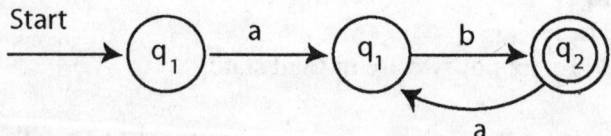

The strings accepted by

> i) $\delta(q_0, ab) = \delta(q_1, b)$
>
> $= q_0$
>
> Hence q_0 is a final state.
>
> ii) $\delta(q_0, ab) = \delta(q_1, b)$
>
> $= q_2$
>
> Here q_2 is final state.

Therefore string 'ab' is accepted by both (i) and (ii)

Problem 13. *Design an NFA to accept set of strings over alphabet set {0,1} and ending with two consecutive 0's.*

Solution :

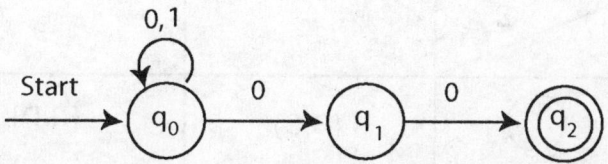

It is an NFA, because from state 'q_0' on seeing '0' we can go to either q_0 or q_1
The above NFA accepts the following strings 00,000,100,0100,1100,11000....01100....

Problem 14. *Design an NFA which accepts set of all strings containing '101' as substring.*

Solution :

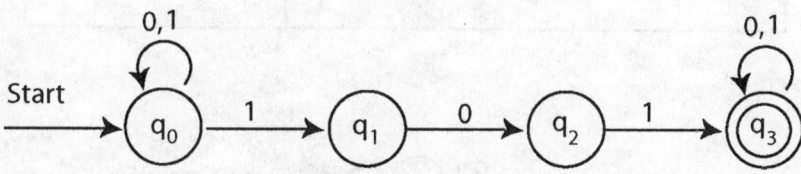

Problem 15. *Draw the transition diagram for the below FA*

$M = (\{A, B, C, D\}, \{0, 1\}, \delta, C, \{A,C\})$ *where*

$\delta(A, 0) = \delta(A, 1) = \{A, B, C\}$

$\delta(B, 0) = \{B\}, \; \delta(B, 1) = \{A, C\}$

$\delta(C, 0) = \{B, C\}, \; \delta(C, 1) = \{B, D\}$

$\delta(D, 0) = \{A, B, C, D\}$

$\delta(D, 1) = \{A\}$

Solution:

Transition table for given NFA:

Σ Q	0	1
→ C	{B,C}	{B,D}
B	{B}	{A,C}
A	{A,B,C}	{A,B,C}
D	{A,B,C,D}	{A}

Transition diagram:

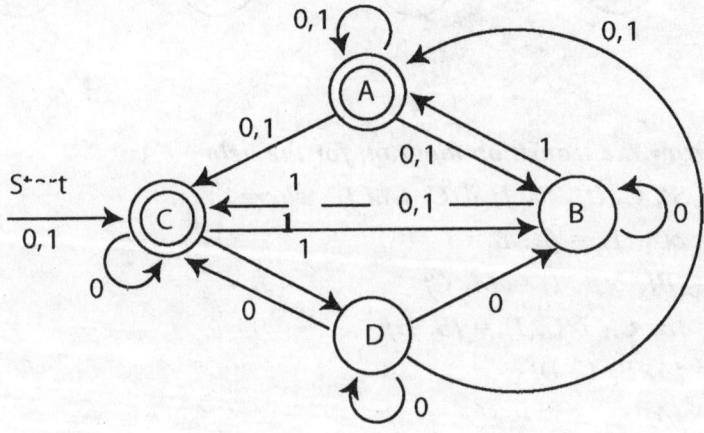

Accepted strings: 0010110, 11100, 111001, ...

Problem 16. *For the NFA given by the following state transition diagram.*
  a) Check whether the string 'abbabba' is accepted or not
  b) Give at least two transition paths.

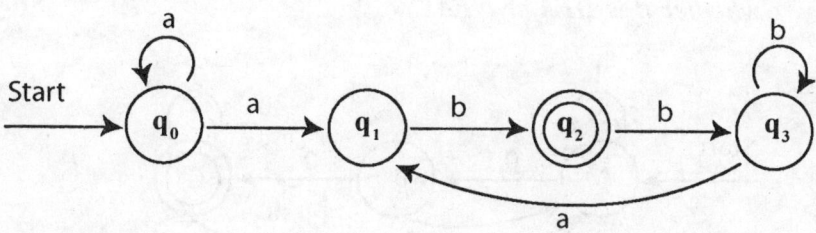

Solution:

a) Acceptance of the string 'abbabba'

$\hat{\delta} (q_0, \varepsilon) = \{q_0\}$

$\hat{\delta} (q_0, a) = \delta(\hat{\delta} (q_0, \varepsilon), a) = \delta(q_0, a) = \{q_0, q_1\}$

$\hat{\delta} (q_0, ab) = \delta(\hat{\delta} (q_0, a), b) = \delta(\{q_0, q_1\}, b) = \{q_2\}$

$\hat{\delta} (q_0, abb) = \delta(\hat{\delta}(q_0, ab), b) = \delta(\{q_2\}, b) = \delta(q_2, b) = \{q_3\}$

$\hat{\delta} (q_0, abba) = \delta(\hat{\delta} (q_0, abb), a) = \delta(q_3, a) = \{q_1\}$

$\hat{\delta} (q_0, abbab) = \delta(\hat{\delta} (q_0, abba), b) = \delta(q_1, b) = \{q_2\}$

$\hat{\delta} (q_0, abbabb) = \delta(\hat{\delta} (q_0, abbab), b) = \delta(q_2, b) = \{q_3\}$

$\hat{\delta} (q_0, abbabba) = \delta(\hat{\delta} (q_0, abbabb), a) = \delta(q_3, a) = \{q_1\}$

$\therefore \quad \hat{\delta} (q_0, abbabba) = \{q_1\}$.

Since the final state is q_2, but we are in q_1 state. Therefore the string 'abbabba' is not accepted by given NFA.

b)   Given at least two transition paths

Several transition paths are possible for the given NFA, two of them are

$$q_0 \rightarrow q_1 \rightarrow q_2$$

$$q_0 \rightarrow q_1 \rightarrow q_2 \rightarrow q_3 \rightarrow q_1 \rightarrow q_2$$

Problem 17. *Represent all 5-tuple for below transition diagram and decide whether it is DFA or NFA*

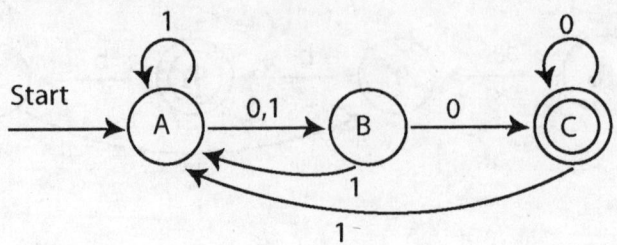

Solution:

The 5-tuple is given by M= $(Q, \Sigma, \delta, q_0, F)$

where   Q= {A, B, C}

Σ= {0, 1}

δ: Q×Σ $\rightarrow$ 2^Q i.e δ:{A,B,C}×{0,1}$\rightarrow 2^Q$

q_0=A

F= {C}

Here 2^Q = {ϕ, {A}, {B}, {C}, {A, B}, {B, C}, {C, A}, {A, B, C} }

Transition table:

Q \ Σ	0	1
$\rightarrow$ A	{B}	{A,B}
B	{C}	{A}
Ⓒ	{C}	{A}

In the above transition table, we can observe that δ(A, 1) = {A, B} i.e. from state 'A', on seeing the input symbol '1', we will go to either state 'A' or state 'B'.

$\therefore$ The given transition diagram is for NFA, whereas for DFA, for each state there should be exactly one transition on each input symbol.

REVIEW QUESTIONS

1. Give a DFA accepting the language over the alphabet {0,1}, the set of all strings starting with '001'.
2. Give a DFA accepting the language over the alphabet {0,1}, the set of all strings such that every block of five consecutive symbols contain at least two 0's.
3. Give a DFA accepting the language over the alphabet {a,b,c}, the set of all strings ending with 'bc'.
4. Design a FSM which can accept strings containing odd number of 0's and any number of 1's.
5. Design a FA which accepts the even binary numbers over {0,1}. Also give the transition table.
6. Design a FA which checks whether a given binary number is divisible by 3 over the alphabet {0,1}.
7. Design a DFA $L(M) = \{ w \mid w \in \{0,1\}^* \}$ and W is a string that does not contain consecutive 1's.
8. Recognize the language given by the following DFA

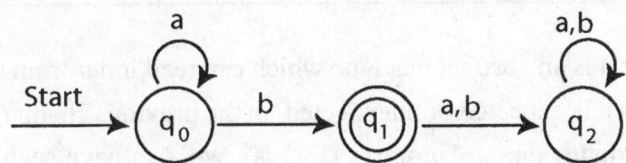

9. Design an NFA over {1,0} where 5th symbol from right side is 1.
10. Design an NFA which accepts the string containing either '01' or '10' over $\Sigma = \{0,1\}$.
11. Design an NFA which accepts the strings containing two 1's followed by two 0's as a substrings over $\Sigma = \{0,1\}$.
12. Design an NFA which accepts the strings containing 'acb' as a substring over $\Sigma = \{a, b, c\}$.
13. Design a DFA to accept the set of integers over the alphabet digit, where digit = $\{0,1,\ldots,9\}$.
14. Design a DFA over $\Sigma = \{a, b, c\}$ to accept language of the form $(abc)^n$, with n>=0.
15. Design a DFA over $\Sigma = \{a, b, c\}$ to accept language of the form $(abc)^n$, with n>=1.
16. Design a DFA over $\Sigma = \{a, b, c\}$ to accept language of the form $a^i b^j c^k$, with i>=0, j>=1 and k>=1.

Chapter 2

Finite Automata

2.1 INTRODUCTION

An automaton is an abstract machine which can read input from the input tape and give the output which is expected or unexpected. In the previous chapter, we have defined an automaton and what a Finite Automata (FA) is? We also have seen types of FA i.e., Deterministic Finite Automata (DFA) and Non- Deterministic Finite Automata (NFA).

In this chapter DFA and NFA are dealt in detail. In addition to NFA, NFA with ε transitions are also discussed. Conversions of NFA to DFA and NFA with ε-transitions to without ε-transitions are explained. Minimization of FSM and also equivalence of two FSM's are discussed. Finally, the machines with output i.e., Moore and Mealy are illustrated with examples.

2.2 SIGNIFICANCE OF NON-DETERMINISTIC FINITE AUTOMATON

As earlier said, an automaton is an abstract model of a computer, which takes certain input and gives output. The output obtained may be desirable or undesirable. DFA is a construction of deterministic model, which is difficult to construct. Hence NFA is first constructed and later this NFA is converted to a DFA. The input which is accepted by NFA is also accepted by the DFA.

2.3   NFA WITH ε-TRANSITIONS

2.3.1 SIGNIFICANCE OF NFA WITH ε

ε-moves play an important role in the construction of NFA. There may be a situation where the input symbol has not been accepted by the machine, but there is a necessity to change the state in order to reach the final state. In such situation, we change the state on ε-move, which helps in reaching final state of the automata.

2.3.2 DEFINITION

A Non-deterministic finite automaton with ε-transitions, **M** is given by **5**-tuple
M = (Q, Σ, δ, q₀, F)

where   **Q** is a finite set of states i.e., $q_0, q_1, q_2.....$

Σ is a finite set of input symbols

δ is a transition function , mapping from $Q \times (\Sigma \cup \{\varepsilon\}) \rightarrow 2^Q$

q_0 is initial state, $q_0 \in Q$

F is set of final states, $F \subseteq Q$.

Here   δ: $Q \times (\Sigma \cup \{\varepsilon\}) \rightarrow 2^Q$.

i.e., $\delta(q_0, a) = \{q_1, q_2\}$

Here **a** may be a symbol in Σ or ε.

Example 2.1

The transition diagram of an NFA accepting the language consisting of any number of 0's, followed by any number of 1's, followed by any number of 2's is given as

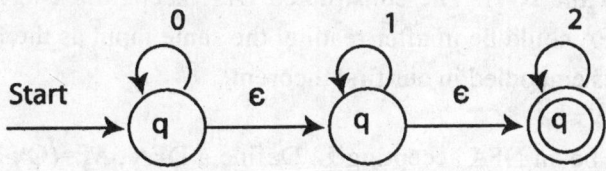

In some situations the machine need to change from the given state to the next state without looking for an input symbol, in order to reach final state. In such case we change the states on input symbol ε, i.e., no symbol is considered to change the states, such machine is said to be NFA with ε.

Extended transition function $\hat{\delta}$:

We shall now extend the transition function δ to a function $\hat{\delta}$ that maps $Q \times (\Sigma^* \cup \{\varepsilon\})$ to 2^Q. i.e., $\hat{\delta} : Q \times (\Sigma^* \cup \{\varepsilon\})$ to 2^Q.

We now define δ as follows:

1) $\hat{\delta}(q, \varepsilon) = \varepsilon\text{-CLOSURE}(q)$

2) $\hat{\delta}(q, a) = \varepsilon\text{-CLOSURE}(\delta(\hat{\delta}(q, \varepsilon), a))$
   $= \varepsilon\text{-CLOSURE}(\delta(\varepsilon\text{-CLOSURE}(q), a)$

3) For **w** in Σ^* and **a** in Σ, $\hat{\delta}(q, wa) = \varepsilon\text{-CLOSURE}(P)$,
   where P={p| for some r in $\delta(q, w)$, p is in $\delta(r, a)$}

4) $\hat{\delta}(S, a) = \cup_{q \text{ in } S} \delta(q, a)$,

5) $\hat{\delta}(S, w) = \cup_{q \text{ in } S} \hat{\delta}(q, w)$, for set of states S.

2.3.3 ACCEPTANCE OF LANGUAGES

We define L (M), the language accepted by $M = (Q, \Sigma, \delta, q_0, F)$ to be
$$L(M) = \{w \mid \hat{\delta}(q_0, w) \text{ contains a state in F, where } w \in \Sigma^*\}$$

2.4 CONVERSIONS AND EQUIVALENCES

2.4.4 THE EQUIVALENCE OF DFA'S AND NFA'S

Since every DFA is a NFA, it is clear that the class of languages accepted by NFA's includes the regular sets (The languages accepted by DFA's). However it turns out that these are only sets accepted by NFA's. The proof hinges on showing that DFA's can simulate DFA's i.e., for every NFA we can construct an equivalent DFA (one which accepts the same language). The way a DFA simulates an NFA is to allow the states of the DFA to correspond to sets of states of the NFA. The constructed DFA keeps track in its finite control of all states that the NFA could be in after reading the same input as the DFA has read. The formal construction is embodied in our first theorem.

Theorem:

Let $M = (Q, \Sigma, \delta, q_0, F)$ be an NFA accepting L. Define a DFA , $M^1 = (Q^1, \Sigma, \delta^1, q_0, F^1)$ as follows:

The states of M^1 are all subsets of the set of states of **M**, i.e., $Q^1 = 2^Q$.

M' will keep track in its state of all the states **M** could be in at any given time.

F' is the set of all states in **Q'** containing a final state of **M**.

An element of **Q'** will be denoted by $[q_1, q_2 \ldots q_i]$, where $q_1, q_2 \ldots q_i$ are in **Q**.

Observe that $[q_1, q_2 \ldots q_i]$ is a single state of the DFA corresponding to a set of states of the NFA. Note that $q_0' = [q_0]$.

We define $\delta'([q_1, q_2 \ldots q_i], a) = [p_1, p_2 \ldots p_j]$

if and only if $\delta(\{q_1, q_2 \ldots q_i\}, a) = \{p_1, p_2 \ldots p_j\}$

i.e., δ' applied to an element $[q_1, q_2 \ldots q_i]$ of **Q'** is computed by applying δ to each state of **Q** represented by $[q_1, q_2 \ldots q_i]$.

On applying δ to each of $q_1, q_2 \ldots q_i$ and taking the union, we get some new set of states $p_1, p_2 \ldots p_j$. This new set of states has a representative $[p_1, p_2 \ldots p_j]$ in **Q'** and that element is the value of $\delta'([q_1, q_2 \ldots q_i], a)$

It is easy to show induction on the length of the input string **x** that

$\delta'(q_0', x) = [q_1, q_2 \ldots q_i]$

if and only if $\delta'(q_0, x) = \{q_1, q_2 \ldots q_i\}$

Basis: The result is trivial for $|x| = 0$, since $[q_0'] = [q_0]$ and **x** must be ε.

Induction: Suppose that the hypothesis is true for inputs of length **m** or less.

Let **xa** be a string of length **m+1** with **a** in Σ. Then

$$\delta'(q_0', xa) = \delta(\delta'(q_0', x), a)$$

By the inductive hypothesis $\delta'(q_0', x) = [p_1, p_2 \ldots p_j]$ if and only if $\delta(q_0, x) = \{p_1, p_2 \ldots p_j\}$

But by definition of δ', $\delta'([p_1, p_2 \ldots p_j], a) = [r_1, r_2 \ldots r_k]$

if and only if $\delta'(\{p_1, p_2 \ldots p_j\}, a) = \{r_1, r_2 \ldots r_k\}$

Thus $\delta'(q_0', xa) = [r_1, r_2 \ldots r_k]$

if and only if $\delta(q_0, xa) = \{r_1, r_2 \ldots r_k\}$

Which establishes the inductive hypothesis.

To complete the proof, we have only to add $\delta'(q_0', x)$ is in **F'** exactly when $\delta(q_0, x)$ contains a state of **Q** that is in **F**.

$$\therefore L(M) = L(M')$$

2.4.1.1 PROCEDURE TO CONVERT NFA TO DFA

1) Write all transitions from initial state on every input symbol in Σ.

2) Repeat (1) for every new state. If a transition on some input symbol resulted in a set of states, then it is also considered as a new single state.

3) Repeat (2) until we do not get a new state.

4) The final states of equivalent DFA are all those states which consists of atleast one accepting state of given NFA.

NOTE: Accessible states: If there is a path (direct or indirect) from the initial state, then that state is called accessible state, otherwise inaccessible state.

Example 2.2   **Construct DFA for the following NFA**

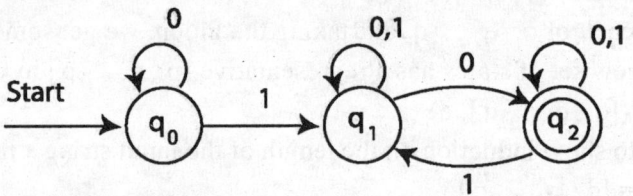

NFA to DFA conversion

Given NFA, $M = (Q, \Sigma, \delta, q_0, F)$

$Q = \{q_0, q_1, q_2\}$

$\Sigma = \{0, 1\}$

$q_0 = q_0$

$F = q_2$, and (δ is given below)

Transition table for NFA:

Q \ Σ	0	1
→ q_0	$\{q_0\}$	$\{q_1\}$
q_1	$\{q_1, q_2\}$	$\{q_1\}$
q_2	$\{q_2\}$	$\{q_1, q_2\}$

Let equivalent DFA be $M^1 = (Q^1, \Sigma, \delta^1, q_0^1, F^1)$

$\delta^1([q_0], 0) = [q_0]$ iff $\delta(q_0, 0) = \{q_0\}$

$\delta^1([q_0], 1) = [q_1]$ iff $\delta(q_0, 1) = \{q_1\}$

$\delta^\mathsf{l}([q_1], 0) = [q_1, q_2]$ iff $\delta(q_1, 0) = \{q_1, q_2\}$

$\delta^\mathsf{l}([q_1], 1) = [q_1]$ iff $\delta(q_1, 1) = \{q_1\}$

$\delta^\mathsf{l}([q_1,q_2], 0) = [q_1,q_2]$ iff $\delta(\{q_1, q_2\}, 0) = \{q_1, q_2\} \cup \{q_2\} = \{q_1, q_2\}$

$\delta^\mathsf{l}([q_1,q_2], 1) = [q_1,q_2]$ iff $\delta(\{q_1, q_2\}, 1) = \{q_1\} \cup \{q_1, q_2\} = \{q_1, q_2\}$

$\therefore$ $Q^\mathsf{l} = \{[q_0], [q_1], [q_1, q_2]\}$

       $q_0^{\;\mathsf{l}} = q_0$

       $F^\mathsf{l} = \{[q_1, q_2]\}$

       δ^l is given below

Transition table for DFA

Q Σ	0	1
$\rightarrow [q_0]$	$[q_0]$	$[q_1]$
$[q_1]$	$[q_1 ,q_2]$	$[q_1]$
$([q_1,q_2])$	$[q_1,q_2]$	$[q_1 ,q_2]$

Transition diagram for DFA:

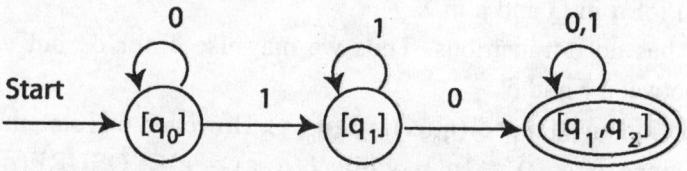

Acceptance of input string by NFA and DFA:

Consider the input string: 11001

Using NFA

$\hat{\delta}(q_0, \varepsilon) = \{q_0\}$

$\hat{\delta}(q_0, 1) = \delta(\hat{\delta}(q_0, \varepsilon), 1) = \delta(q_0, 1) = \{q_1\}$

$\hat{\delta}(q_0, 11) = \delta(\hat{\delta}(q_0, 1), 1) = \delta(q_1, 1) = \{q_1\}$

$\hat{\delta}(q_0, 110) = \delta(\hat{\delta}(q_0, 11), 0) = \delta(q_1, 0) = \{q_1, q_2\}$

$\hat{\delta}(q_0, 1100) = \delta(\hat{\delta}(q_0, 110), 0) = \delta(\{q_1, q_2\}, 0) = \{q_1, q_2\} \cup \{q_2\} = \{q_1,q_2\}$

$\hat{\delta}(q_0, 11001) = \delta(\hat{\delta}(q_0, 1100), 1) = \delta(\{q_1, q_2\}, 1) = \{q_1, q_2\}$ $\rightarrow$ Accepted.

Using DFA:

$\hat{\delta}(q_0, \varepsilon) = \{q_0\}$

$\hat{\delta}(q_0, 1) = \delta(\hat{\delta}(q_0, \varepsilon), 1) = \delta(q_0, 1) = \{[q_1]\}$

$\hat{\delta}(q_0, 11) = \delta(\hat{\delta}(q_0, 1), 1) = \delta([q_1], 1) = \{[q_1]\}$

$\hat{\delta}(q_0, 110) = \delta(\hat{\delta}(q_0, 11), 0) = \delta([q_1], 0) = \{[q_1, q_2]\}$

$\hat{\delta}(q_0, 1100) = \delta(\hat{\delta}(q_0, 110), 0) = \delta([q_1, q_2], 0) = \{[q_1, q_2]\}$

$\hat{\delta}(q_0, 11001) = \delta(\hat{\delta}(q_0, 1100), 1) = \delta([q_1, q_2], 1) = \{[q_1, q_2]\}$ → Accepted.

The string "11001" is accepted by both NFA and DFA.

2.4.2 EQUIVALENCE OF NFA'S WITH AND WITHOUT ε-MOVES

Like nondeterministic, the ability to make transitions on ε does not allow the NFA to accept non regular sets. We show this by simulating an NFA with ε-transitions by a NFA without such transitions.

Theorem: If **L** is accepted by an NFA with ε-transitions, then **L** is accepted by an NFA without ε-transitions.

Proof :Let $M = (Q, , \delta, q_0, F)$ be an NFA with ε-transitions. Construct $M^1 = (Q, \Sigma, \delta^1, q_0, F^1)$ where

$$F^1 = \begin{cases} F \cup \{q_0\} & \text{if } \varepsilon\text{-CLOSURE}(q_0) \text{ contains a state of F,} \\ F & \text{otherwise,} \end{cases}$$

and $\delta^1(q, a)$ is $\hat{\delta}(q, a)$ for **q** in **Q** and **a** in Σ.

Note that M^1 has no ε-transitions. Thus we may use δ^1 for $\hat{\delta}$, but we must continue to distinguish between δ and δ^1.

We wish to show by induction on $|x|$ that $\delta^1(q_0, x) = \hat{\delta}(q_0, x)$. However, this statement may not be true for $x = \varepsilon$, since $\delta^1(q_0, \varepsilon) = \{q_0\}$, while $\delta(q_0, \varepsilon) = \varepsilon\text{-CLOSURE}(q_0)$. We therefore begin our induction at 1.

Basis: $|x| = 1$. Then x is a symbol a and $\delta^1(q_0, a) = \hat{\delta}(q_0, a)$ by definition of δ^1.

Induction: $|x| > 1$. Let x = wa for symbol a in Σ. Then $\delta^1(q_0, wa) = \delta(\delta^1(q_0, w), a)$.

By the induction hypothesis, $\delta^1(q_0, w) = \hat{\delta}(q_0, w)$.

Let $\hat{\delta}(q_0, w) = P$. We must show that

$\delta^1(P, a) = \hat{\delta}(q_0, wa)$.

But $\delta^1(p, a) = \cup_{q \text{ in } P} \delta^1(q, a) = \cup_{q \text{ in } P} \delta(q, a)$.

Then as $P = \hat{\delta}(q_0, w)$ we have $\cup_{q \text{ in } P} \hat{\delta}(q, a) = \hat{\delta}(q_0, wa)$

By rule (2) in the definition of $\hat{\delta}$. Thus $\delta^1(q_0, wa) = \hat{\delta}(q_0, wa)$

To complete the proof we shall show that $\delta^l(q_0, x)$ contain a state of F^l if and only if $\hat{\delta}(q_0, x)$ contains a state of F. if $x = \varepsilon$, this statement is immediate from the definition F^l. That is, $\delta^l(q_0, \varepsilon) = \{q_0\}$, and q_0 is placed in F^l whenever $\hat{\delta}(q_0, \varepsilon)$, which is ε-CLOSURE(q_0), contains a state(possibly q_0) in F. If $x \neq \varepsilon$, then $x = wa$ for some symbol a. If $\delta^l(q_0, x)$ contains a state of F, then surely $\delta^l(q_0, x)$ contains the same state in F^l. Conversely, if $\delta^l(q_0, x)$ contains a state in F^l other than q_0, then $\hat{\delta}(q_0, x)$ contains a state in F. If $\delta^l(q_0, x)$ contains q_0, and q_0 is not in F, then as $\hat{\delta}(q_0, x) = \varepsilon$-CLOSURE$(\delta(\hat{\delta}(q_0, w), a))$, the state in ε-CLOSURE(q_0) an F must be in $\hat{\delta}(q_0, x)$.

2.4.2.1 CONVERSION FROM NFA WITH ε-TRANSITION TO NFA WITHOUT ε-TRANSITIONS

1) Find ε-CLOSURE(q) of all states of M ε-CLOSURE(q) function returns a set of NFA states including q, that are reachable from input symbol ε alone.

2) Determine the extended transition function $\hat{\delta}$ as follows:
   a) $\hat{\delta}(q, \varepsilon) = \varepsilon$-CLOSURE$(q)$
   b) $\hat{\delta}(q, a) = \varepsilon$-CLOSURE$(\delta(\hat{\delta}(q, \varepsilon), a))$

3) Final states of M^l includes all states whose ε-CLOSURE contains a final state of M.

Example 2.3

Convert NFA-ε to ordinary NFA for the following figure

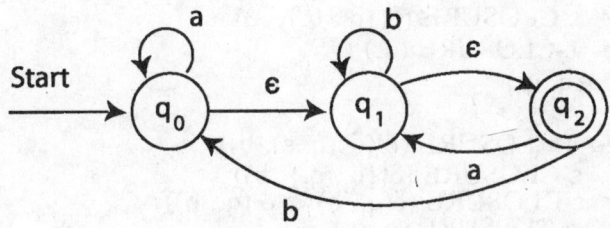

Solution:

Let NFA with ε-transitions be $M = (Q, \Sigma, \delta, q_0, F)$ where $Q = \{q_0, q_1, q_2\}$
$\Sigma = \{a, b\}$
$F = q_2$

Step 1 : ε-CLOSURE $(q_0) = \{q_0, q_1, q_2\}$
ε-CLOSURE $(q_1) = \{q_1, q_2\}$
ε-CLOSURE $(q_2) = \{q_2\}$

Transition table for NFA- ε

Q \ Σ	a	b	ε
→ q_0	q_0	-	q_1
q_1	-	q_1	q_2
q_2	q_1	q_0	-

Step 2: Use the transition function properties

a) $\hat{\delta}(q, \varepsilon) = \varepsilon\text{-CLOSURE}(q)$

b) $\hat{\delta}(q, a) = \varepsilon\text{-CLOSURE}(\delta(\hat{\delta}(q_0, \varepsilon), a))$

1) $\hat{\delta}(q_0, a) = \varepsilon\text{-CLOSURE}(\delta(\hat{\delta}(q_0, \varepsilon), a))$
$= \varepsilon\text{-CLOSURE}(\delta(\{q_0, q_1, q_2\}, a))$
$= \varepsilon\text{-CLOSURE}(q_0, q_1)$
$= \{q_0, q_1, q_2\}$

2) $\hat{\delta}(q_0, b) = \varepsilon\text{-CLOSURE}(\delta(\hat{\delta}(q_0, \varepsilon), b))$
$= \varepsilon\text{-CLOSURE}(\delta(\{q_0, q_1, q_2\}, b))$
$= \varepsilon\text{-CLOSURE}(q_0, q_1)$
$= \{q_0, q_1, q_2\}$

3) $\hat{\delta}(q_1, a) = \varepsilon\text{-CLOSURE}(\delta(\hat{\delta}(q_1, \varepsilon), a))$
$= \varepsilon\text{-CLOSURE}(\delta(\{q_1, q_2\}, a))$
$= \varepsilon\text{-CLOSURE}(q_1)$
$= \{q_1, q_2\}$

4) $\hat{\delta}(q_1, b) = \varepsilon\text{-CLOSURE}(\delta(\hat{\delta}(q_1, \varepsilon), b))$
$= \varepsilon\text{-CLOSURE}(\delta(\{q_1, q_2\}, b))$
$= \varepsilon\text{-CLOSURE}(\delta(q_1, b) \cup \delta(q_2, b))$
$= \varepsilon\text{-CLOSURE}(q_0, q_1)$
$= \{q_0, q_1, q_2\}$

5) $\hat{\delta}(q_2, a) = \varepsilon\text{-CLOSURE}(\delta(\hat{\delta}(q_2, \varepsilon), a))$
$= \varepsilon\text{-CLOSURE}(\delta(q_2, a))$
$= \varepsilon\text{-CLOSURE}(q_1)$
$= \{q_1, q_2\}$

6) $\hat{\delta}(q_2, b) = \varepsilon\text{-CLOSURE}(\delta(\hat{\delta}(q_2, \varepsilon), b))$
$= \varepsilon\text{-CLOSURE}(\delta(q_2, b)$

$$= \varepsilon-\text{CLOSURE }(q_0)$$
$$= \{q_0, q_1, q_2\}$$

Step 3 : The final states of M^1 are q_0, q_1, q_2 because the final state of M is q_2 and is included in ε–CLOSUREs of q_0, q_1, q_2

$$\therefore F^1 = \{q_0, q_1, q_2\}$$

Transition table for NFA without ε–moves

Q \\ Σ	a	b
→ q_0	$\{q_0, q_1, q_2\}$	$\{q_0, q_1, q_2\}$
q_1	$\{q_1, q_2\}$	$\{q_0, q_1, q_2\}$
q_2	$\{q_1, q_2\}$	$\{q_0, q_1, q_2\}$

Transition diagram without ε– moves

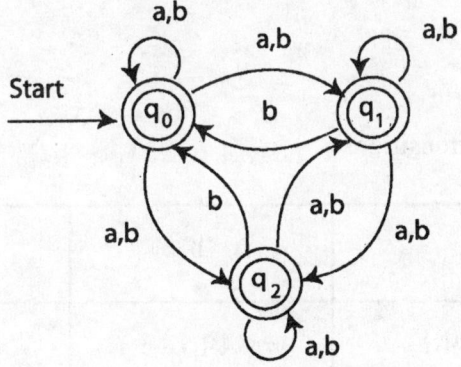

2.5 NFA TO DFA CONVERSION

2.5.1 INDIRECT METHOD

Example 2.3
For the NFA–ε given, check whether the string "aabbabab" is accepted or not, if

accepted write the transition path. Find equivalent NFA without ε−transitions, explain the procedure used and check the string given on your new NFA.

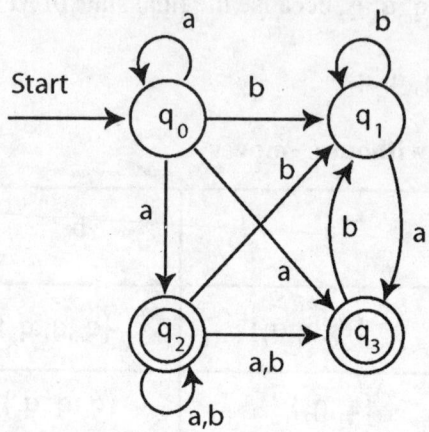

Solution:

Let the given NFA− ε be M=(Q, Σ,δ,q_0,F)

where   Q = $\{q_0, q_1, q_2, q_3\}$

Σ = $\{a,b\}$,

q_0 = q_0 ,

F = $\{q_3\}$

Transition table for NFA− ε transitions

Q \ Σ	a	b	ε
→ q_0	$\{q_0, q_2\}$	$\{q_1\}$	∅
q_1	$\{q_3\}$	$\{q_1\}$	∅
q_2	$\{q_2\}$	$\{q_2\}$	$\{q_3\}$
q_3	∅	$\{q_1\}$	∅

Acceptance of the string by NFA with ε–moves

i) **aabbabab** : The given NFA accepts set of strings consisting of **a**'s and **b**'s only. But the string "aabbabab" is not accepted because the transition path does not exist. **aabbabab** : The given NFA accepts set of strings consisting of **a**'s and **b**'s only. But the string "aabbabab" is not accepted because the transition path does not exist.

ii) **Acceptance of the input string "aabbabab" using NFA– ε**

We know that

i) $\hat{\delta}$ (q_0, ε) = ε –CLOSURE(q_0)

ii) $\hat{\delta}$ (q_0,a) = ε –CLOSURE($\delta(\hat{\delta}$ $(q_0, \varepsilon),a))$

iii) $\hat{\delta}$ (q_0,xa) = ε –CLOSURE($\delta(\hat{\delta}$ $(q_0,x),a))$

$\qquad$ ε –CLOSURE(q_0)={q_0}

$\qquad$ ε–CLOSURE(q_1)={q_1}

$\qquad$ ε –CLOSURE(q_2)={q_2,q_3}

$\qquad$ ε –CLOSURE(q_3)={q_3}

$\hat{\delta}$ (q_0, ε) = ε –CLOSURE(q_0) = {q_0}

$\hat{\delta}$ (q_0, a) = ε –CLOSURE($\delta(\hat{\delta}$ $(q_0, \varepsilon), a))$

$\qquad$ = ε –CLOSURE($\delta(q_0, a)$)

$\qquad$ = ε –CLOSURE({q_0, q_2})

$\qquad$ = ε –CLOSURE(q_0) ∪ ε –CLOSURE(q_2)

$\qquad$ = {q_0}∪{q_2, q_3}

$\qquad$ = {q_0, q_2, q_3}

$\hat{\delta}$ (q_0, aa) = ε –CLOSURE($\delta(\{q_0, q_2, q_3\}, a)$)

$\qquad$ = ε –CLOSURE({q_0, q_2})

$\qquad$ = {q_0, q_2, q_3}

$\hat{\delta}$ (q_0, aab) = ε –CLOSURE($\delta(\{q_0, q_2, q_3\}, b)$)

$\qquad$ = ε –CLOSURE({q_1, q_2})

$\qquad$ = {q_1, q_2, q_3}

$\hat{\delta}$ $(q_0, aabb)$ = ε –CLOSURE($\delta(\{q_1, q_2, q_3\}, b)$

$\qquad$ = ε –CLOSURE({q_1, q_2})

$\qquad$ = {q_1, q_2, q_3}

$\hat{\delta}$ $(q_0, aabba)$ = ε –CLOSURE($\delta(\{q_1, q_2, q_3\}, a)$)
        = ε –CLOSURE($\{q_2, q_3\}$)
        = $\{q_2, q_3\}$

$\hat{\delta}$ $(q_0, aabbab)$ = ε –CLOSURE($\delta(\{q_2, q_3\}, b)$)
        = ε – CLOSURE($\{q_1, q_2\}$)
        = $\{q_1, q_2, q_3\}$

$\hat{\delta}$ $(q_0, aabbaba)$ = ε –CLOSURE($\delta(\{q_1, q_2, q_3\}, a)$)
        = ε –CLOSURE($\{q_2, q_3\}$)
        = $\{q_2, q_3\}$

$\hat{\delta}$ $(q_0, aabbabab)$ = ε –CLOSURE($\delta(\{q_2, q_3\}, b)$)
        = ε –CLOSURE$\{q_1, q_2\} = \{q_1, q_2, q_3\}$

$\therefore$ The string "aabbabab" is accepted by NFA with ε –moves.

The transition path for "aabbabab" is as follows.

$q_0 \xrightarrow{a} q_2 \xrightarrow{a} q_2 \xrightarrow{b} q_2 \xrightarrow{b} q_2 \xrightarrow{a} q_2 \xrightarrow{b} q_2 \xrightarrow{a} q_2 \xrightarrow{b} q_2 \xrightarrow{\in} q_3 \xrightarrow{} FS$

iii)Construction of NFA without ε –moves

Let NFA without ε –transitions be $M^1 = (Q^1, \Sigma, \delta^1, q_0^1, F^1)$

Step 1 : ε –CLOSURE(q_0) = $\{q_0\}$
        ε –CLOSURE(q_1) = $\{q_1\}$
        ε –CLOSURE(q_2) = $\{q_2, q_3\}$
        ε –CLOSURE(q_3) = $\{q_3\}$

Step 2 : Transition function δ^1 is calculated as follows.

$\hat{\delta}$ $(q, \varepsilon)= \varepsilon$ –CLOSURE(q)
$\hat{\delta}$ $(q, a) = \varepsilon$ –CLOSURE($\delta(\hat{\delta}(q, \varepsilon), a)$)

1) $\hat{\delta}$ $(q_0, a) = \varepsilon$ –CLOSURE($\delta(\hat{\delta}(q_0, \varepsilon), a)$)
        = ε –CLOSURE($\delta(q_0, a)$)

$$= \varepsilon -\text{CLOSURE}(\{q_0, q_2\})$$
$$= \{q_0, q_2, q_3\}$$

2) $\hat{\delta}$ $(q_0, b) = \varepsilon -\text{CLOSURE}(\delta\ (\hat{\delta}\ (q_0, \varepsilon), b))$
$$= \varepsilon -\text{CLOSURE}(\delta(q_0, b))$$
$$= \varepsilon -\text{CLOSURE}(q_1)$$
$$= \{q_1\}$$

3) $\hat{\delta}$ $(q_1, a) = \varepsilon -\text{CLOSURE}(\delta(\hat{\delta}\ (q_1, \varepsilon), a))$
$$= \varepsilon -\text{CLOSURE}(\delta(q_1, a))$$
$$= \varepsilon -\text{CLOSURE}(\{q_3\})$$
$$= \{q_3\}$$

4) $\hat{\delta}$ $(q_1, b) = \varepsilon -\text{CLOSURE}(\delta\ (\hat{\delta}\ (q_1, \varepsilon), b))$
$$= \varepsilon -\text{CLOSURE}(\delta(q_1, b))$$
$$= \varepsilon -\text{CLOSURE}(\{q_1\}$$
$$= \{q_1\}$$

5) $\hat{\delta}$ $(q_2, a) = \varepsilon -\text{CLOSURE}(\delta\ (\hat{\delta}\ (q_2, \varepsilon), a))$
$$= \varepsilon -\text{CLOSURE}(\delta(\{q_2, q_3\}, a))$$
$$= \varepsilon -\text{CLOSURE}(q_2)$$
$$= \{q_2, q_3\}$$

6) $\hat{\delta}$ $(q_2, b) = \varepsilon -\text{CLOSURE}(\delta\ (\hat{\delta}\ (q_2, \varepsilon), b))$
$$= \varepsilon -\text{CLOSURE}(\delta(\{q_2, q_3\}, b))$$
$$= \varepsilon -\text{CLOSURE}(\{q_1, q_2\})$$
$$= \{q_1, q_2, q_3\}$$

7) $\hat{\delta}$ $(q_3, a) = \varepsilon -\text{CLOSURE}(\delta(\hat{\delta}\ (q_3, \varepsilon), a))$
$$= \varepsilon -\text{CLOSURE}(\delta(q_3, a))$$
$$= \varepsilon -\text{CLOSURE}(\emptyset)$$
$$= \emptyset$$

8) $\hat{\delta}$ $(q_3, b) = \varepsilon -\text{CLOSURE}(\delta(\hat{\delta}\ (q_3, \varepsilon), b))$
$$= \varepsilon -\text{CLOSURE}(\delta(q_3, b))$$

$$= \varepsilon -\text{CLOSURE}(q_1)$$
$$= \{q_1\}$$

Step 3 : Final states of M^1

The final state of M is q_3 and ε –CLOSUREs of q_2 and q_3 contain q_3. Therefore final states of M^1 are q_2 and q_3.

Q \ Σ	a	b
→ q_0	$\{q_0, q_2, q_3\}$	$\{q_1\}$
q_1	$\{q_3\}$	$\{q_1\}$
q_2	$\{q_2, q_3\}$	$\{q_1, q_2, q_3\}$
⊙ q_3	$\varnothing$	$\{q_1\}$

Transition diagram for NFA without ε–moves

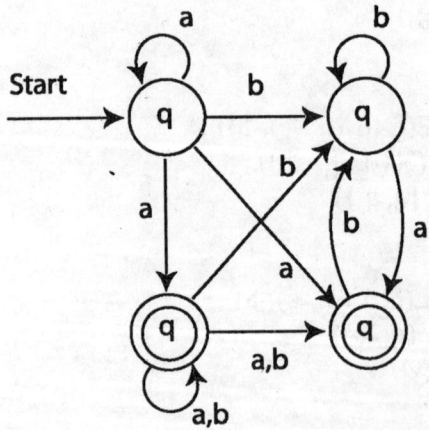

Acceptance of string "aabbabab" by NFA without ε–transitions

$$\hat{\delta}(q_0, \varepsilon) = q_0 \qquad \text{(since there are no other } \varepsilon \text{ –moves from } q_0\text{)}$$

$\hat{\delta}$ $(q_0, a) = \delta(\hat{\delta} (q_0, \varepsilon), a) = \delta(q_0, a) = \{q_0, q_2, q_3\}$

$\hat{\delta}$ $(q_0, aa) = \delta(\hat{\delta} (q_0, a), a) = \delta(\{q_0, q_2, q_3\}, a) = \{q_0, q_2, q_3\}$

$\hat{\delta}$ $(q_0, aab) = \delta(\hat{\delta} (q_0, aa), b) = \delta(\{q_0, q_2, q_3\}, b) = \{q_1, q_2, q_3\}$

$\hat{\delta}$ $(q_0, aabb) = \delta(\hat{\delta} (q_0, aab), b) = \delta(\{q_1, q_2, q_3\}, b) = \{q_1, q_2, q_3\}$

$\hat{\delta}$ $(q_0, aabba) = \delta(\hat{\delta} (q_0, aabb), a) = \delta(\{q_1, q_2, q_3\}, a) = \{q_2, q_3\}$

$\hat{\delta}$ $(q_0, aabbab) = \delta(\{q_2, q_3\}, b) = \{q_1, q_2, q_3\}$

$\hat{\delta}$ $(q_0, aabbaba) = \delta(\{q_1, q_2, q_3\}, a) = \{q_2, q_3\}$

$\hat{\delta}$ $(q_0, aabbabab) = \delta(\{q_2, q_3\}, b) = \{q_1, q_2, q_3\}$

since $\{q_1, q_2, q_3\}$ consists of final states q_2 and q_3.

∴ The string is accepted by both NFA with and without ε –moves

transition path for NFA without ε–moves is

Transition path for NFA without ε moves is

$q_0 \xrightarrow{a} q_2 \xrightarrow{a} q_2 \xrightarrow{b} q_2 \xrightarrow{b} q_2 \xrightarrow{a} q_2 \xrightarrow{b} q_2 \xrightarrow{a} q_2 \xrightarrow{b} q_2$ or q_3

Example 2.4

For the NFA– ε given, check whether the string "aabbabab" is accepted or not, if accepted write the transition path. Find equivalent NFA without ε–transitions, explain the procedure used and check the string given on your new NFA.

Take the string as "aabbabab" and verify whether it is accepted or not

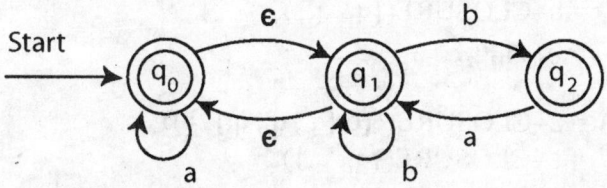

Solution : Transition table for NFA with ε moves

Q \ Σ	a	b	ε
→ (q₀)	{q₀}	-	{q₁}
(q₁)	-	{q₁,q₂}	{q₀}
(q₂)	{q₁}	-	-

Acceptance of a string by NFA with ε-transitions

Consider the string "**aabbabab**"

We have

       i) $\hat{\delta}(q_0, \varepsilon)$ = ε–CLOSURE(q_0)

      ii) $\delta(q_0, a)$ = ε–CLOSURE($\delta(\hat{\delta}(q_0, \varepsilon), a)$)

ε–CLOSURE(q_0) = {q_0, q_1}

ε–CLOSURE(q_1) = {q_0, q_1}

ε–CLOSURE(q_2) = {q_2}

$\hat{\delta}(q_0, \varepsilon)$ = ε–CLOSURE(q_0) = {q_0, q_1}

$\hat{\delta}(q_0, a)$ = ε–CLOSURE($\delta(\hat{\delta}(q_0, \varepsilon), a)$)

              = ε–CLOSURE($\delta(\hat{\delta}\{q_0, q_1\}, a)$)

              = ε–CLOSURE({q_0})

              = {q_0, q_1}

$\hat{\delta}(q_0, aa)$ = ε–CLOSURE($\delta(\hat{\delta}(q_0, a), a)$)

            = ε–CLOSURE(($\delta\{q_0, q_1\}, a$))

            = ε–CLOSURE({q_0})

            = {q_0, q_1}

$\hat{\delta}(q_0, aab)$ = ε–CLOSURE($\delta(\{q_0, q_1\}, b)$)

             = ε–CLOSURE({q_1, q_2})

             = {q_0, q_1, q_2}

$\hat{\delta}(q_0, aabb)$ = ε–CLOSURE($\delta(\hat{\delta}\{q_0, q_1, q_2\}, b)$)

              = ε–CLOSURE({q_1, q_2})

              = {q_0, q_1, q_2}

$$\hat{\delta}(q_0, aabba) = \varepsilon\text{-CLOSURE}(\delta(\hat{\delta}\{q_0, q_1, q_2\}, a))$$
$$= \varepsilon\text{-CLOSURE}(\{q_0, q_1\})$$
$$= \{q_0, q_1\}$$

$$\hat{\delta}(q_0, aabbab) = \varepsilon\text{-CLOSURE}(\{q_1, q_2\})$$
$$= \{q_0, q_1, q_2\}$$

$$\hat{\delta}(q_0, aabbaba) = \varepsilon\text{-CLOSURE}(\{q_0, q_1\})$$
$$= \{q_0, q_1\}$$

$$\hat{\delta}(q_0, aabbabab) = \varepsilon\text{-CLOSURE}(\{q_1, q_2\})$$
$$= \{q_0, q_1, q_2\}$$

The set $\{q_0, q_1, q_2\}$ consists of final states q_0, q_1, q_2.

$\therefore$ The string "**aabbabab**" is accepted by NFA–ε

Conversion of NFA–ε to NFA without ε

Step 1 : Find ε–CLOSURE of all states of NFA–ε

$$\varepsilon\text{-CLOSURE}(q_0) = \{q_0, q_1\}$$
$$\varepsilon\text{-CLOSURE}(q_1) = \{q_0, q_1\}$$
$$\varepsilon\text{-CLOSURE}(q_2) = \{q_2\}$$

Step 2 : Calculation δ–transitions

i) $\hat{\delta}(q, \varepsilon) = \varepsilon$–CLOSURE(q)

ii) $\hat{\delta}(q, a) = \varepsilon$–CLOSURE($\delta(\hat{\delta}(q, \varepsilon), a)$)

1) $\hat{\delta}(q_0, a) = \varepsilon$–CLOSURE($\delta(\hat{\delta}(q_0, \varepsilon), a)$)
$$= \varepsilon\text{-CLOSURE}(\delta\{q_0, q_1\}, a))$$
$$= \varepsilon\text{-CLOSURE}(\{q_0\})$$
$$= \{q_0, q_1\}$$

2) $\hat{\delta}(q_0, b) = \varepsilon$–CLOSURE($\delta(\hat{\delta}(q_0, \varepsilon), b)$)
$$= \varepsilon\text{-CLOSURE}(\delta(\{q_0, q_1\}, b))$$
$$= \varepsilon\text{-CLOSURE}(\{q_1, q_2\})$$
$$= \{q_0, q_1, q_2\}$$

3) $\hat{\delta}\,(q_1, a)$ = ε –CLOSURE($\delta(\hat{\delta}\,(q_1, \varepsilon), a)$)

               = ε –CLOSURE($\delta(\{q_0, q_1\}, a)$)

               = ε –CLOSURE(q_0)

               = $\{q_0, q_1\}$

4) $\hat{\delta}\,(q_1, b)$ = ε –CLOSURE($\delta(\hat{\delta}\,(q_1, \varepsilon), b)$)

               = ε –CLOSURE($\delta(\{q_0, q_1\}, b)$)

               = ε –CLOSURE($\{q_1, q_2\}$)

               = $\{q_0, q_1, q_2\}$

5) $\hat{\delta}\,(q_2, a)$ = ε –CLOSURE($\delta\,(\hat{\delta}\,(q_2, \varepsilon), a)$)

               = ε –CLOSURE($\delta(q_2, a)$)

               = ε –CLOSURE(q_1)

               = $\{q_0, q_1\}$

6) $\hat{\delta}\,(q_2, b)$ = ε –CLOSURE($\delta\,(\hat{\delta}\,(q_2, \varepsilon), b)$)

               = ε–CLOSURE($\delta(q_2, b)$)

               = ε –CLOSURE($\emptyset$)

               = $\emptyset$

Step 3 : calculate the final state of M^1

The final state of **M** are q_0, q_1, q_2, these q_0, q_1, q_2 are in ε –CLOSUREs of q_0, q_1, q_2 . Therefore the final states of M^1 are q_0, q_1, q_2 .

Q \ Σ	a	b
→ q_0	$\{q_0, q_1\}$	$\{q_0, q_1\}$
q_1	$\{q_0, q_1\}$	$\{q_0, q_1\}$
q_2	$\{q_0, q_1\}$	$\emptyset$

Transition diagram

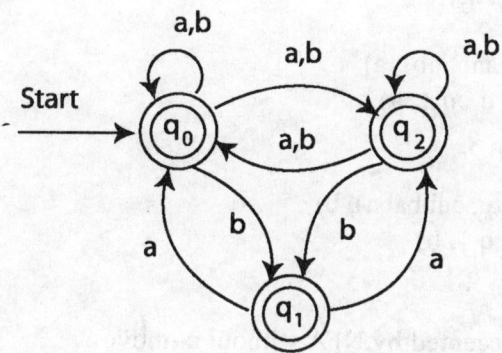

Acceptance of the string 'aabbabab'

$$\hat{\delta}(q_0, \varepsilon) = q_0$$

$$\begin{aligned}
\hat{\delta}(q_0, a) &= \delta(\hat{\delta}(q_0, \varepsilon), a)) \\
&= \delta(q_0, a) \\
&= \{q_0, q_1\}
\end{aligned}$$

$$\begin{aligned}
\hat{\delta}(q_0, aa) &= \delta(\hat{\delta}(q_0, a), a) \\
&= \delta(\{q_0, q_1\}, a) \\
&= \{q_0, q_1\}
\end{aligned}$$

$$\begin{aligned}
\hat{\delta}(q_0, aab) &= \delta(\hat{\delta}(q_0, aa), b) \\
&= \delta(\{q_0, q_1\}, b) \\
&= \{q_0, q_1, q_2\}
\end{aligned}$$

$$\begin{aligned}
\hat{\delta}(q_0, aabb) &= \delta(\hat{\delta}(q_0, aab), b) \\
&= \delta(\{q_0, q_1, q_2\}, b) \\
&= \{q_0, q_1, q_2\}
\end{aligned}$$

$$\begin{aligned}
\hat{\delta}(q_0, aabba) &= \delta(\hat{\delta}(q_0, aabb), a) \\
&= \delta(\{q_0, q_1, q_2\}, a) \\
&= \{q_0, q_1\}
\end{aligned}$$

$$\hat{\delta}\ (q_0, \text{aabbab})\ = \delta(\hat{\delta}\ (q_0, \text{aabba}), b)$$
$$= \delta(\{q_0, q_1\}, b)$$
$$= \{q_0, q_1, q_2\}$$

$$\hat{\delta}\ (q_0, \text{aabbaba}) = \delta((q_0, \text{aabbab}), a)$$
$$= \delta(\{q_0, q_1, q_2\}, a)$$
$$= \{q_0, q_1, q_2\}$$

$$\hat{\delta}\ (q_0, \text{aabbabab}) = \delta(\hat{\delta}\ (q_0, \text{aabbaba}), b)$$
$$= \delta(\{q_0, q_1\}, b)$$
$$= \{q_0, q_1, q_2\}$$

$\therefore$ The string '**aabbabab**' is accepted by NFA without ε–moves.

2.5 CONVERSION OF NFA ε TO DFA

2.5.1 CONVERSION OF NFA WITH ε -TRANSITIONS TO DFA(INDIRECT METHOD) PROCEDURE FOR HERE INCLUDE STEPI AND STEP II

Step I: Convert NFA ε to NFA
StepII: Convert NFA to DFA

2.5.2 PROCEDURE TO CONVERT NFA WITH ε-TRANSITIONS TO DFA (DIRECT METHOD)

1) Find ε-CLOSURE(q) for all states of NFA with ε-moves

2) $q_0{}^!= \varepsilon$-CLOSURE(q_0) where q_0 is the initial state of **M**.

3) $\delta^!(q_0{}^!, a) = \varepsilon$-CLOSURE $(\delta\ (q_0{}^!, \infty))$

   where $q_0{}^!$ is the set of states

   i.e., $\delta^!([q_0, q_1........q_i], a) = \varepsilon$- CLOSURE($\delta\ (\{q_0, q_1........q_i\}, a)$)

4) Repeat step 3 for each new state found in step 3

5) $F^!$ is the set of all states of DFA, containing at least one final state of NFA - ε.

NOTE: We can observe that $[q_0, q_1........q_i]$ is a single state corresponding to a set of states of NFA.

Example 2.5
Convert the following NFA- ε to an equivalent DFA.

Present state	Next state		Epsilon ε
	0	1	
→a	b	{ b,d}	–
b	–	d	c
c	e	–	–
d	{ b,c}	–	e
ⓔ	e	e	–

Solution : Let the given NFA with ε- transitions be **M = (Q, Σ, δ, q₀, F)**

where $Q = \{a, b, c, d, e\}$,

$\Sigma = \{0,1\}$

$\delta : Q \times (\Sigma \cup \{\varepsilon\}) \rightarrow 2^Q$, $q_0 = a$, $F = \{e\}$

Converting NFA with ε-transitions to NFA without ε-transitions:

Transition diagram for NFA

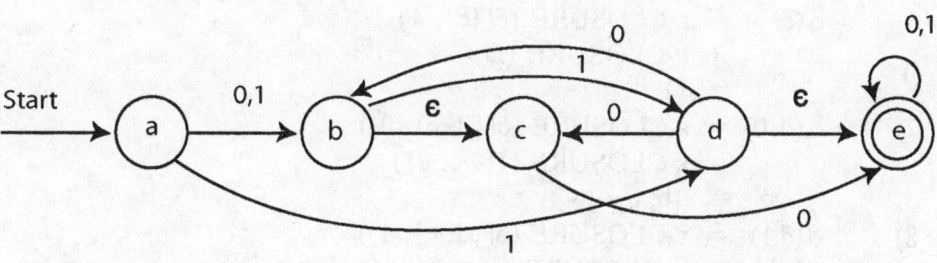

ε-CLOSURE of all states

ε-CLOSURE (a) = {a}

ε-CLOSURE (b) = {b, c}

ε-CLOSURE (c) = {c}

ε-CLOSURE (d) = {d, e}

ε-CLOSURE (e) = {e}

(Using ε-CLOSURE of the states and transition table , we can find the transitions for new NFA)

Transition function is determined as follows:

1) $\hat{\delta}(a,0)$ = ε-CLOSURE ($\delta(\hat{\delta}((a, \ \varepsilon), 0)$

            = ε-CLOSURE ($\delta($ ε-CLOSURE (a), 0))

            = ε-CLOSURE ($\delta(a, \ 0)$)

            = ε-CLOSURE (b)

            = {b, c}   (Since $\hat{\delta}(a, \ \varepsilon)$= ε-CLOSURE (a))

2) $\hat{\delta}(a,1)$ = ε-CLOSURE ($\delta(\hat{\delta}(a, \ \varepsilon), 1)$

            = ε-CLOSURE ($\delta($ ε-CLOSURE (a), 1))

            = ε-CLOSURE ($\delta(a, 1)$)

            = ε-CLOSURE (b, d)

            = ε-CLOSURE (b) $\cup$ ε-CLOSURE(d)

            = {b, c} $\cup$ {d, e}

            = {b, c, d, e}

3) $\hat{\delta}(b,0)$ = ε-CLOSURE ($\delta(\{b, c\}, 0)$)

            = ε-CLOSURE (e)

            = {e}

4) $\hat{\delta}(b,1)$ = ε-CLOSURE ($\delta(\{b, c\}, 1)$)

            = ε-CLOSURE (d)

            = {d, e}

5) $\hat{\delta}(c,0)$ = ε-CLOSURE ($\delta(\{c\}, 0)$)

            = ε-CLOSURE (e)

            = {e}

6) $\hat{\delta}(c,1)$ = ε-CLOSURE ($\delta(\{c\}, 1)$)

            = ε-CLOSURE ($\emptyset$)

            = $\emptyset$

7) $\hat{\delta}(d,0)$ = ε-CLOSURE ($\delta(\{d, e\}, 0)$)

            = ε-CLOSURE ({b, c, e})

            = {b, c, e}

8) $\hat{\delta}(d,1)$ = ε-CLOSURE ($\delta(\{d, e\}, 1)$)

            = ε-CLOSURE (e)

            = {e}

9) $\hat{\delta}(e,0)$ = ε-CLOSURE ($\delta(e, 0)$)

            = ε-CLOSURE (e)

            = {e}

10) $\hat{\delta}(e,1)$ = ε-CLOSURE ($\delta($ ε-CLOSURE(e)), 1))

            = ε-CLOSURE ($\delta(e, 1)$)

$$= \text{ε-CLOSURE (e)}$$
$$= \{e\}$$

Final states for new NFA are d and e

$$\therefore \text{F}^1 = \{d, e\}$$

∴ Transition table for NFA without ε-moves:

Q \\ Σ	0	1
→a	{b,c}	{b,c,d,e}
b	{e}	{d,e}
c	{e}	∅
(d)	{b,c,e}	{e}
(e)	{e}	{e}

NFA to DFA conversion:

Q \\ Σ	0	1
→[a]	[b,c]	[b,c,d,e]
[b,c]	[e]	[d,e]
([b,c,d,e])	[b,c,e]	[d,e]
([e])	[e]	[e]
([d,e])	[b,c,e]	[e]
([b,c,d])	[e]	[d,e]

∴ [b, c, e] = [b, c]

Transition diagram for DFA

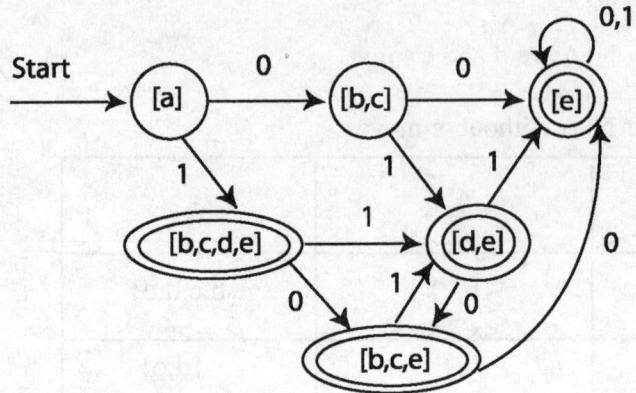

2.6 MINIMIZATION OF FINITE AUTOMATON

Procedure for minimization of FA:

1) Prepare the transition table for the given finite automata **M**.

2) Consider **Q** of **M** and partition **Q** into sets of final and non final states.

3) Take the sets of final and non-final states as group1 and group2.

4) Consider each group and consider the transitions for each group on every input symbol.

5) Partition the group when on an input symbol, if the state from a group goes to a different group.

6) Do not partition the group, when on an input symbol, if the state from a group goes to the same group.

7) Repeat steps 5 and 6, till there is no further partitioning.

8) Finally we get the minimized DFA **M¹**,

   $M^1 = (Q^1, \Sigma, \delta^1, q_0^{\ 1}, F^1)$ which is equivalent to the given DFA.

Example 2.6

Minimize the finite automaton given below and show both given and reduced are equivalent.

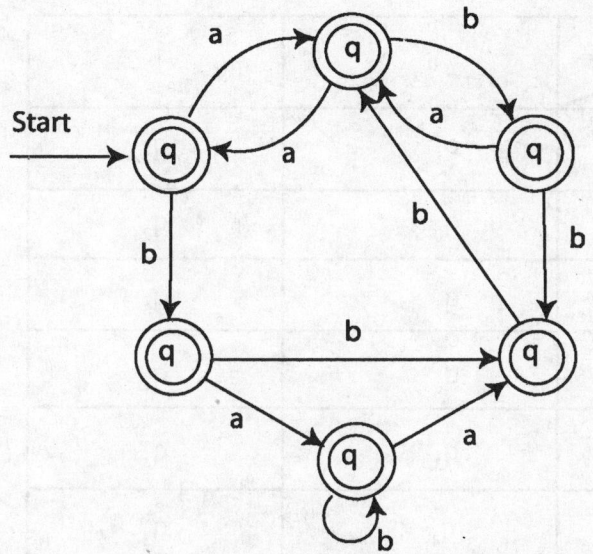

Solution : The transition table for the given DFA is shown below.

Q \ Σ	a	b
→ q_0	q_1	q_5
q_1	q_0	q_2
q_2	q_1	q_3
q_3	–	q_1
q_4	q_3	q_4
q_5	q_4	q_3

From the above table we observe that there is no transition from state q_3 on input **a**. In DFA there is a transition from each state on every input symbol.

To overcome this situation add a dead state q_6 which has transitions to itself on both input symbols **a** and **b**, and we add a transition from q_3 to q_6 on input symbol **a**.

The transition table for the new state is shown below.

Q \ Σ	a	b
→ q_0	q_1	q_5
q_1	q_0	q_2
q_2	q_1	q_3
q_3	q_6	q_1
q_4	q_3	q_4
ⓠ$_5$	q_4	q_3
q_6	q_6	q_6

∴ $Q = \{q_0, q_1, q_2, q_3, q_4, q_5, q_6\}$

Partition **Q** into sets of final and non-final states.

$q_0, q_1, q_2, q_3, q_4, q_6$	q_5
group-1	group-2

Consider transition for each group on every input symbol.

Consider group-1:

$\delta(q_0, a) = q$, $\delta(q_0, b) = q_5$
$\delta(q_1, a) = q_0$, $\delta(q_1, b) = q_2$
$\delta(q_2, a) = q_1$, $\delta(q_2, b) = q_3$
$\delta(q_3, a) = q_6$, $\delta(q_3, b) = q_1$
$\delta(q_4, a) = q_3$, $\delta(q_4, b) = q_4$
$\delta(q_5, a) = q_4$, $\delta(q_5, b) = q_3$
$\delta(q_6, a) = q_6$, $\delta(q_6, b) = q_6$

Partition group-1, since q_0 on input **b** goes to q_5 i.e., different group

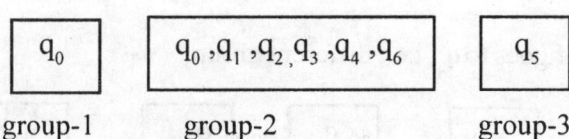

group-1   group-2   group-3

Consider group-2:

$\delta(q_1, a) = q_0$
$\delta(q_2, a) = q_1$
$\delta(q_3, a) = q_6$
$\delta(q_4, a) = q_3$
$\delta(q_6, a) = q_6$

Partition group-2, since q_1 on input **a** goes to q_0 i.e., different group.

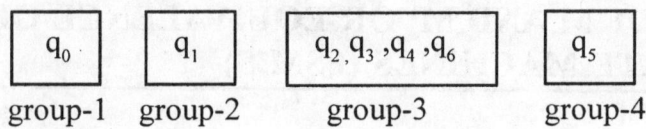

group-1   group-2   group-3   group-4

Consider group-3:

$\delta(q_2, a) = q_1$
$\delta(q_3, a) = q_6$
$\delta(q_4, a) = q_3$
$\delta(q_6, a) = q_6$

Partition group-3, since on input **a**, **q2** goes to **q1** i.e., different group

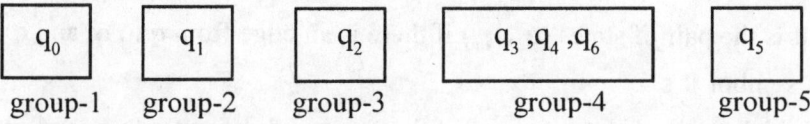

group-1   group-2   group-3   group-4   group-5

Consider group-4 :

$\delta(q_3, a) = q_6,$ $\delta(q_3, b) = q_1$
$\delta(q_4, a) = q_3,$ $\delta(q_4, b) = q_4$
$\delta(q_6, a) = q_6,$ $\delta(q_6, b) = q_6$

Partition group-4, since on input **b**, q_3 goes to **q1** i.e., different group

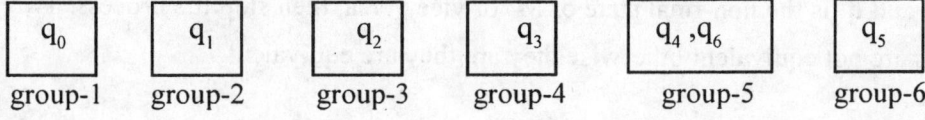

group-1   group-2   group-3   group-4   group-5   group-6

Consider group-5:

$$\delta(q_4, a) = q_3$$
$$\delta(q_6, a) = q_6$$

Partition group-5, since on input **a**, **q4** goes to q_3, i.e., different group

q_0	q_1	q_2	q_3	q_4	q_5	q_6
group-1	group-2	group-3	group-4	group-5	group-6	group-7

The set Q^1 for the reduced DFA is

$$Q^1 = \{q_0, q_1, q_2, q_3, q_4, q_5, q_6\}.$$

∴ The given DFA is itself in reduced form, because both the sets Q and Q^1 consist of same number of states.

2.7 EQUIVALENCE OF M AND M¹ OR EQUIVALENCE OF TWO FINITE STATE MACHINES (FSM'S)

Procedure:

1) Construct comparison table as follows

   a) Comparison table consists of **n+1** columns where **n** is the number of input symbols in Σ. Where $\Sigma = \{0,1\}$.

   b) **Column 1**: It is the ordered pair of states (q, q^1) where $q \in M$, $q^1 \varepsilon M^1$. The first pair in this column is the pair of initial states of two FSM's.

   c) **Column 2**: It is the pair of states (q_0, q_0^1) if there is an edge from **q** to q_0 and q^1 to q_0^1 on input symbol $0 \varepsilon \Sigma$.

   d) **Column 3**: It is the pair of states (q_1, q_1^1) if there is an edge from **q** to q_1 and q^1 to q_1^1 on input symbol $1 \varepsilon \Sigma$.

2) Repeat step 1 for every new pair got in Column 2 and Column 3.

3) Stop the process when there are no new pairs in Column 2 and Column 3.

4) In the construction of comparison table, if we get a pair (q, q^1) where **q** is the final state of **M** and q^1 is the non-final state of M^1 or vice versa, then stop the process. Two FSM's are not equivalent otherwise they are they are equivalent.

2.8 MOORE AND MEALY MACHINES

2.8.1 MOORE MACHINE

A moore machine is a 6-tuple $M = (Q, \Sigma, \Delta, \delta, \lambda, q_0)$

where   Q - number of states

Σ - set of input symbols

Δ - set of output symbols

δ - indicates the transition

q_0 - initial state

λ - a mapping from Q to Δ (Δ is the output alphabet)

The output of **M** in response to input $a_1, a_2, \dots\dots a_n$ where $\mathbf{n \geq 0}$ is $\lambda(q_0), \lambda(q_1), \dots\dots\lambda(q_n)$

where $q_0, q_1, \dots q_n$ are sequence of states such that $\delta(q_{i-1}, a_i) = q_i$ for $\mathbf{1 \leq i \leq n}$

Note: Any Moore machine gives output $\lambda(q_0)$ in response to input ε.

Example 2.8

Check whether the string "0111" is accepted by the Moore machine whose

transition table is given as below. What is the output string?

Present state	Next state		O/p λ
	a = 0	b = 1	
→ q_0	q_3	q_1	0
q_1	q_1	q_2	1
q_2	q_2	q_3	0
q_3	q_3	q_0	0

$$q_0 \xrightarrow{\ 0\ } q_3 \xrightarrow{\ 1\ } q_0 \xrightarrow{\ 1\ } q_1 \xrightarrow{\ 1\ } q_2$$

The transition path is $q_0 \, q_3 \, q_0 \, q_1 \, q_2$.

Since the output is associated with states, the output for the given string is "00010".

Note: If number of inputs is in **n**, then the number of Output symbols in a Moore machine will be **n+1**.

Example 2.8

For the following Moore machine if the input string is "1010" give the output string.

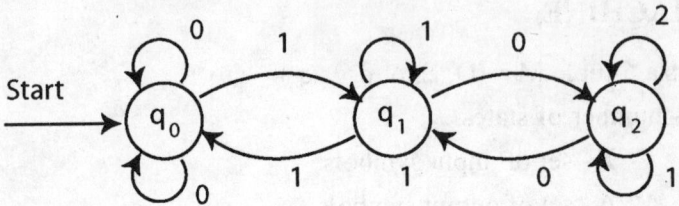

Solution : Transition table:

Present state	Next state		O/p λ
	a = 0	b = 1	
→ q_0	q_0	q_1	0
q_1	q_2	q_0	1
q_2	q_1	q_2	2

For input string "1010"q_0

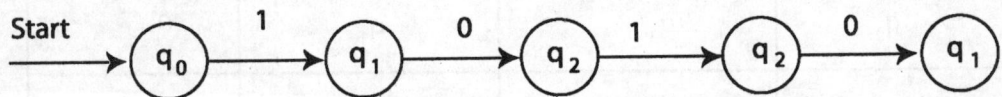

Output String: "01221"

2.8.2 MEALY MACHINE

A mealy machine is a $M = (Q, \Sigma, \Delta, \delta, \lambda, q_0)$

where   **Q** - number of states

Σ - set of input symbols

δ - indicates the transition function

q_0 - initial state

Δ - set of output symbols

λ - a mapping from Q×Σ to Δ

Example 2.9

That is, $\lambda(q, a)$ gives the output associated with the transition from state **q** on input **a**.

The output of **M** in response to input $a_1, a_2, \ldots a_n$ is $\lambda(q_0, a_1), \lambda(q_0, a_2) \ldots \ldots \lambda(q_n, a_n)$ where $q_0, q_1, q_2, \ldots q_n$ is the sequence of states such that $\delta(q_{i-1}, a_i) = q_i$ for $1 \le i \le n$

Example 2.10

Check whether the string **01100** is accepted by the Mealy machine give the output string where $M = (\{q_0, p_0, p_1\}, \{0, 1\}, \{y, n\}, \delta, \lambda, q_0)$. The label **a/b** on an arc from state **p** to **q** indicates that $\delta(p, a) = q$ and $\lambda(p, a) = b$

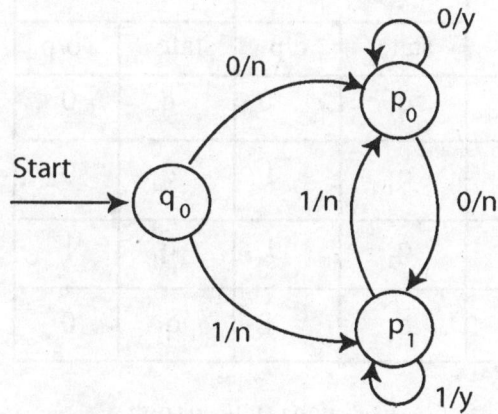

Solution : The transition table for the above diagram is

Present state	Next state δ			
	a = 0		a = 1	
	state	o/p	state	o/p
→q_0	p_0	n	p_1	n
p_0	p_0	y	p_1	n
p_1	p_0	n	p_1	y

The transition path for the input string "01100" is

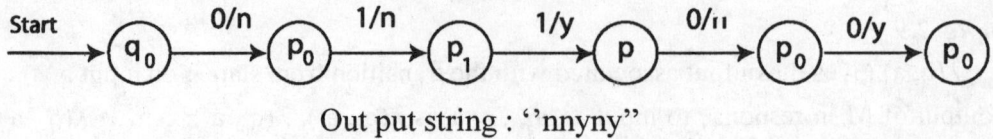

Out put string : "nnyny"

Example 2.11

Write the O/P for the following string "0011", with a mealy machine whose transition table is given as below

Present state	Next state δ			
	$a = 0$		$a = 1$	
	state	o/p	state	o/p
$\rightarrow$ q_1	q_3	0	q_2	0
q_2	q_1	1	q_4	0
q_3	q_2	1	q_1	1
q_4	q_4	1	q_3	0

Solution: The output for the input string "0011" is "0100"

Note: In mealy machine, if the length of the input string is 'n' then the length of the output string is also 'n'.

2.8.3 PROCEDURE FOR CONVERTING MEALY MACHINE TO MOORE MACHINE

1) For any state q_i, determine the number of outputs **n** associated with q_i in the next state column.
2) Split q_0 into **n** different states.
3) Construct the transition table for the new states.

4) Arrange the pair of states and outputs in the next state column. The resulting state is the Moore machine.

5) In the Moore machine, if the output of the initial state, say q_0 is **1** then add a new state say **q** whose transitions are identical to q_0, but whose output is **0**. The transformed table is the Moore machine.

Example 2.12
Construct the Moore machine for the given Mealy machine.

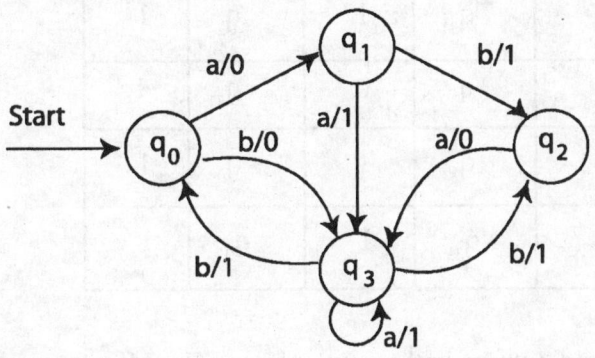

Transition table:

Present state	Next state			
	a		**b**	
	state	o/p	state	o/p
$\rightarrow q_0$	q_1	0	q_3	0
q_1	q_3	1	q_2	1
q_2	q_3	0	q_3	1
q_3	q_3	1	q_0	1

Examine the next state column:
1. q_0 is associated with one output **1**. No need to split it.
2. q_1 is associated with one output **0**. No need to split it.
3. q_2 is associated with one output **1**. No need to split it.
4. q_3 associated with two outputs **0** and **1** so split q_3 into q_{30} and q_{31}.

The transition table constructed for the new states is shown below

Present state	Next state			
	a		b	
	state	o/p	state	o/p
→q_0	q_1	0	q_{30}	0
q_1	q_{31}	1	q_2	1
q_2	q_{30}	0	q_{31}	1
q_{30}	q_{31}	1	q_0	1
q_{30}	q_{31}	1	q_0	1

Arrange the pair of states and outputs in the next column. The Moore machine is shown below.

Present state	Next state		Output
	a	b	
→q_0	q_1	q_{30}	1
q_1	q_{31}	q_2	0
q_2	q_{30}	q_{31}	1
q_{30}	q_{31}	q_0	0
q_{31}	q_{31}	q_0	1

We observe that the output of the initial state q_0 is **1**. This means the Moore machine accepts the null input string (ε), which is not accepted by the Mealy machine.

To overcome this situation, we add a new state say **q** whose transitions are identical to q_0, but whose output is **0**.

The resulting Moore machine is shown below.

Present state	Next state		Output
	a	b	
→ q	q_1	q_{30}	0
q_0	q_1	q_{30}	1
q_1	q_{31}	q_2	0
q_2	q_{30}	q_{31}	1
q_{30}	q_{31}	q_0	0
q_{31}	q_{31}	q_0	1

Transition diagram

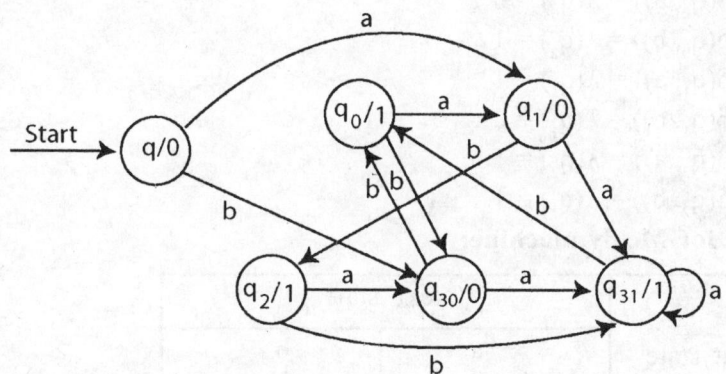

2.8.4 PROCEDURE FOR CONVERTING MOORE MACHINE TO MEALY MACHINE

1) Define the output function λ^1 for the Mealy machine as follows:

$\lambda^1(q, a) = \lambda(\delta(q, a))$ for all states **q** and input symbol **a**.

2) The transition function δ for Mealy machine is same as that of a given Moore machine.

NOTE:

1) In Mealy machine, in next state column some states are associated with different outputs. So split the states which are associated with different outputs.

2) In Moore machine, in present state each state is associated with only one output.

Example 2.13

Construct the Mealy machine for the given Moore machine

Q \ Σ	a	b	Output
q_0	q_0	q_2	0
q_1	q_1	q_0	1
q_2	q_2	q_1	1

Solution:

In the equivalent Mealy machine only the output function λ^1 changes and the transition function δ remains same .

Output function λ^1 is calculated as follows.

$\lambda^1(q_0, a) = \lambda(\delta(q_0, a)) = \lambda(q_0) = 0$

$\lambda^1(q_0, b) = \lambda(\delta(q_0, b)) = \lambda(q_2) = 1$

$\lambda^1(q_1, a) = \lambda(\delta(q_1, a)) = \lambda(q_1) = 1$

$\lambda^1(q_1, b) = \lambda(\delta(q_1, b)) = \lambda(q_1) = 0$

$\lambda^1(q_2, a) = \lambda(\delta(q_2, a)) = \lambda(q_2) = 1$

$\lambda^1(q_2, b) = \lambda(\delta(q_2, b)) = \lambda(q_1) = 1$

The transition table for Mealy machine:

Present state	Next state			
	a		b	
	state	o/p	state	o/p
→ q_0	q_0	0	q_2	1
q_1	q_1	1	q_0	0
q_2	q_2	1	q_1	1

Transition diagram

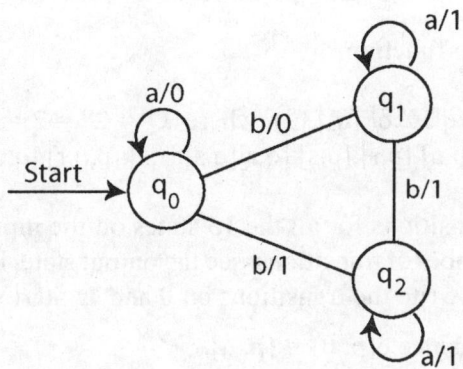

NOTE:

    1) In Mealy machine, in next state column some states are associated with different outputs. So split the states which are associated with different outputs.

    2) In Moore machine, in present state each state is associated with only one output.

SOLVED PROBLEMS

Problem 1. *Construct a DFA equivalent to the NFA, M = ({p, q, r, s},{0,1},δ,p,{s}), where transition table is given below.*

Q ＼ Σ	0	1
→ p	{p,q}	p
q	r	r
r	s	-
ⓢ	s	s

Solution:

    NFA to DFA conversion:

    Given NFA, $M = (Q, \Sigma, \delta, q_0, F)$

where $Q = \{p, q, r, s\}$,

$\Sigma = \{0,1\}$,

$q_0 = p$,

δ = transition function,

$F = \{s\}$.

we construct a DFA $M^1 = (q^1, \Sigma, \delta^1, [q_0], F^1)$, where $Q^1 = 2^Q = 2^4 = 16$

$\therefore Q^1 = \{\emptyset, [p], [q], [r], [s], [p,q], [p,r], [p,s], [q,r][q,s], [r,s], [p,q,r], [q,r,s], [r, s, p], [s, p, q], [p, q, r, s]\}$

If we write the Q^1-transitions for all the **15** states on the inputs **0** and **1**, we get **30** transitions. To reduce the number of transitions, see the output state, if it is old state leave it, otherwise for that new state write the transitions on **0** and **1**. Start with start state **p**.

Consider $\delta^1([p], 0) = [p, q]$, since $\delta(p, 0) = \{p, q\}$

$\delta^1([p], 1) = [p]$, since $\delta(p, 1) = \{p\}$ Since c $\delta^1 : Q^1 \times \Sigma \rightarrow 2^Q$

Here [p, q] is new state, [p] is old state. So write the transitions for [p, q] on 0 and 1

$\delta^1([p, q], 0) = [p, q, r]$, since $\delta (\{p, q\}, 0) = \delta (p, 0) \cup \delta(q, 0) = \{p, q, r\}$

$\delta^1([p, q], 1) = [p, r]$, since $\delta (\{p, q\}, 1) = \{p, q\} \cup \{r\} = \{p, q, r\}$

The final state for M is $F = \{s\}$

The final state for M^1 are $F^1 = \{[s], [p, s], [q, s], [r, s], [q, r, s], [r, s, p], [s, p, q], [p, q, r, s]\}$

Transition table for DFA:

Q ＼ Σ	0	1
→[p]	[p,q]	[p]
[p,q]	[p,q,r]	[p,r]
[p,q,r]	[p,q,r,s]	[p,r]
[p,r]	[p,q,s]	[p]
⬭[p,q,r,s]	[p,q,r,s]	[p,r,s]
⬭[p,q,s]	[p,q,r,s]	[p,r,s]
⬭[p,r,s]	[p,q,s]	[p,s]
⬭[p,s]	[p,q,s]	[p,s]

Transition diagram for DFA:

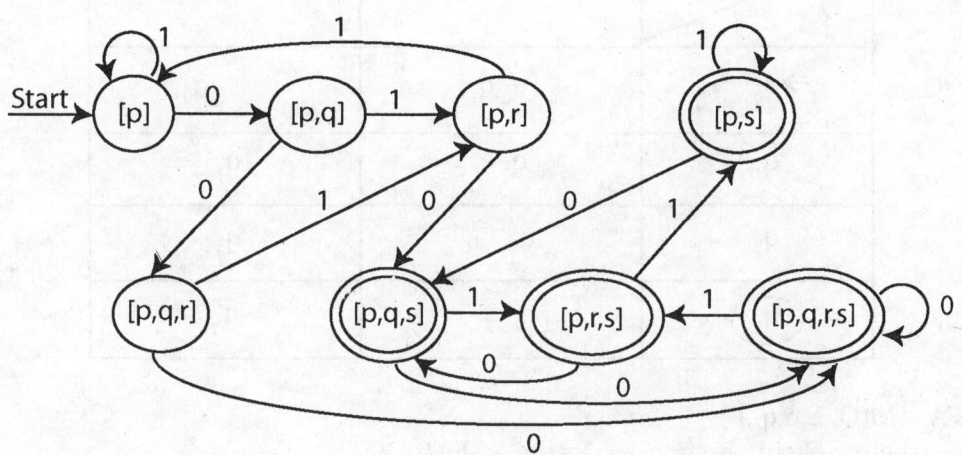

Reduce the number of states: If two rows are identical in the transition diagram then if the corresponding states are both non -final we can remove one row. Otherwise we cannot remove.

NOTE: If two rows are identical, we can remove one row. As the rows corresponding to [p, r, s] and [p, s] are identical and [p, q, r, s] and [p, q, s] are equal and all are final states.

Transition diagram for the reduced states:

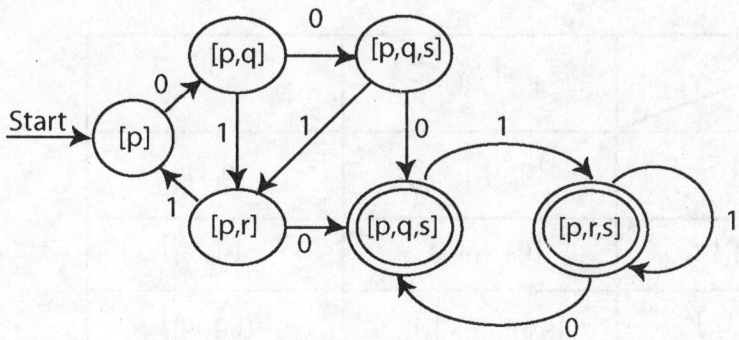

Problem 2. Construct a DFA equivalent to M= ({q₀, q₁, q₂,q₃},{0,1}, , {q₃}) where is given by the fol¹owing table

Q \ Σ	a	b
→q_0	{q_0,q_1}	q_0
q_1	q_2	q_1
q_2	q_3	q_3
q_3	-	q_2

NFA = M(Q, Σ,δ,q_0,F)

    where   Q={q_0, q_1, q_2,q_3},   Σ={0,1},   F={q_3}

    δ:Q× Σ→Q

DFA=M^1(Q^1, Σ,$δ^1$,q_0^1,F^1)

    where Q^1=2^Q=2^4=16.

Q^1={Ø, [q_0], [q_1], [q_2], [q_3], [q_0, q_1,], [q_0, q_2], [q_0, q_3], [q_1, q_2], [q_1, q_2], [q_2,q_3], [q_0, q_1, q_2], [q_0, q_2, q_3], [q_2, q_3, q_1], [q_3, q_0, q_1], [q_0, q_1, q_2, q_3]}

    $δ^1$: Q^1× Σ→2^Q

    q_0→Initial state.

    F^1= {[q_3], [q_0, q_3], [q_1, q_3], [q_1, q_2, q_3], [q_0, q_1, q_2, q_3]

Transition table for $δ^1$ is given below.

Q \ Σ	a	b
→[q_0]	[q_0,q_1]	[q_0]
[q_0,q_1]	[q_0,q_1,q_2]	[q_0,q_1]
[q_0,q_1,q_2]	[q_0,q_1,q_2,q_3]	[q_0,q_1,q_3]
[q_0,q_1,q_3]	[q_0,q_1,q_2]	[q_0,q_1,q_2]
[q_0,q_1,q_2,q_3]	[q_0,q_1,q_2,q_3]	[q_0,q_1,q_2,q_3]

Transition diagram is as follows:

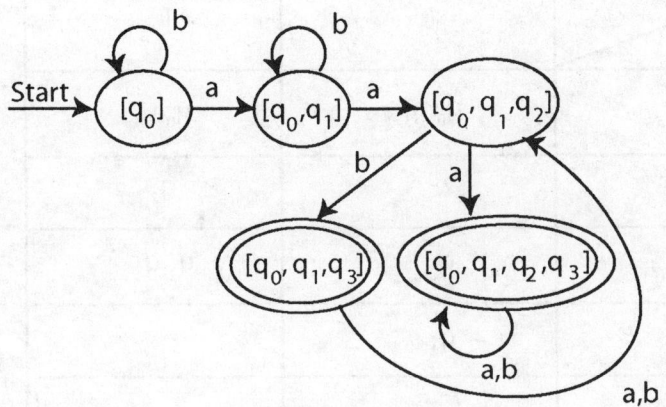

Problem 3. Construct a DFA equivalent to NFA given in the following diagram.

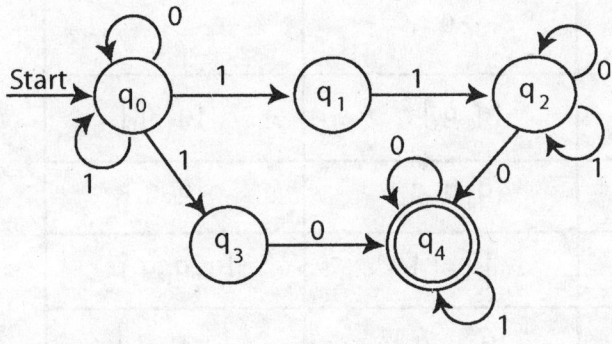

Solution :

Given that NFA=M=($\{q_0, q_1, q_2, q_3, q_4\}$, $\{0, 1\}$, q_0, $\{q_4\}$)
construct DFA, i.e., M^1=(Q^1, Σ, δ^1, q_0, F^1)
where $Q^1=2^Q=2^5=32$ states.

Transition table for NFA is:

Q \ Σ	0	1
→q_0	q_0, q_3	q_0, q_1
q_1	-	q_2
q_2	q_2	q_2, q_4
q_3	q_4	-
$\boxed{q_4}$	q_4	q_4

$\quad$ F= $\{q_4\}$

Transition table for DFA is:

Q \ Σ	0	1
→$[q_0]$	$[q_0, q_3]$	$[q_0, q_1]$
$[q_0, q_3]$	$[q_0, q_3, q_4]$	$[q_0, q_1]$
$[q_0, q_1]$	$[q_0, q_3]$	$[q_0, q_1, q_2]$
$[q_0, q_3, q_4]$	$[q_0, q_3, q_4]$	$[q_0, q_1, q_4]$
$[q_0, q_1, q_2]$	$[q_0, q_2, q_3]$	$[q_0, q_1, q_2, q_4]$
$[q_0, q_1, q_4]$	$[q_0, q_3, q_4]$	$[q_0, q_1, q_2, q_4]$
$[q_0, q_2, q_3]$	$[q_0, q_2, q_3, q_4]$	$[q_0, q_1, q_2, q_4]$
$[q_0, q_1, q_2, q_4]$	$[q_0, q_2, q_3, q_4]$	$[q_0, q_1, q_2, q_4]$
$[q_0, q_2, q_3, q_4]$	$[q_0, q_2, q_3, q_4]$	$[q_0, q_1, q_2, q_4]$

From theorem we have $\delta^1([q_0, q_3], 0) = [q_0, q_3, q_4]$
if $\delta(\{q_0, q_3\}, 0) = \delta(q_0, 0) \cup \delta(q_3, 0)$
$$= \{q_0, q_3\} \cup \{q_4\}$$
$$= \{q_0, q_3, q_4\}$$

Transition diagram for DFA is as follows:

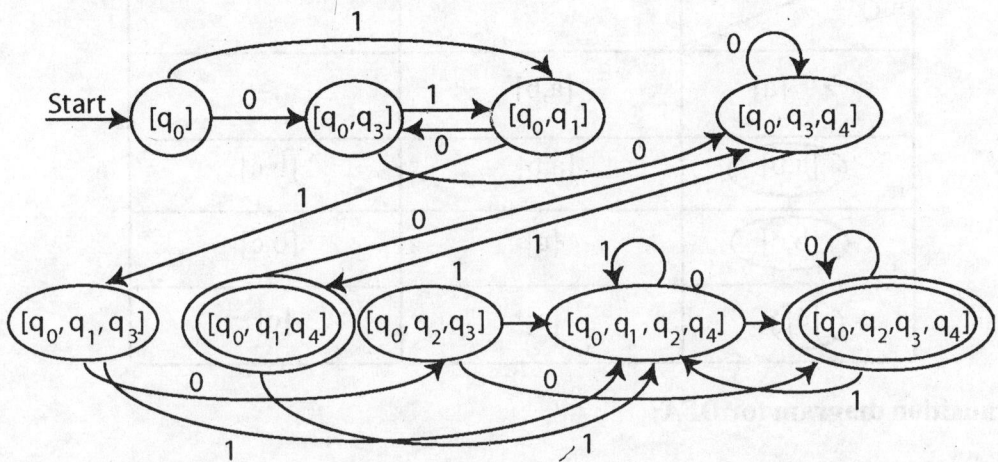

Problem 4. For the following NFA, construct the equivalent DFA.

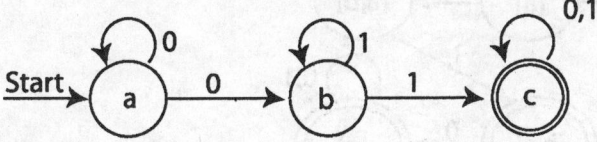

Solution : **The transition table for the given NFA is**

Q \ Σ	0	1
→a	{a,b}	-
b	-	{b,c}
ⓒ	c	c

$M=\{Q, \Sigma, \delta, q_0, F\}$ where $Q=\{a, b, c\}$ $\Sigma=\{0,1\}$ $q_0=a$, $F=\{c\}$
construct DFA $M=\{Q^1, \Sigma, \delta^1, q_0^1, F^1\}$ as follows.
$Q^1 = 2^Q = \{\emptyset, [a], [b], [c], [a, b], [b, c], [c, a], [a, b, c]\}$

Transition table for DFA:

Q Σ	0	1
→ [a]	[a,b]	-
[a,b]	[a,b]	[b,c]
[b,c]	[c]	[b,c]
[c]	[c]	[c]

Transition diagram for DFA:

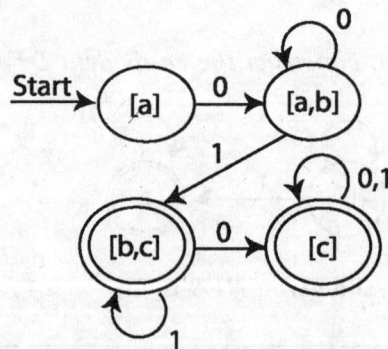

Problem 5. Construct DFA for the following NFA

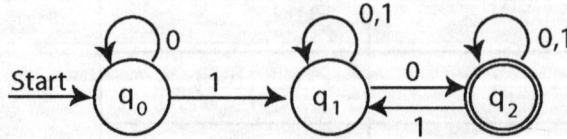

Solution:

Transition table for NFA:

Q \ Σ	0	1
→ q_0	$\{q_0\}$	$\{q_1\}$
q_1	$\{q_1, q_2\}$	$\{q_2\}$
q_2 (final)	$\{q_2\}$	$\{q_1, q_2\}$

Transition table for DFA:

Q \ Σ	0	1
→ $[q_0]$	$[q_0]$	$[q_1]$
$[q_1]$	$[q_1, q_2]$	$[q_1]$
$[q_1, q_2]$ (final)	$[q_1, q_2]$	$[q_1, q_2]$

Transition diagram for DFA:

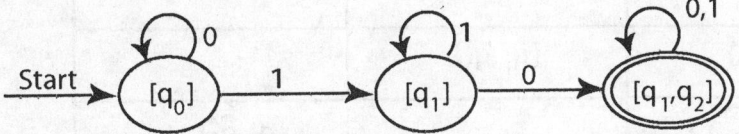

Acceptance of input string by NFA and DFA:

Consider the input string: 11001 using NFA.

$\hat{\delta}(q_0, \varepsilon) = \{q_0\}$

$\hat{\delta}(q_0, 1) = \delta(\hat{\delta}(q_0, \varepsilon), 1) = \delta(q_0, 1) = \{q_1\}$

$\hat{\delta}(q_0, 11) = \delta(\hat{\delta}(q_0, 1), 1) = \delta(q_1, 1) = \{q_1\}$

$\hat{\delta}(q_0, 110) = \delta(\hat{\delta}(q_0, 11), 0) = \delta(\{q_1\}, 0) = \{q_1, q_2\}$

$\hat{\delta}(q_0, 1100) = \delta(\hat{\delta}(q_0, 110), 0) = \delta(\{q_1, q_2\}, 0) = \delta(q_1, 0) \cup \delta(q_2, 0)$

$$= \{q_1, q_2\} \cup \{q_2\} = \{q_1, q_2\}$$
$$\hat{\delta}(q_0, 11001) = \delta(\hat{\delta}(q_0, 1100), 1) = \delta(\{q_1, q_2\}, 1) = \delta(q_1, 1) \cup \delta(q_2, 1) = \{q_1, q_2\} \rightarrow \text{Accepted.}$$

Using DFA:

$$\hat{\delta}(q_0, \varepsilon) = \{q_0\}$$
$$\hat{\delta}(q_0, 1) = \delta(\hat{\delta}(q_0, \varepsilon), 1) = \delta(q_0, 1) = [q_1]$$
$$\hat{\delta}(q_0, 11) = \delta(\hat{\delta}(q_0, 1), 1) = \delta([q_1], 1) = [q_1]$$
$$\hat{\delta}(q_0, 110) = \delta(\hat{\delta}(q_0, 11), 0) = \delta([q_1], 0) = [q_1, q_2]$$
$$\hat{\delta}(q_0, 1100) = \delta(\hat{\delta}(q_0, 110), 0) = \delta([q_1, q_2], 0) = \delta([q_1, q_2], 0) = [q_1, q_2]$$
$$\hat{\delta}(q_0, 11001) = \delta(\hat{\delta}(q_0, 1100), 1) = \delta([q_1, q_2], 1) = [q_1, q_2] \rightarrow \text{Accepted.}$$

The string '11001' is accepted by both NFA and DFA.

Problem 6. Construct DFA for the following NFA

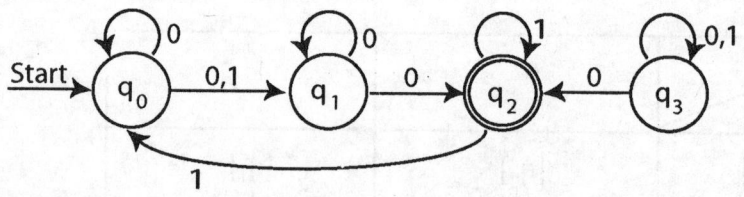

Solution :

Transition table for NFA:

Q \ Σ	0	1
→ q_0	$\{q_0, q_1\}$	$\{q_1\}$
q_1	$\{q_1, q_2\}$	-
(q_2)	-	$\{q_0, q_2\}$
q_3	$\{q_2, q_3\}$	$\{q_3\}$

Transition table for DFA:

Q \ Σ	0	1
→[q_0]	[q_0, q_1]	[q_1]
[q_1]	[q_1, q_2]	-
[q_0, q_1]	[q_0, q_1, q_2]	[q_1]
([q_1, q_2])	[q_1, q_2]	[q_0, q_2]
([q_0, q_1, q_2])	[q_0, q_1, q_2]	[q_0, q_1, q_2]
([q_0, q_2])	[q_0, q_1]	[q_0, q_1, q_2]

Transition diagram for DFA:

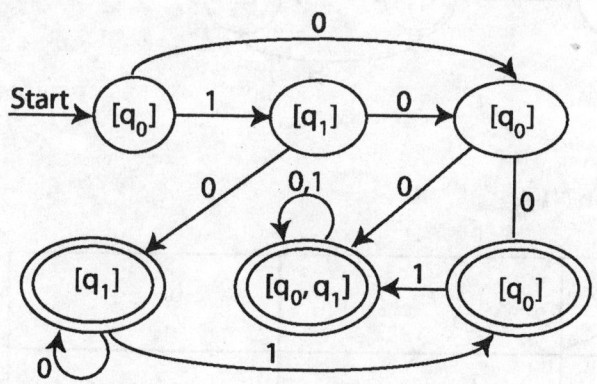

Acceptance of input string by NFA and DFA:
Consider the following **String – 10111.**
Using NFA -10111.
$\hat{\delta}(q_0, \ \varepsilon) = \{q_0\}$
$\hat{\delta}(q_0, 1) = \delta(\hat{\delta}(q_0, \ \varepsilon), 1) = \delta(q_0, 1) = \{q_1\}$

$\hat{\delta}(q_0, 10) = \delta(\hat{\delta}(q_0, 1), 1) = \delta(q_1, 0) = \{q_1, q_2\}$

$\hat{\delta}(q_0, 101) = \delta(\hat{\delta}(q_0, 10), 1) = \delta(\{q_1, q_2\}, 1) = \delta(q_1, 1) \cup \delta(q_2, 1) = \{q_0, q_2\}$

$\hat{\delta}(q_0, 1011) = \delta(\hat{\delta}(q_0, 101), 1) = \delta(\{q_0, q_2\}, 1) = \delta(q_1, 1) \cup \delta(q_2, 1) = \{q_0, q_1, q_2\}$

$\hat{\delta}(q_0, 10111) = \delta(\hat{\delta}(q_0, 1011)) = \delta(\{q_0, q_1, q_2\}, 1)) = \delta(q_0, 1) \cup \delta(q_1, 1) \cup \delta(q_2, 1)$
$$= \{q_1\} \cup \varnothing \cup \{q_0, q_2\}$$
$$= \{q_0, q_1, q_2\} \quad \rightarrow \text{Accepted}$$

Using DFA-10111

$\hat{\delta}(q_0, \varepsilon) = [q_0]$

$\hat{\delta}(q_0, 1) = \delta(\hat{\delta}(q_0, \varepsilon), 1) = \delta([q_0], 1) = [q_1]$

$\hat{\delta}(q_0, 10) = \delta(\hat{\delta}(q_0, 1), 0) = \delta([q_1], 0) = [q_1, q_2]$

$\hat{\delta}(q_0, 101) = \delta(\hat{\delta}(q_0, 10), 1) = \delta([q_1, q_2], 1) = [q_0, q_2]$

$\hat{\delta}(q_0, 1011) = \delta(\hat{\delta}(q_0, 101), 1) = \delta([q_0, q_2], 1) = [q_0, q_1, q_2]$

$\hat{\delta}(q_0, 10111) = \delta(\hat{\delta}(q_0, 1011), 1) = \delta([q_0, q_1, q_2], 1) = [q_0, q_1, q_2] \quad \rightarrow \text{Accepted}$

$\therefore$ The string '10111' is accepted by NFA and DFA

Problem 7. Construct DFA equivalent to the finite state machine.

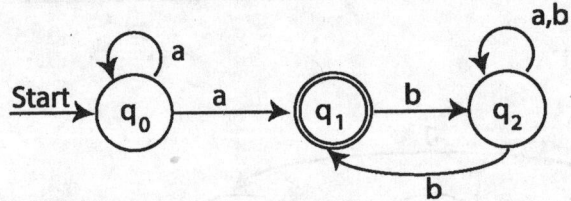

Solution :

Transition table for NFA

Q \ Σ	a	b
→ q_0	$[q_0, q_1]$	-
(q_1)	-	$\{q_2\}$
$\{q_2\}$	$\{q_2\}$	$\{q_1, q_2\}$

Transition table for DFA

Q \ Σ	a	b
→ [q_0]	[q_0,q_1]	-
([q_0,q_1])	[q_0,q_1]	[q_2]
[q_2]	[q_2]	[q_1,q_2]
([q_1,q_2])	[q_2]	[q_1,q_2]

Transition diagram for DFA:

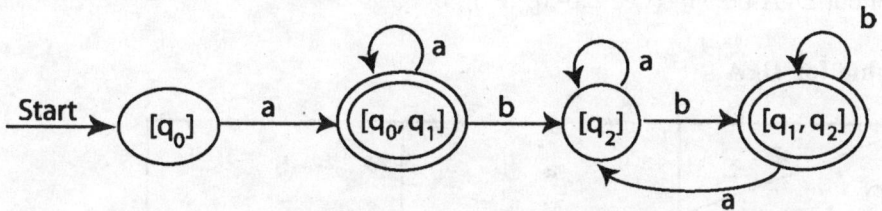

The strings which are accepted by NFA and DFA are aabab, abbbab....
The strings which are not accepted by NFA and DFA are aab, ababba......

Problem 8. Construct DFA equivalent to the following NFA

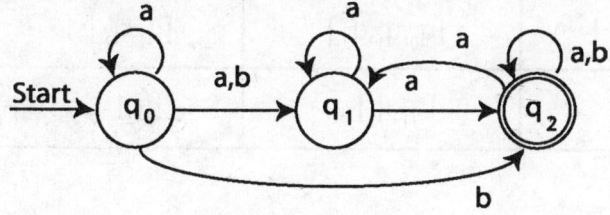

Solution

Given   NFA=M=(Q, Σ,δ , q_0, F)

where   Q={q_0, q_1, q_2} Σ={a,b}, q_0= q_0, F={q_2} δ:Q× Σ→Q

δ is given by the following transition table

Q＼Σ	a	b
→ q_0	{q_0,q_1}	{q_1,q_1}
q_1	{q_1,q_2}	$\varnothing$
q_2	{q_1,q_2}	{q_2}

Let the equivalent DFA be M'=(Q', Σ,δ',q_0',F') , δ':Q'× Σ→2^Q

Transition table for DFA

Q＼Σ	a	b
→[q_0]	[q_0,q_1]	[q_1,q_2]
[q_0,q_1]	[q_0,q_1,q_2]	[q_1,q_2]
[q_1,q_2]	[q_1,q_2]	[q_2]
[q_0,q_1,q_2]	[q_0,q_1,q_2]	[q_1,q_2]
[q_2]	[q_1,q_2]	[q_2]

Transition diagram for DFA:

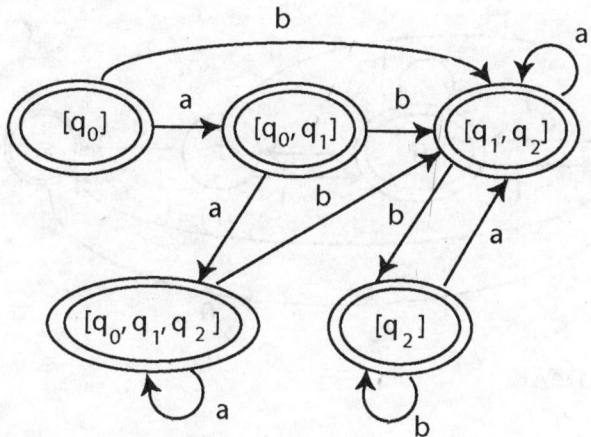

Accepted strings are baab, aaaa, baba....

Problem 9. Find DFA equivalent to NFA, described by the following state transition table. I.S=p, F.S={q, s}

Q \ Σ	0	1
→ p	{q,s}	q
(q)	r	{q,r}
r	s	p
(s)	-	p

Q \ Σ	0	1
→ [q₀]	[q₀]	[q₁]
[q₀,q₁]	[q₁]	[q₀,q₁]
([q₀,qᵣ])	[q₀,q₁]	[q₀,q₁]

Transition diagram for NFA

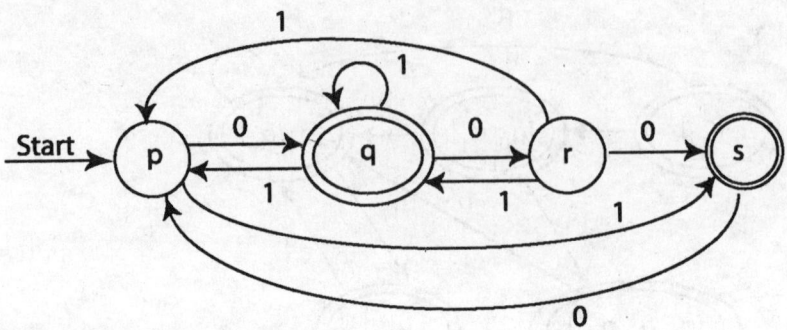

Transition table for DFA

Q \ Σ	0	1
→[p]	[q,s]	[q]
([q])	[r]	[q,r]
[r]	[s]	[p]
([s])	[∅]	[p]
([q,s])	[r]	[p, q, r]
([q,r])	[r,s]	[p, q, r]
([r,s])	[s]	[p]
([p,q,r])	[q,r,s]	[p, q, r]
([q, r,s])	[r,s]	[p, q, r]

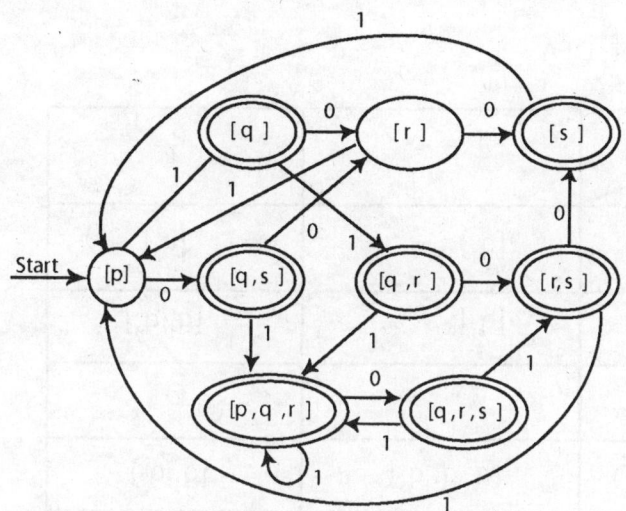

Problem 10. For the following state transition table draw the state transition diagram. Find its equivalent machine. q_0 is the initial state and q_3 is the final state. a=0, b=1.

Q \ Σ	0	1
→ q_0	q_1	q_2
q_1	q_1	$\{q_1, q_3\}$
q_2	$\varnothing$	$\varnothing$
(q_3)	$\{q_0, q_3\}$	q_3

Solution :

The state transition diagram for NFA

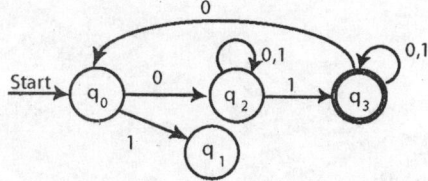

NFA to DFA conversion:

The state transition table for DFA

Q \ Σ	0	1
→[q_0]	[q_1]	[q_2]
[q_1]	[q_1]	[q_1,q_3]
[q_2]	∅	∅
([q_1,q_3])	[q_0,q_1,q_3]	[q_1,q_3]
([q_0,q_1,q_3])	[q_0,q_1,q_3]	[q_1,q_2,q_3]
([q_1,q_2,q_3])	[q_0,q_1,q_3]	[q_1,q_3]

Transition diagram for DFA

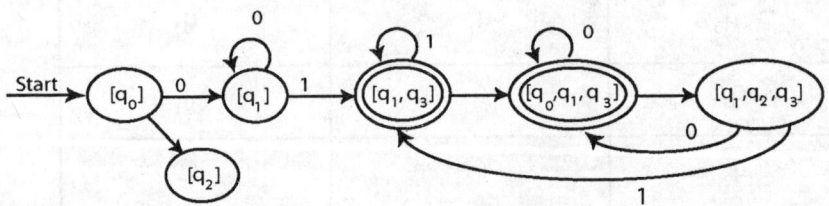

Reduce the number of states:

In the above transition table, rows corresponding to [q_1, q_3] and [q_1, q_2, q_3] are identical.

∴ Remove [q_1, q_2, q_3] consider [q_1, q_3] and substitute [q_1, q_2, q_3]= [q_1, q_3] in the above table.

Transition table: for the reduced DFA

Q \ Σ	0	1
→ [q_0]	[q_1]	[q_2]
[q_1]	[q_1]	[q_1, q_3]
[q_2]	∅	∅
([q_1, q_3])	[q_1, q_3]	[q_1, q_3]

Probelm 11. For the transition table given find the state transition diagram. Check whether the string '00101011' is accepted or not. Determine equivalent DFA and check for the same string and prove that both are equivalent, q_0 is the initial state and q_2 is the final state.

Solution :

Q \ Σ	0	1
→ q_0	{q_0, q_1}	-
q_1	∅	{q_1, q_2}
(q_2)	q_1	∅

State transition diagram for NFA:

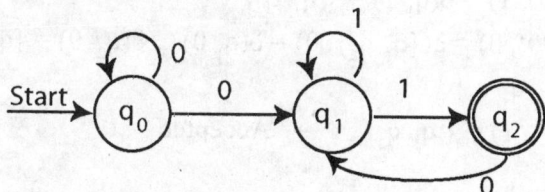

Transition table for DFA:

Σ Q	0	1
→ $[q_0]$	$[q_0,q_1]$	$\varnothing$
$[q_0,q_1]$	$[q_0,q_1]$	$[q_1,q_2]$
$([q_1,q_2])$	$[q_1]$	$[q_1,q_2]$
$[q_1]$	$\varnothing$	$[q_1,q_2]$

Transition diagram for DFA:

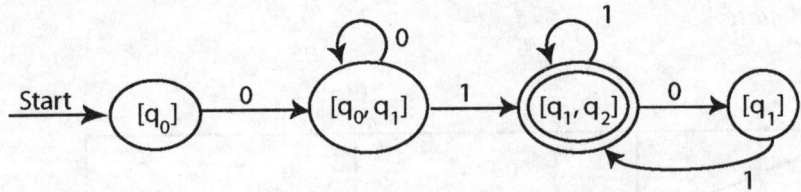

Acceptance of input string '00101011' using NFA:

$\hat{\delta}(q_0, \ \varepsilon) = q_0$

$\hat{\delta}(q_0, \ 0) = \delta(\hat{\delta}(q_0, \ \varepsilon), 0) = \delta(q_0, 0) = \{q_0, q_1\}$

$\hat{\delta}(q_0, \ 00) = \delta(\hat{\delta}(q_0, \ 0), 0) = \delta(\{q_0, q_1\}, 0) = \delta(q_0,0) \cup \delta(q_1, 0) = \{q_0, q_1\}$

$\hat{\delta}(q_0, \ 001) = \delta(\hat{\delta}(q_0, \ 00), 1) = \delta(\{q_0, q_1\}, 1) = \delta(q_0,1) \cup \delta(q_1, 1) = \{q_1, q_2\}$

$\hat{\delta}(q_0, \ 0010) = \delta(\hat{\delta}(q_0, \ 001), 0) = \delta(\{q_1, q_2\}, 0) = \delta(q_1,0) \cup \delta(q_2, 0) = \{q_1\}$

$\hat{\delta}(q_0, \ 00101) = \delta(\hat{\delta}(q_0, \ 0010), 1) = \delta(q_1, 1) = \{q_1, q_2\}$

$\hat{\delta}(q_0, \ 001010) = \delta(\hat{\delta}(q_0, \ 0010), 0) = \delta(\{q_1, q_2\}, 0) = \delta(q_1,0) \cup \delta(q_2, 0) = \{q_1\}$

$\hat{\delta}(q_0, \ 0010101) = \delta(q_1, 1) = \{q_1, q_2\}$

$\hat{\delta}(q_0, \ 00101011) = \delta(\{q_1, q_2\}, 1) = \{q_1, q_2\} \quad \rightarrow$ Accepted

Using DFA

$\hat{\delta}(q_0, \varepsilon) = q_0$

$\hat{\delta}(q_0, 0) = \delta(\hat{\delta}(q_0, \varepsilon), 0) = \delta(q_0, 0) = [q_0, q_1]$

$\hat{\delta}(q_0, 00) = \delta(\hat{\delta}(q_0, 0), 0) = \delta([q_0, q_1], 0) = [q_0, q_1]$

$\hat{\delta}(q_0, 001) = \delta(\hat{\delta}(q_0, 00), 1) = \delta([q_0, q_1], 1) = [q_1, q_2]$

$\hat{\delta}(q_0, 0010) = \delta(\hat{\delta}(q_0, 001), 0) = \delta([q_0, q_1], 0) = [q_1]$

$\hat{\delta}(q_0, 00101) = \delta([q_1], 1) = [q_1, q_2]$

$\hat{\delta}(q_0, 001010) = \delta([q_1, q_2], 0) = [q_1]$

$\hat{\delta}(q_0, 0010101) = \delta(\hat{\delta}(q_0, 001010), 1) = \delta([q_1], 1) = [q_1, q_2]$

$\hat{\delta}(q_0, 00101011) = \delta([q_1, q_2], 1) = [q_1, q_2] \quad \rightarrow$ Accepted

$\therefore$ The string 00101011 is accepted by NFA and DFA.

Problem 12. Construct DFA equivalent to the following finite automaton

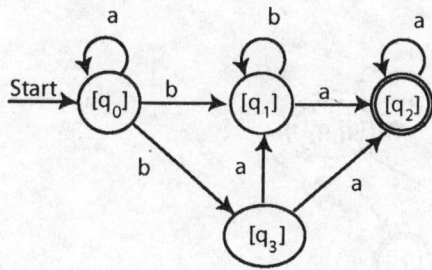

Solution :
Transition table for NFA:

Q \ Σ	A	b
→ q_0	q_0	$\{q_1, q_3\}$
q_1	q_2	q_1
q_2	q_2	$\varnothing$
q_3	$\{q_1, q_2\}$	$\varnothing$

Transition table for DFA :

Q ＼ Σ	a	b
→ q_0	$[q_0]$	$[q_1,q_3]$
$[q_1,q_3]$	$[q_1,q_2]$	$[q_1]$
$([q_1,q_2])$	$[q_2]$	$[q_1]$
$[q_1]$	$[q_2]$	$[q_1]$
$([q_2])$	$[q_2]$	$\varnothing$

Transition diagram for DFA :

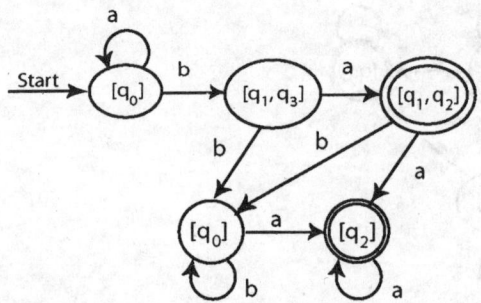

Problem 13. Define NFA- ε transitions and write the differences between NFA- ε and ordinary NFA.

Solution:

NFA with ε- transitions:

A non-deterministic finite automaton, NFA with ε-transitions, M is given by 5-tuple,

M= (Q, Σ, δ , q_0, F)

where　Q = A finite set of states i.e., q_0, q_1, q_2...... .

　　　　Σ = A finite set of input symbols,

　　　　δ = A transition function that defines the rules for change of state

It maps from Q× (Σ∪ ε) to 2^Q.

　　　　q_0 = initial state, $q_0 \in Q$

　　　　F = the set of final states, F⊆Q

NFA without ε-moves(NFA)	NFA with ε-moves
1. Transition function of NFA maps from Q× Σ to 2^Q i.e., for a given input symbol there would be at least one transition from a state.	1. Transition function of NFA with ε-moves maps from Q×(Σ ∪ ε) to 2^Q, i.e., for a given input there would be more than one transition including 'ε' from a state.
2. It accepts an input string if there exists at least one path from start state to final state.	2. It is same as NFA. It defines a function ε-CLOSURE (q) to determine a path.
3. It is slower than DFA.	3. It is much slower when compared to DFA and NFA.
4. It is smaller than DFA.	4. It is also smaller than DFA
5. Not all NFA's are equivalent to DFA's	5. All NFA with ε-moves are equivalent to NFA without ε-moves.

Problem 14. Define epsilon CLOSURE? Find epsilon (ε) CLOSUREs for all the states of given NFA- ε. Remove epsilons without changing the acceptance.

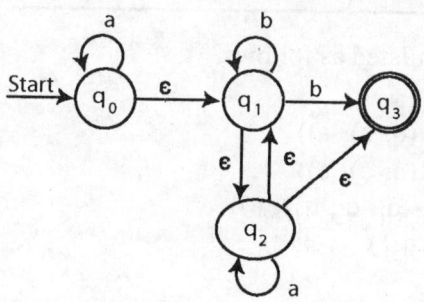

Solution :
Epsilon CLOSURE or ε-CLOSURE:
Epsilon CLOSURE of state 'q', represented as ε-CLOSURE (q) is a set of NFA states including **'q'**, that are reachable from q on input symbol **'ε'** alone.

ε-CLOSURE(q) cannot be an empty set, because 'q' is always reachable from itself irrespective of an input symbol.

Therefore ε-CLOSURE of 'q' always includes **'q'**.

If P is a set of states, then

$$\varepsilon\text{-CLOSURE (q)} =\cup_{q\varepsilon p}\ \varepsilon\text{-CLOSURE (q) where q } \varepsilon\ P$$

Finding ε-CLOSURE:

ε-CLOSURE of all states of given NFA with ε transitions is given below.

Step 1 :

ε-CLOSURE (q_0) = $\{q_0, q_1, q_2, q_3\}$

ε-CLOSURE (q_1) = $\{q_1, q_2, q_3\}$

ε-CLOSURE (q_2) = $\{q_1, q_2, q_3\}$

ε-CLOSURE (q_3) = $\{q_3\}$,

Final state of M= $\{q_3\}$

∴ Final states of M^1 are q_0, q_1, q_2, q_3 .

Conversion of NFA with ε-moves to ordinary NFA

Transition tabe for NFA-ε moves

Q \ Σ	a	b	ε
→q_0	q_0	-	q_1
q_1	-	$\{q_1, q_3\}$	q_2
q_2	q_2	-	$\{q_1, q_3\}$
q_3	-	-	-

Step 2: The transition function is calculated as follows.

i) $\hat{\delta}(q, \varepsilon)$ = ε-CLOSURE(q)

ii) $\hat{\delta}(q, a)$ = ε-CLOSURE $(\delta(\hat{\delta}(q, \varepsilon), a))$

1) $\hat{\delta}(q_0, a)$ = ε-CLOSURE $(\delta(\hat{\delta}(q, \varepsilon), a))$

= ε-CLOSURE $(\delta(\{q_0, q_1, q_2, q_3\}, a))$

= ε-CLOSURE $(\{q_0, q_2\})$

= $\{q_0, q_1, q_2, q_3\}$

2) $\hat{\delta}(q_0, b)$ = ε-CLOSURE $(\delta(\hat{\delta}(q_0, \varepsilon), b))$

= ε-CLOSURE $(\delta(\{q_0, q_1, q_2, q_3\}, b))$

= ε-CLOSURE $(\{q_1, q_3\})$

= $\{q_1, q_2, q_3\}$

3) $\hat{\delta}(q_1, a)$ = ε-CLOSURE $(\delta(\hat{\delta}(q_1, \varepsilon), a))$

= ε-CLOSURE $(\delta(\{q_1, q_2, q_3\}, a))$

= ε-CLOSURE (q_2)

$$= \{q_1, q_2, q_3\}$$

4) $\hat{\delta}(q_{1,} b) = \varepsilon\text{-CLOSURE } (\delta(\hat{\delta}(q_1, \varepsilon), b))$

$\qquad\qquad = \varepsilon\text{-CLOSURE } (\delta(\{q_1, q_2, q_3\}, b))$

$\qquad\qquad = \varepsilon\text{-CLOSURE } (\{q_1, q_3\})$

$\qquad\qquad = \{q_{1,} q_{2,} q_3\}$

5) $\hat{\delta}(q_{2,} a) = \varepsilon\text{-CLOSURE } (\delta(\hat{\delta}(q_2, \varepsilon), a))$

$\qquad\qquad = \varepsilon\text{-CLOSURE } (\delta(\{q_1, q_2, q_3\},a))$

$\qquad\qquad = \varepsilon\text{-CLOSURE } (q_2)$

$\qquad\qquad = \{q_1, q_2, q_3\}$

6) $\hat{\delta}(q_{2,} b) = \varepsilon\text{-CLOSURE } (\delta(\hat{\delta}(q_2, \varepsilon), b))$

$\qquad\qquad = \varepsilon\text{-CLOSURE } (\delta(\{q_1, q_2, q_3\}, b)$

$\qquad\qquad = \varepsilon\text{-CLOSURE } (\{q_1, q_3\})$

$\qquad\qquad = \{q_{1,} q_{2,} q_3\}$

7) $\hat{\delta}(q_{3,} a) = \varepsilon\text{-CLOSURE } (\delta(\hat{\delta}(q_3, \varepsilon), a))$

$\qquad\qquad = \varepsilon\text{-CLOSURE } (\delta(q_3, a))$

$\qquad\qquad = \varepsilon\text{-CLOSURE } (\emptyset)$

$\qquad\qquad = \emptyset$

8) $\hat{\delta}(q_{3,} b) = \varepsilon\text{-CLOSURE } (\delta(\hat{\delta}(q_3, \varepsilon), b))$

$\qquad\qquad = \varepsilon\text{-CLOSURE } (\delta(q_3, b))$

$\qquad\qquad = \varepsilon\text{-CLOSURE } (\emptyset)$

$\qquad\qquad = \emptyset$

Step 3 :

Final states of M^1 are $q_0, q_1, q_2, q_3.$

Transition table for M^1

Q $\diagdown$ Σ	a	b
$\rightarrow$ q_0	$\{q_0, q_1, q_2, q_3\}$	$\{q_1, q_2, q_3\}$
q_1	$\{q_1, q_2, q_3\}$	$\{q_1, q_2, q_3\}$
q_2	$\{q_1, q_2, q_3\}$	$\{q_1, q_2, q_3\}$
q_3	$\emptyset$	$\emptyset$

Transition table for NFA without ε-moves:

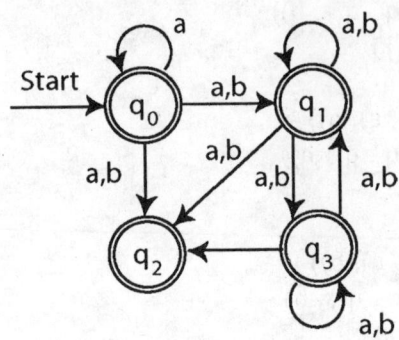

Problem 15. Compare three finite automata critically

Deterministic Finite Automata(DFA)	Non-Deterministic Finite Automata(NFA)	NFA with ε moves
1. Transition function of DFA maps from $Q \times \Sigma$ to Q i.e. for a given input symbol there would be at most one transition from a state.	1. Transition function of NFA maps from $Q \times \Sigma$ to 2^Q i.e. for a given input symbol there would be at least one transition from a state.	1. Transition function of NFA with ε-moves maps from $Q \times (\Sigma \cup \varepsilon)$ to 2^Q, i.e. for a given input there would be more than one transition including 'ε' from a state.
2. It accepts an input string if it is in one of the final state at the end of the string	2. It accepts an input string if there exists at least one path from start state to final state.	2. It is same as NFA. It defines a function ε-CLOSURE (q) to determine a path.
3. It is faster than NFA.	3. It is slower than DFA.	3. It is more slower when compared to DFA and NFA.
4. It is larger than NFA.	4. It is smaller than DFA.	4. It is also smaller.
5. All DFA's are equivalent to NFA.	5. Not all NFA are equivalent to DFA	5. All NFA's with ε-move are equivalent to NFA without ε-moves.

Problem 16. Minimize the following finite automaton and prove given and resultants are equivalent. Prove that given and resultant are equivalent.(model- 5, type-1)

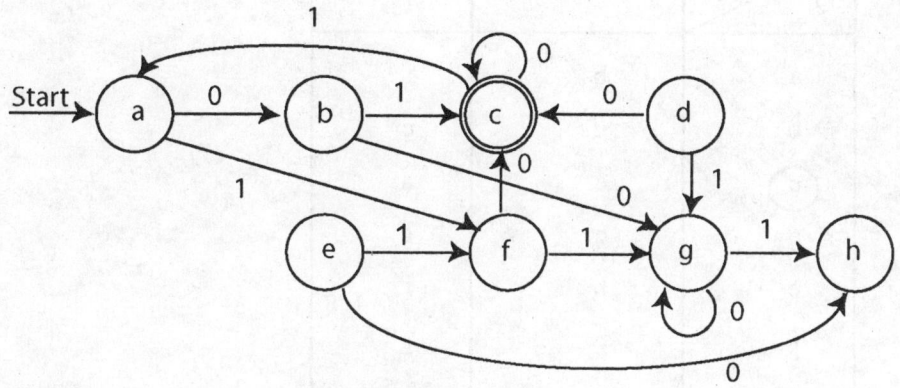

Solution:

Transition table for the given DFA

Q \ Σ	0	1
→ a	b	f
b	g	c
ⓒ	c	a
d	c	g
e	h	f
f	c	g
g	g	h
h	-	-

Given that there is no transition from state 'h' on any input symbol. But a DFA has transitions from each state on every input symbol.

So we add a dead state 'i' which transitions to itself on any input symbol and we add transitions from **h** to **i** on both input symbols **0** and **1**

∴ Q={a,b,c,d,e,f,g,h,i}

The transition table is shown below.

Q \ Σ	0	1
a	b	f
→ b	g	c
(c)	c	a
d	c	g
e	h	f
f	c	g
g	g	h
h	i	i
i	i	i

Q= {a, b, c, d, e, f, g, h, i}

Partition Q into sets of final and non-final states

a, b, d, e, f, g, h, i	c
Group 1	Group 2

Consider transitions for each group on every input symbol.

Consider group 1:

$\delta (a, 0) = b$

$\delta (b, 0) = g$

$\delta (d, 0) = c$

$\delta (e, 0) = h$

$\delta (f, 0) = c$

$\delta (g, 0) = g$

$\delta (h, 0) = i$

$\delta (i, 0) = i$

Partition group 1, since on input '0', d and f go to different group

a, b, e, g, h, i	d,f	c
Group 1	Group 2	Group 3

Consider group 1:

$\delta(a, 0) = b$ $\delta (a,1) = f$
$\delta(b, 0) = g$ $\delta (b, 1) = c$
$\delta(e, 0) = h$ $\delta (e, 1) = f$
$\delta(g, 0) = g$ $\delta (g, 1) = h$
$\delta(h, 0) = i$ $\delta (h,1) = i$
$\delta(i, 0) = i$ $\delta (i, 1) = i$

Partition group 1, since on input 1, a, b, e go to different groups.
But make a group of a and e, because they go to the same group.

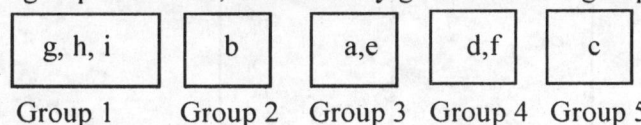

  Group 1   Group 2   Group 3   Group 4   Group 5

Consider group 1:

$\delta(g, 0) = g$ $\delta (g, 1) = h$
$\delta(h, 0) = i$ $\delta (h,1) = i$
$\delta(i, 0) = i$ $\delta (i, 1) = i$

Group 1 cannot be partitioned, because all members of this group go to the same group on given input.

Consider group 3:

$\delta (a, 0) = b$
$\delta (e, 0) = h$

Partition group 3, since on input **0**, **a** and **e** go to different groups

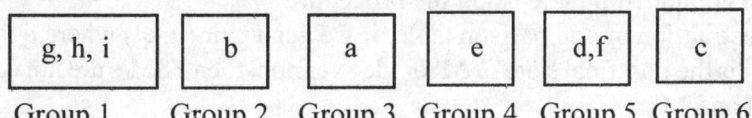

  Group 1   Group 2   Group 3   Group 4   Group 5   Group 6

Consider group 5:

    $\delta(d, 0) = c$ $\delta (d, 1) = g$
    $\delta(f, 0) = c$ $\delta (f, 1) = g$

Group 5 cannot be partitioned, since all members of this group go to the same group on given input.

Consider group 6:

It contains only single member. Therefore it can not be partitioned.

Let $a = a^1$
    $b = b^1$
    $c = c^1$ **Equivalent states**
   $\{d, f\} = d^1$
    $e = e^1$
   $\{g, h, i\} = g^1$

Let the minimized DFA be $M^1 = (Q^1, \Sigma, \delta^1, q_0^{\ 1}, F^1)$

where $Q^1 = \{a^1, b^1, c^1, d^1, e^1, g^1\}$
   $q_0{}^1 = a^1$
   $F^1 = c^1$

The minimized DFA, M^1 is shown below

Q $\diagdown$ Σ	0	1
→ a^1	b^1	d^1
b^1	g^1	c^1
$\textcircled{$c^1$}$	c^1	a^1
d^1	c^1	g^1
e^1	g^1	d^1
g^1	g^1	g^1

In the original DFA, the no. of states = 9
In the minimized DFA, the no. of states = 6.

Equivalence of M and M^1:

Step 1: Construct comparison table using the procedure

Step 2: In the construction of comparison table, if we get a pair (q, q^1) where q is the final state of '**M**' and q^1 is the non-final state of **M^1** or vice version, then FSM's are not equivalent.

Equivalence of M and M^1

The comparison table is shown below

(q, q^1)	$(q_0, q_0{}^1)$	$(q_1, q_1{}^1)$
(a, a^1)	(b, b^1)	(f, d^1)
(b, b^1)	(g, g^1)	(c, c^1)
(f, d^1)	(c, c^1)	(g, g^1)
(c, c^1)	(g, g^1)	(h, g^1)
(h, g^1)	(i, g^1)	(i, g^1)
(i, g^1)	(i, g^1)	(i, g^1)

In the above comparison table, we don't have a pair (q, q^1), where q is the **final state** of M and q^1 is **non-final** state of M^1 or vice versa. All pairs in column 2 and column 3 are also occur in column 1

$\therefore$ M and M^1 are equivalent.

Problem 17. *Distinguish Moore and Mealy finite state machines.*

Solution :

Moore machine: The Moore machine is a six-tuple $(Q, \Sigma, \Delta, \delta, \lambda, q_0)$ where

i)   **Q** is a finite set of states.

ii) Σ is an input alphabet.

iii) Δ is the output alphabet.

iv) δ is the transition function $Q \times \Sigma$ into Q, i.e., $\delta: Q \times \Sigma \rightarrow Q$

v) λ is the output function mapping Q into Δ ,i.e., $\lambda : Q \rightarrow \Delta$

vi) q_0 is the initial state.

Mealy machine: Mealy machine is a six-tuple $(Q, \Sigma, \Delta, \delta, \lambda, q_0)$ where

i)   **Q** is a finite set of states.

ii) Σ is an input alphabet.

iii) Δ is the output alphabet.

iv) δ is the transition function $Q \times \Sigma$ into Q. i.e., $\delta: Q \times \Sigma \rightarrow Q$

v) λ is the output function mapping $Q \times \Sigma$ into Δ, i.e., $\lambda : Q \times \Sigma \rightarrow \Delta$

vi) q_0 is the initial state.

Differences between Moore and Mealy machines:

1)   In Moore machine output is associated with state. In mealy machine output is associated with transition.

2)   In mealy machine: $\lambda: Q \times \Sigma \rightarrow \Delta$.
       In Moore machine: $\lambda: Q \rightarrow \Delta$.

Problem 18. *Construct the Moore machine to the given Mealy machine.*

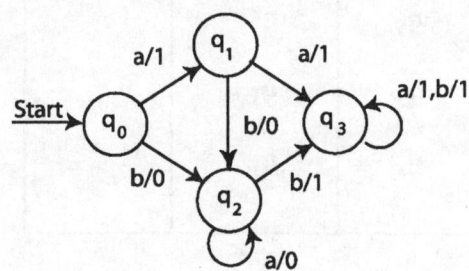

Solution :

The transition table for given Mealy machine is shown below

Present state	Next state			
	Input = a		Input = b	
	State	output	State	output
→ q_0	q_1	1	q_2	0
q_1	q_3	1	q_2	0
q_2	q_2	0	q_3	1
q_3	q_3	1	q_3	1

Examine the next state column

- State 'q_0' is associated with output 'ε'
- State 'q_1' is associated with output '1'
- State 'q_2' is associated with output '0'
- State 'q_3' is associated with output '1'

Since none of the states is splitted, we construct the Moore machine directly as shown below.

Present state	Next state		Output
	Input = a	Input = b	
→ q_0	q_1	q_2	ε
q_1	q_3	q_2	1
q_2	q_2	q_3	0
q_3	q_3	q_3	1

Problem 20. Convert the Mealy machine shown below into equivalent Moore machine.

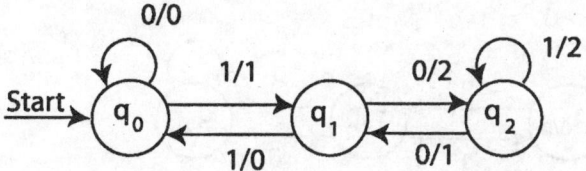

Solution:

Transition table for Mealy machine

Present state	Next state			
	Input = 0		Input = 1	
	State	output	State	output
→ q_0	q_0	0	q_1	1
q_1	q_2	2	q_0	0
q_2	q_1	1	q_2	2

Examine the next state column

- q_0 is associated with **one** output '**0**'. No need to split it.
- q_1 is associated with **one** output '**1**'. No need to split it.
- q_2 is associated with **one** output '**2**'. No need to split it.

Arrange the pairs of states and outputs in the next state column as shown below

Present state	Next state		output
	a = 0	a = 1	
→ q_0	q_0	q_1	0
q_1	q_2	q_0	1
q_2	q_1	q_2	2

Transition diagram for Moore machine

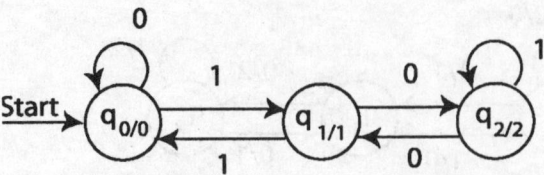

Problem 21. Construct the Mealy machine for the given Moore machine.

Q \ Σ	a	b	Output
→ q_0	q_0	q_2	0
q_1	q_1	q_0	1
q_2	q_2	q_1	1

Solution :

In the equivalent Mealy machine only the output function 'λ^1' changes and the transition function 'δ' remains same.

Output function λ^1 is calculated as follows

$\lambda^1(q_0, a) = \lambda(\delta(q_0, a)) = \lambda(q_0) = 0$
$\lambda^1(q_0, b) = \lambda(\delta(q_0, b)) = \lambda(q_2) = 1$
$\lambda^1(q_1, a) = \lambda(\delta(q_1, a)) = \lambda(q_1) = 1$
$\lambda^1(q_1, b) = \lambda(\delta(q_1, b)) = \lambda(q_0) = 0$
$\lambda^1(q_2, a) = \lambda(\delta(q_2, a)) = \lambda(q_2) = 1$
$\lambda^1(q_2, b) = \lambda(\delta(q_2, b)) = \lambda(q_1) = 1$

The transition table for Mealy machine:

Present state	Input = a		Input = b	
	State	output	State	output
→ q_0	q_0	0	q_2	1
q_1	q_1	1	q_0	0
q_2	q_2	1	q_1	1

Transition diagram

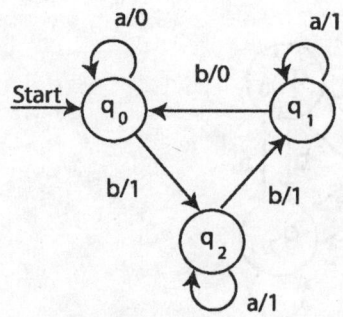

EXERCISE PROBLEMS

1) Show that the automaton M_1 and M_2 are not equivalent.

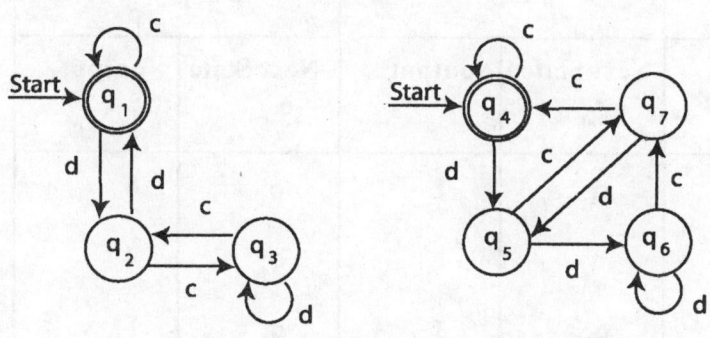

$$M_1 \qquad\qquad M_2$$

2) Check whether the following two finite state machines are equivalent

P.S	N.S.	
	0	1
→ A	A	B
B	C	B
Ⓒ	B	D
Ⓓ	D	D

P.S	N.S.	
	0	1
→ P	P	R
Q	R	P
Ⓡ	Q	Q

$$M_1 = (Q, \Sigma, \delta, q_0, F) \qquad\qquad M_2 = (Q_2, \Sigma, \delta^1, q_0^1, F^1)$$

3) **Construct state transition table for the following Moore machine.**

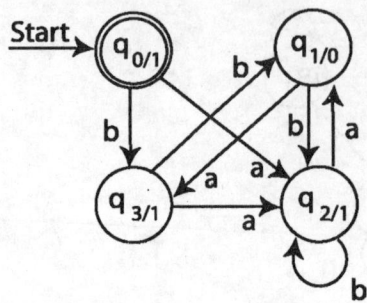

4) **Construct Moore machine equivalent to the Mealy machine described below.**

Present state	a = 0		a = 1	
q_i	Next State q_{i+1}	output	Next State q_{i+1}	output
→ q_1	q_1	1	q_2	0
q_2	q_4	1	q_4	1
q_3	q_2	1	q_3	1
q_4	q_3	0	q_1	1

5) **Construct Mealy machine equivalent to the Moore machine given by the following table.**

q_i	a = 0 $q_i + 1$	a = 1 $q_i + 1$	Output
→ q_0	q_1	q_2	1
q_1	q_3	q_2	0
q_2	q_2	q_1	1
q_3	q_0	q_3	1

6) Construct Moore machine to the given Mealy machine.

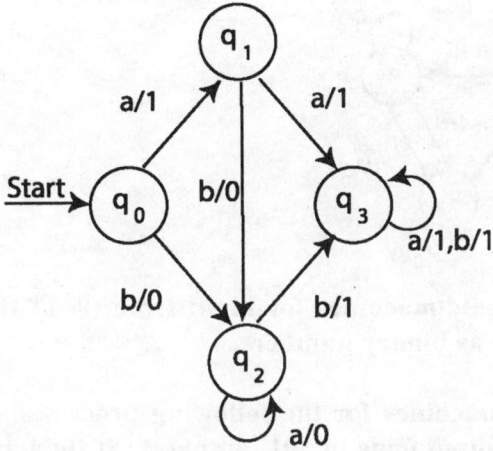

7) Construct Moore machine to the given Mealy machine

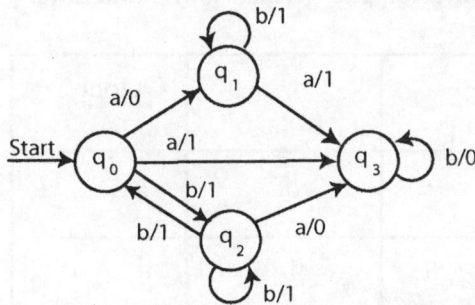

8) construct the Moore machine for given Mealy machine

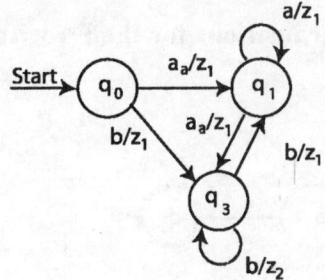

9)   Construct the following Mealy machine into equivalent Moore machine as shown in the following figure.

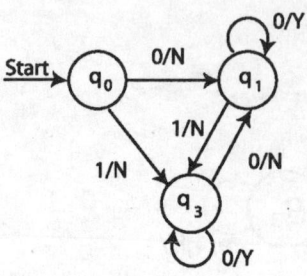

10) Construct Moore and Mealy machines for input from $(0+1)^*$ that gives reside modulo 5 of the input treated as binary number.

11) Give Mealy and Moore machines for the following processes.
For input from $(0+1)^*$, if the input ends in 101, output A; if the input ends in 110, output B; otherwise output C.

12) Construct the Mealy machine for the given Moore machine.

State \ Input	A	b	Output
→ q_0	q_1	q_2	1
q_1	q_1	q_1	0
q_2	q_1	q_0	1

13) Find an equivalent NFA without °-transitions for the FA with °-transition shown below. Give also transition table

Start → q_0 (0 loop) — ε → q_1 (1 loop) — ε → q_2 (2 loop)

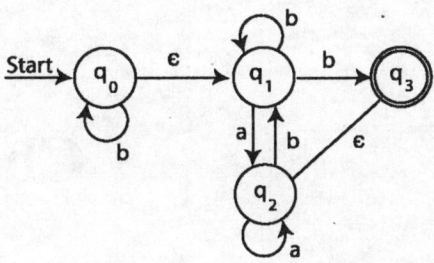

14) Construct equivalent DFA for the following

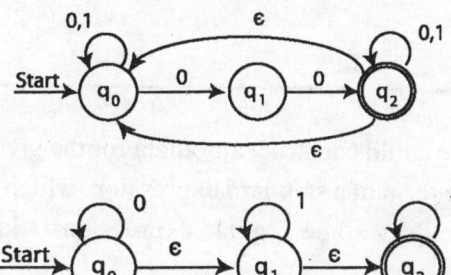

Chapter 3

Regular Languages

3.1 INTRODUCTION

In the previous chapter, we could construct automata for the given language. Every automata can be represented in the form of a standard expression, which is called as regular expression. The language accepted by some regular expression is known as a regular language.

The given chapter focuses on the construction of a regular expression for a given model or automata. The method to construct automata for a given regular expression and vice-versa is discussed. All languages need not be regular. To prove certain languages are not regular, pumping lemma is used and finally the applications of regular expressions and the closure properties of a regular language are discussed.

3.2 REGULAR SETS

3.2.1 DEFINITIONS

Definition: A language is a regular set, if it is a set accepted by some finite automata. If the language contains ε, then the start state itself is the final state.

Example 3.1

Show that whether the following language is regular or not
$L = \{\varepsilon\ 0,00,000,0000,....\}$ i.e, the set of zero's

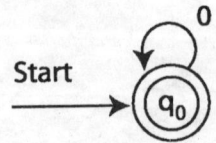

Here, the language L is accepted by finite automata. Hence L is a regular set.

Example 3.2

Show that L = {ε, 00, 0000....} i.e., the set of even number of zero's is regular or not.

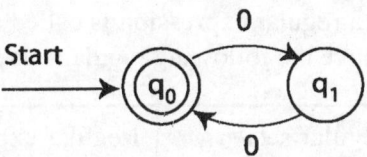

Here, the language L is accepted by finite automata. Hence L is a regular set.

3.2.2 REGULAR EXPRESSIONS AND REGULAR LANGUAGES

Regular expressions: The languages accepted by finite automata are described by expressions called regular expressions.

Regular expressions are useful for representing some sets of strings in an algebraic fashion.

A formal recursive definition of regular expression over Σ is as follows

1. Any terminal symbol (i.e., an element of Σ), ε and $\emptyset$ are regular expressions. When we view **a** in Σ as a regular expression, we denote it by **a**.

2. The union of two regular expressions **R1** and **R2** written as **R1+R2** is also a regular expression.

3. The concatenation of two regular expressions **R1** and **R2** written as **R1R2** is also a regular expression.

4. The iteration (or closure) of a regular expression **R**, written as **R*** is also a regular expression.

5. If **R** is a regular expression then **(R)** is also a regular expression.

6. The regular expressions over Σ are precisely those obtained recursively by the application of the rules 1-5 once or several times.

NOTE:

(a) The parentheses used in rule-5 influence the order of evaluation of a regular expression.

(b) In the absence of parentheses, we have the hierarchy of operations as follows: iteration (closure), concatenation and union. That is, in evaluating a regular expression involving various operations, we perform iteration first, the concatenation and finally union. This hierarchy is similar to that followed for arithmetic expressions (exponentiation, multiplication and addition).

Definition: Any set represented by a regular expression is called a regular set.

For example, if $\{a, b\} \in \Sigma$ then we have the following regular sets and regular expressions

	Regular sets	Regular expressions
1	$\{a\}$	a
2	$\{a, b\}$	$a + b$
3	$\{ab\}$	ab
4	$\{\varepsilon, a, aa, aaa, ---\}$	a^*
5	$\{a, b\}^*$	$(a + b)^*$

Example 3.3

Describe the following sets by regular expressions

(a) $\{101\}$ (b) $\{abba\}$ (c) $\{01, 10\}$ (d) $\{\in, ab\}$

(e) $\{abb, a, b, bba\}$ (f) $\{\in, 0, 00, 000, ... \}$ (g) $\{1, 11, 111, ...\}$

Solution:

(a) $\{1\}$, $\{0\}$ are represented by 1 and 0 respectively. '101' is obtained by concatenating   1, 0 and 1. $\{101\}$ is represented by regular expression '**101**'.

(b) $\{abba\}$ is represented by regular expression '**abba**'

(c)   As $\{01, 10\}$ is the union of $\{01\}$ and $\{10\}$.

         $\{01, 10\}$ is represented by '**01+10**'

(d) $\{\varepsilon, ab\}$ is represented by '$\varepsilon + ab$'.

(e) The set $\{abb, a, b, bba\}$ is represented by the regular expression '**abb+a+b+bba**'

(f) As $\{\varepsilon, 0, 00, 000,....\}$ is simply$\{0\}*$, is represented by 0^*

(g) Any element in $\{1, 11, 111, ...\}$ can be obtained by concatenating 1 and any element of $\{1\}*$. Hence $1(1)^*$ represents $\{1+\}$

Example 3.4

Describe the following sets by regular expressions

(a) L_1 = the set of all strings of 0's and 1's ending in '00'.

(b) L_2 = the set of all strings of 0's and 1's beginning with '0' and ending with '1'.

(c) $L_3 = \{\varepsilon, 11, 1111, 111111,....\}$.

Solution:

(a) Any string of L_1 is obtained by concatenating any string over $\{0, 1\}$ and the string 00. $\{0, 1\}$ is represented by $0+1$

Hence L_1 is represented by $(0+1)^*00$

i.e., $L_1 = \{00, 000, 100, 0100, 0000, 1100, 1000 ...\}$

(b) As any element of L_2 if obtained by concatenating 0 and string over $\{0, 1\}$ and 1. L_2 can be represented by $0(0+1)^*1$

i.e., $L_2 = \{01, 001, 011, 0001, 0111, 0011, 0101, ...\}$

(c) Any element of L_3 is either ε or a string of even numbers of 1's, i.e., a string of the form $(11)^n$, $n \geq 0$

Hence L_3 is represented by $(11)^*$

i.e., $L_3 = \{\varepsilon, 11, 1111, 111111,.......\}$

3.2.3 IDENTITY RULES FOR REGULAR EXPRESSIONS

Two regular expressions **P** and **Q** are equivalent (we write $P = Q$), if **P** and **Q** represent the same set of strings.

The following are the identities for regular expressions; these are useful for simplifying regular expressions

I_1 : $\emptyset + R = R$

I_2 : $\emptyset R = R\emptyset = \emptyset$

I_3 : $\varepsilon R = R\varepsilon = R$

I_4 : $\varepsilon^* = \varepsilon$ and $\emptyset^* = \varepsilon$

I_5 : $R + R = R$

I_6 : $R^* R^* = R^*$

I_7 : $RR^* = R^*R$

I_8 : $(R^*)^* = R^*$

I_9 : $\varepsilon + RR^* = R^* = \varepsilon + R^*R$

I_{10}: $(PQ)^*P = P\,(QP)^*$

I_{11}: $(P + Q)^* = (P^*Q^*)^* = (P^* + Q^*)^*$

I_{12}: $(P + Q)\,R = PR + QR$ and $R\,(P + Q) = RP + RQ$

3.2.4 EQUIVALENCE OF REGULAR EXPRESSIONS

Let us see one important theorem, named **Arden's Theorem,** which helps in checking the equivalence of two regular expressions.

Arden's Theorem: Let **P** and **Q** be the two regular expressions over the input set Σ. The regular expression **R** is given as **R = Q + RP,** which has a unique solution as **R = QP***

Proof: Let **P** and **Q** are two regular expressions over the input alphabet Σ.
If **P** does not contain ε then there exists **R** such that

$$\mathbf{R = Q + RP} \quad\text{..........................} \quad (1)$$

We will replace **R** by **QP*** in equation (1).

Consider R.H.S of equation (1).
$$= Q + QP^* P$$
$$= Q(\varepsilon + P^*P)$$
$$= QP^* \qquad\qquad\qquad \text{(Since } \varepsilon + R^* R = R^*)$$

Thus **R=QP*** is proved.

To prove that $R = QP^*$ is a unique solution, we will now replace L.H.S. of equation (1) by $Q + RP$. Then it becomes $Q + RP$

But again R can be replaced by $Q + R\,P$

$$\therefore \qquad Q + RP \;=\; Q + (Q + RP)\,P$$
$$=\; Q + QP + RP^2$$

Again replace R by $Q + RP$
$$=\; Q + QP + (Q + RP)\,P^2$$
$$=\; Q + QP + QP^2 + RP^3$$

Thus if we go on replacing R by $Q + R\,P$, then we get,

$$Q + R\,P = Q + QP + QP^2 + \ldots + QP^i + RP^{i+1}$$

$$= Q(\varepsilon + P + P^2 + \ldots P^i) + RP^{i+1}$$

From equation (1),

$$R = Q\,(\varepsilon + P + P^2 + \ldots\, P^i) + RP^{i+1} \quad \ldots(2)$$

where $i \geq 0$

Consider equation (2),

$$R = Q\,(\varepsilon + P + P^2 + \ldots + P^i) + RP^{i+1}$$

$$\text{since } P^* = \varepsilon + P + P^2 + \ldots + P^i$$

$$\therefore \quad R = QP^* + RP^{i+1}$$

Let **w** be a string of length **i**.

In RP^{i+1} we have no string of less than **i+1** length. Hence **w** is not in set RP^{i+1}.

Hence **R** and **QP*** represent the same set. Hence it is proved that

$$R = Q + RP \text{ has a unique solution.}$$

$$\therefore R = QP^*.$$

Example 3.5

Show that $(0^*\ 1^*)^* = (0 + 1)^*$

Solution: Consider L.H.S. $= (0^*\ 1^*)^*$

$$= \{\varepsilon, 0, 00, 1, 11, 111, 01, 10, \ldots\}$$

$$= \{\varepsilon, \text{ any number of 0's, any number of 1's, any combination}$$
$$\text{of 0's and 1's}\}$$

Consider R.H.S $= (0 + 1)^*$

$$= \{\varepsilon, 0, 00, 1, 11, 111, 01, 10, \ldots\}$$

$$= \{\varepsilon, \text{ any number of 0's, any number of 1's, any combination of}$$
$$\text{0's and 1's}\}$$

Hence, L.H.S. = R.H.S. is proved.

Example 3.6

Show that $(ab)^* \neq (a^*b^*)$

Solution: Consider

$$\text{L.H.S.} = (ab)^* = \{\varepsilon, ab, abab, ababab, \ldots\}$$

and $\text{R.H.S.} = (a^*b^*) = \{\varepsilon, a, aa, aaa, b, bb, bbb, ab, abb, aab \ldots\}$

Note that in the L.H.S, the **R.E.** will give the strings of **ab** combination i.e., ab, abab, ababab and so on. But in R.H.S. there is possibility of **ab** but not of **abab** and so on.

Thus L.H.S.≠ R.H.S. is proved.

Example 3.7

Show that $(r + s)^* \neq r^* + s^*$

Solution: Let us consider the L.H.S. first,
$$\text{L.H.S.} = (r + s)^*$$
$$= \{\varepsilon, r, rr, s, ss, rs, sr, rsrs, ... \}$$
$$= \{\varepsilon, \text{any combination of r's and s's}\}$$
$$\text{R.H.S} = r^* + s^*$$
$$\{\varepsilon, r, rr, rrr, s, ss, ...\}$$
$$= \{\varepsilon, \text{any number of r's or any number s's}\}$$
Note that in R.H.S. there is no combination of r's and s's together.

Hence   L.H.S.≠ R.H.S. is proved.

3.3 CONSTRUCTING FINITE AUTOMATA FOR A GIVEN REGULAR EXPRESSION.

Construction of FA from a given regular expression can be diagramatically illustrated as follows

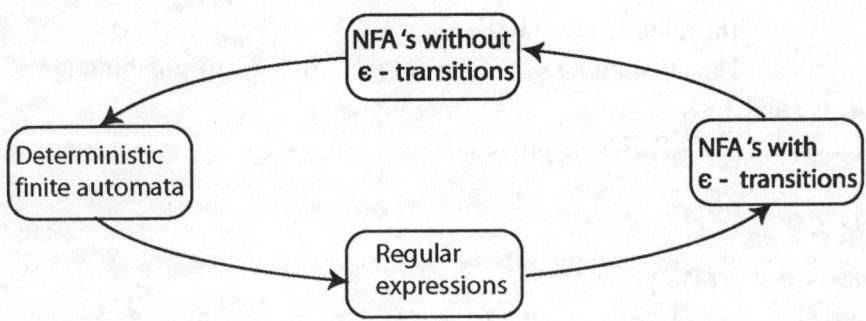

Every DFA can be represented as a regular expression. Every regular expression can be converted to NFA with ε -transitions (or ε-moves). Every NFA with e-transitions can be converted to NFA without ε-transitions. Finally every NFA can be converted to a DFA.

3.3.1 EQUIVALENCE OF NFA AND REGULAR EXPRESSION

Theorem 1: Let **r** be a regular expression, and then there exists a NFA with ε-transitions that accepts **L(r)**.

Proof: This theorem can be proved by induction method.

The basis of induction will be by considering **r** has zero operators.

Basis (zero operators) – Now, since **r** has zero operators, means **r** can be either ε or Ø or **a** for some **a** in input set Σ.

The finite automata for the same can be written as

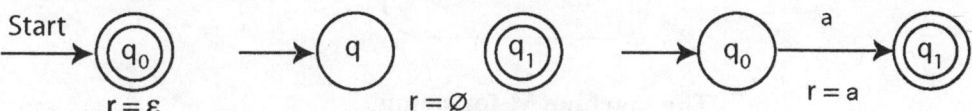

(no path to final state)

Finite automata for given regular expression

Induction: This theorem can be true for **n** number of operators. where **n** is greater than or equal to **1**. The regular expression contains equal to or more than one operator.

In any type of regular expression there are only three cases possible.

      1. Union   2. Concatenation   3. Closure.

Let us see each case,

Case 1: Union

      Let $r = r_1 + r_2$ where r_1 and r_2 are the regular expressions. There exists two NFA's $M_1 = (Q_1, \Sigma_1, \delta_1, q_1 \{f_1\})$ and $M_2 = (Q_2, \Sigma_2, \delta_2, q_2 \{f_2\})$. $L(M_1) = L(r_1)$ means the language states by regular expression r_1 is same which is represented by M_1. Similarly $L(M_2) = L(r_2)$. Q_1 represents the set of all the states in machine M_1. Q_2 represents the set of all the states in machine M_2.

      We assume that Q_1 and Q_2 are disjoint.

Let q_0 be new initial state and f_0 be the new final state, we will form $M = ((Q_1 \cup Q_2 \cup \{q_0, f_0\}), (\Sigma_1 \cup \Sigma_2), \delta, q_0 \{f_0\})$.

The δ is denoted by,

    i) $\delta (q_0, \varepsilon) = \{q_1, q_2\}$

    ii) $\delta (q, a) = \delta_1 (q, a)$ for q in $Q - \{f_1\}$ and a in $\Sigma_1 \cup \{\varepsilon\}$.

    iii) $\delta (q, a) = \delta_2 (q, a)$ for q in $Q_2 - \{f_2\}$ and **a** in $\Sigma_2 \cup \{\varepsilon\}$.

    iv) $\delta_1 (f_1, \varepsilon) = \delta_2 (f_2, \varepsilon) = \{f_0\}$

All the moves are now present in machine M which is as shown in the figure.

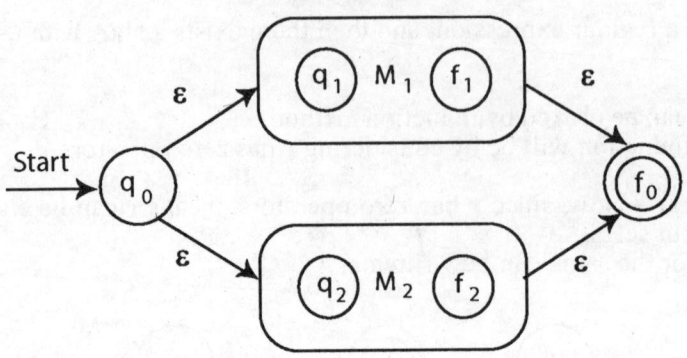

The machine M for union

The construction of machine **M** shows the transition from q_0 to f_0 must begin by going to q_1 or q_2 on ε. If the path goes to q_1, then it follows the path in machine M_1 and goes to the state f_1 and then to f_0 on ε. Similarly, if the path goes to q_2, then it follows the path in machine M_2 and goes to state f_2 and then to f_0 on ε. Thus the $L(M_1) \cup L(M_2)$. That means either the path in machine M_1 or M_2 will be followed.

Case 2: Concatenation

Let $r = r_1 r_2$, where r_1 and r_2 are two regular expressions. The M_1 and M_2 denote the two machines such that $L(M_1) = L(r_1)$ and $L(M_2) = L(r_2)$.
The construction of machine **M** will be

$$M = (Q_1 \cup Q_2, \Sigma_1 \cup \Sigma_2, \delta, \{q_1\}, \{f_2\})$$

The mapping function δ will be given as

i) $\delta(q, a) = \delta_1(q, a)$ for q in $Q_1 - \{f_1\}$ and **a** in $\Sigma_1 \cup \{\varepsilon\}$

ii) $\delta(f_1, \varepsilon) = \{q_2\}$

iii) $\delta(q, a) = \delta_2(q, a)$ for q in Q_2 and **a** in $\Sigma_2 \cup \{\varepsilon\}$

The machine **M** is shown in the figure.

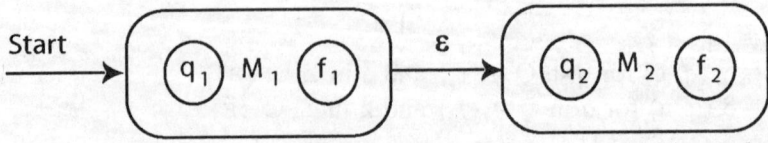

Machine M for concatenation

The initial state is q_1 by some input a, the next state will be f_1. And on receiving ε the transition will be from f_1 to q_2 and the final state will be f_2. The transition from q_2 to f_2 will be on receiving some input b.

Thus $L(M) = ab$

i.e. a is in $L(M_1)$ and b is in $L(M_2)$.

Hence we can prove $L(M) = L(M_1) L(M_2)$.

Case 3: Closure

Let $r = r_1^*$ where r_1 be a regular expression.

The machine M_1 is such that $L(M_1) = L(r_1)$.

Then construct $M = (Q_1 \cup \{q_0, f_0\}, \Sigma_1, \delta_0, \{q_0\}, \{f_0\})$

The mapping function δ is given by,

i) $\delta(q_0, \varepsilon) = \delta(f_1, \varepsilon) = \{q_1, f_0\}$

ii) $\delta(q, a) = \delta_1(q, a)$ for q in $Q_1 - \{f\}$ and a in $\Sigma_1 \cup \{\varepsilon\}$

The machine M will be

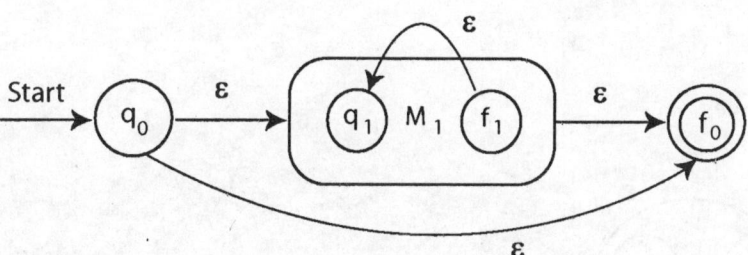

The machine M_1 shows that from q_0 to q_1 there is a transition on receiving ε. Similarly, from q_0 to f_0 on ε there is a path. The path exists from f_1 to q_1 on ε i.e., a back path. Similarly a transition from f_1 to f_0 final state, on receiving ε. The total recursion is possible. Thus one can derive, ε, a, aa, aaa,..... for the input a

Thus $L(M) = L(M_1)^*$ is proved.

Example 3.8

Design a FSA for the language 0*1*.

Solution: Let $r = r_1 r_2$

where $r_1 = 0^*$

$r_2 = 1^*$

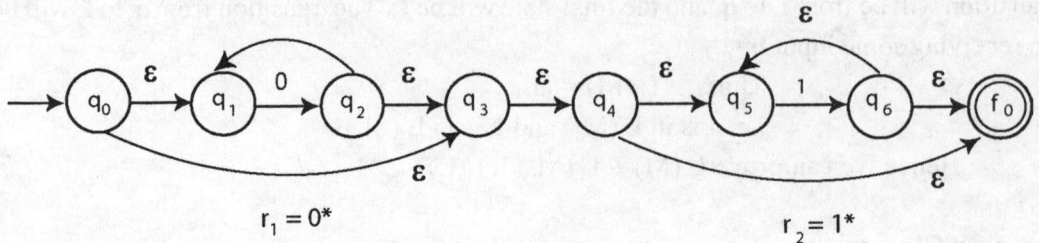

$$r_1 = 0^*$$ $$r_2 = 1^*$$

Example 3.9

Design a FSA for the language $(0 + 1)^* \ 111^*$

Solution: Let $\mathbf{r} = \mathbf{r_1 r_2 r_3 r_4}$

where $r_1 = (0 + 1)^*$

$r_2 = 1$

$r_3 = 1$

$r_4 = 1^*$

NFA with ε – moves

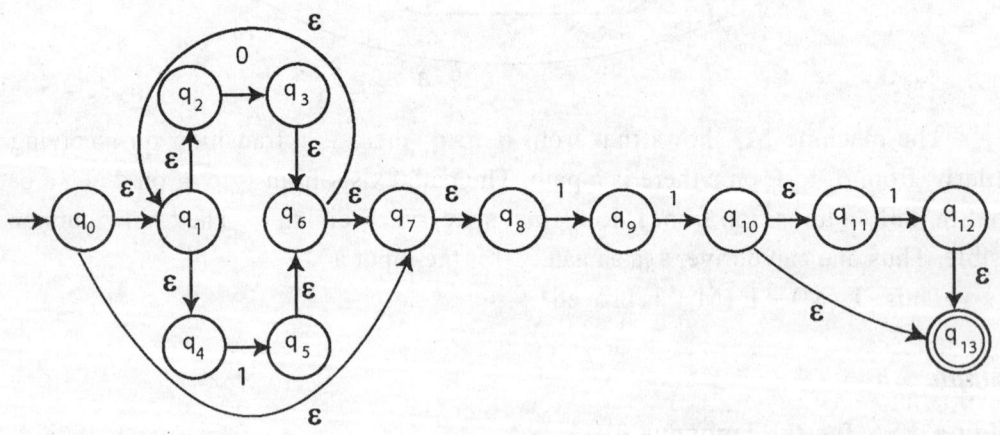

$$(0 + 1)^* \ 111^*$$

NFA without ε– moves

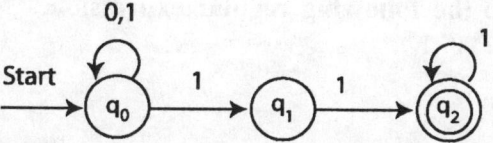

Example 3.10

Design a FSA for the language $(0^*1\ 1^* + 101)$

Solution:

Let $r = r_1 + r_2$
where $r_1 = 0^*1\ 1^*$
$\qquad r_2 = 101$

NFA corresponding to the given R.E is as follows

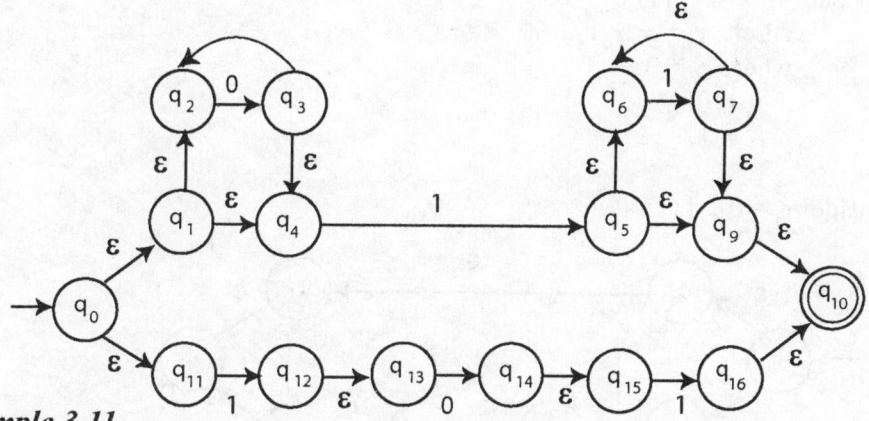

Example 3.11

Design DFA for the regular expression $(010)^*$

Solution:

The language generated by RE - $(010)^*$
$\qquad L = \{\ , 010, 010010, 010010010, ...\}$

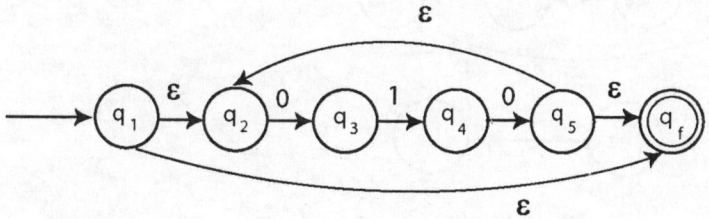

Example 3.12

Construct an NFA equivalent to the following regular expression.

$$r = 10 + (0 + 11) \, 0^* \, 1$$

Solution:

Let $r = r_1 + r_2$

   where $r_1 = 10$

         $r_2 = (0 + 11) \, 0^* \, 1$

   Consider $r_1 = 10$

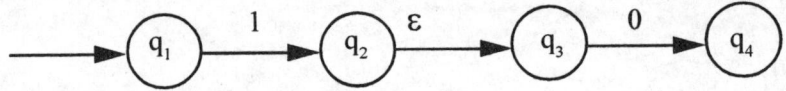

   Consider $r_2 = (0 + 11) \, 0^* \, 1$

       Let $r_2 = r_3 \, r_4 \, r_5$

     where $r_3 = 0 + 11$

         $r_4 = 0^*$

         $r_5 = 1$

   Consider $r_3 = 0 + 11$

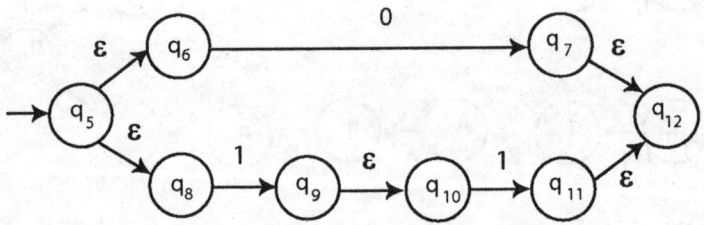

   Consider $r_4 = 0^*$

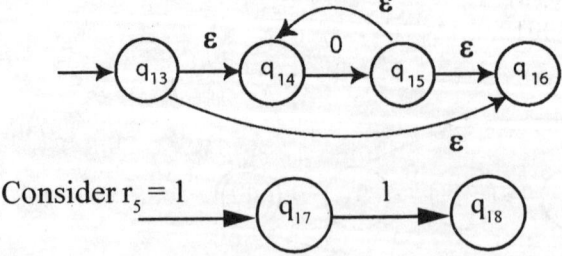

   Consider $r_5 = 1$

Transition diagram for $r_2 = (0 + 11) 0^* 1$

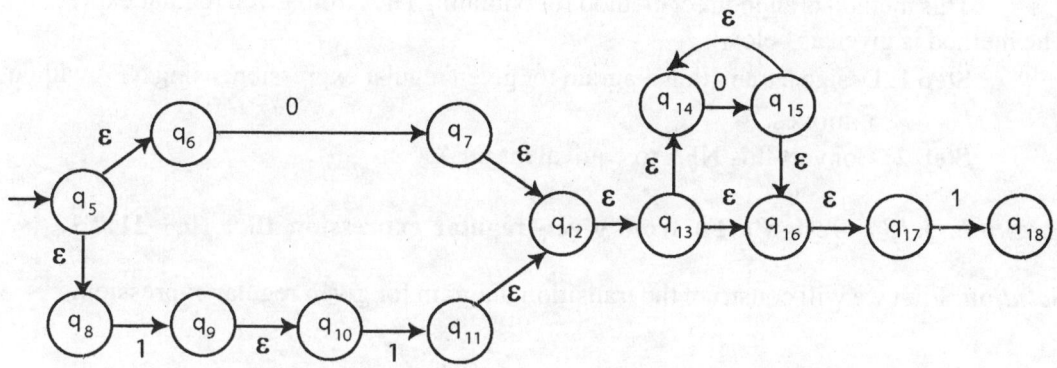

∴ The transition diagram for $r = r_1 + r_2$ i.e., union of r_1 and r_2 is

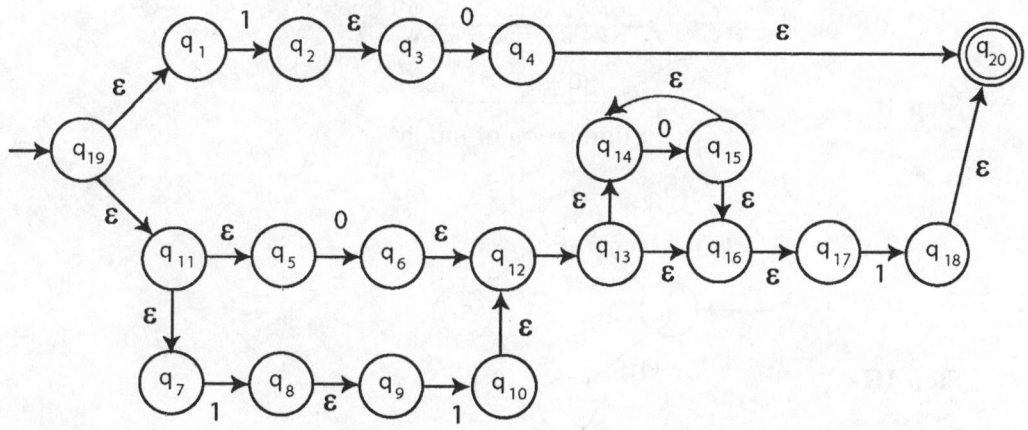

∴ Minimized automata for the above is

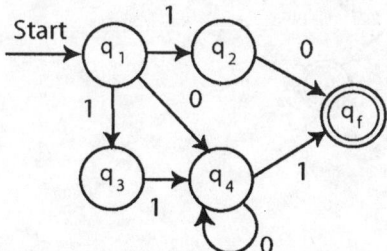

3.3.2   DIRECT METHOD FOR CONVERSION OF RE TO DFA

This method is an indirect method for obtaining DFA from given regular expression. The method is given as below:

Step 1: Design a transition diagram for given regular expression, using NFA without ε-moves.

Step 2: Convert this NFA to equivalent DFA.

Example 3.13   **Design a FA from given regular expression $10 + (0 + 11)0^*1$.**

Solution: First we will construct the transition diagram for given regular expression.

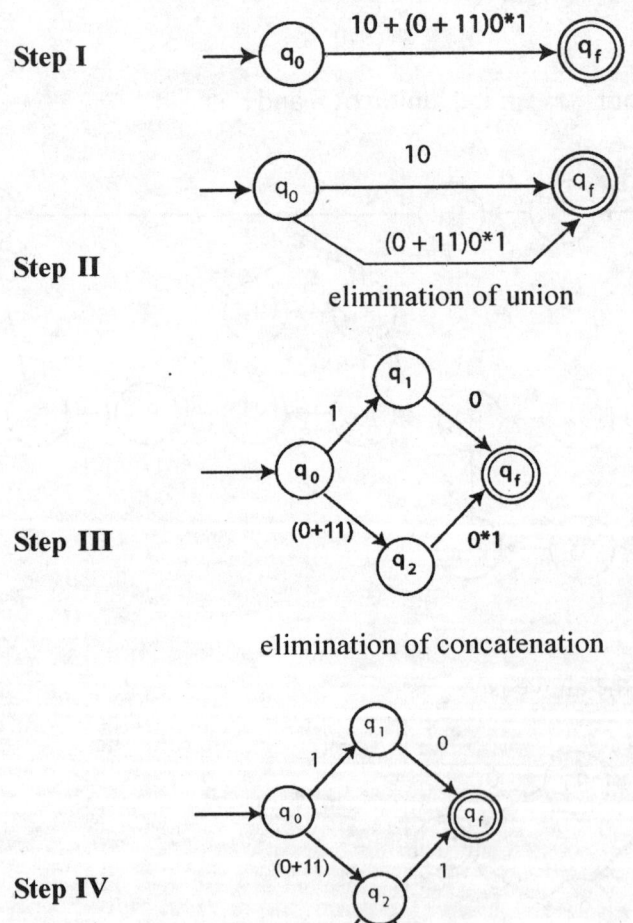

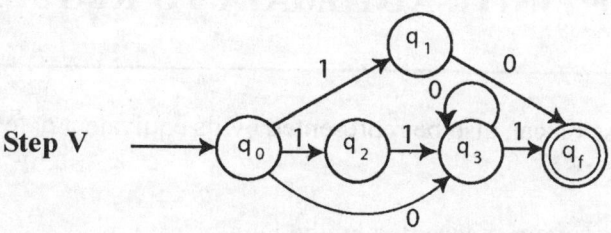

Step V

NFA for   10+(0+11) 0*1

Now we have got NFA without ε. Now we will convert it to required DFA for that, we will first write a transition table for this NFA.

State \ Input	0	1
→ q_0	q_3	$\{q_1, q_2\}$
q_1	q_f	∅
q_2	∅	q_3
q_3	q_3	q_f
$\;q_f$	∅	∅

The equivalent DFA will be

State \ Input	0	1
→ $[q_0]$	$[q_3]$	$[q_1, q_2]$
$[q_3]$	$[q_3]$	$[q_f]$
$[q_1, q_2]$	$[q_f]$	$[q_3]$
$\;[q_f]$	∅	∅

3.4 CONVERSION OF FINITE AUTOMATA TO REGULAR EXPRESSIONS

Theorem: The regular expression can also be represented by its equivalent deterministic finite automata.

Proof: Let, **L** be the set of the language accepted by the DFA.
The DFA can be denoted by the

$$M = (\{q_1, q_2, \ldots q_n\}, \Sigma, \delta, q_1, F)$$

Let $r_{ij}^{\ k}$ denotes the set of the strings **x** such that $\delta(q_i, x) = q_j$. The q_i and q_j indicates source state to target state respectively. The inputs are going through the states of finite automata means that with some input entering into the states and coming out of it. The value of **k** is always less than **i** or **j**.
The $r_{ij}^{\ k}$ is denoted by,

$$r_{ij}^{\ k} = r_{ik}^{\ k-1} (r_{kk}^{\ k-1})^* r_{kj}^{\ k-1} \cup r_{ij}^{\ k-1}$$

$$r_{ij}^{\ 0} = \begin{cases} \{\alpha \mid \delta(q_i,\alpha) = q_j\} & \text{if } i \neq j \\ \{\alpha \mid \delta(q_i,\alpha) = q_j\} \cup \{\epsilon\} & \text{if } i = j \end{cases}$$

We have to show that for each i, j, k there exists a regular expression $r_{ij}^{\ k}$ denoting the language $r_{ij}^{\ k}$. We will put the induction on **k** basis (k = 0). The $r_{ij}^{\ 0}$ is a set of strings each of which is either ϵ or a single symbol. The $r_{ij}^{\ 0}$ is based on $\delta(q_i, a) = q_j$. The $r_{ij}^{\ 0}$ denotes the set of such a's, if there is no such **a** then it will be taken a's $\emptyset$. If i = j the all the a's + ϵ will be set.

Induction: The formula for getting the language $r_{ij}^{\ k}$ is given by regular expression.

$$r_{ij}^{\ k} = (r_{ik}^{\ k-1}) (r_{kk}^{\ k-1})^* (r_{kj}^{\ k-1}) + r_{ij}^{\ k-1}$$

which completes the induction

To get the final regular expression, we have to simply get the language $r_{ij}^{\ n}$, where **i** indicates start state, and **j** indicates final state and **n** will be number of items. If there are **p** number of paths, leading to final state, the

$$r_{ij}{}^n = r_{ij1}{}^n + r_{ij2}{}^n + \ldots + r_{ijp}{}^n$$

where $\mathbf{F}$ is a set of final states $F = \{q_{j1}, q_{j2,} q_{jp}\}$.

Example 3.14
Construct the regular expression for the finite automata given in figure.

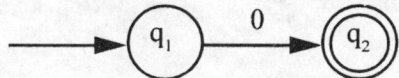

Finite Automaton

Solution:
Let us build the table when k=0

$r_{ij}{}^k$	regular expression
$r_{11}{}^0$	ε
$r_{12}{}^0$	0
$r_{21}{}^0$	$\varnothing$
$r_{22}{}^0$	ε

In the above table we have calculated the values as $\mathbf{r}_{ij}$ will indicate the set of all the input strings from $\mathbf{q}_i$ to $\mathbf{q}_j$, If $\mathbf{i} = \mathbf{j}$ then we add ε with the input string. If $\mathbf{i} \neq \mathbf{j}$ and there is no path from $\mathbf{q}_i$ to $\mathbf{q}_j$ then we add $\varnothing$.

Let us compute $\mathbf{r}_{11}{}^0$

$\mathbf{r}_{11}{}^0$ where $i = 1, j = 1, k = 0$. There is no path from $\mathbf{q}_i$ to $\mathbf{q}_j$ but $\mathbf{i} = \mathbf{j}$. So we add ε in the $\mathbf{k} = \mathbf{0}$ column at $\mathbf{r}_{11}{}^0$ row.

Similarly

$\quad\quad r_{12}{}^0 = $ The input from q_1 to q_2

$r_{12}{}^0 = 0$

$r_{21}{}^0 =$ No input from q_2 to q_1 + since $i \neq j$.

So we add $\emptyset$ over there.

$r_{22}{}^0 =$ No input from q_2 to q_2 and since $i = j$, We will add ε.

$r_{ij}{}^k = r_{ik}{}^{k-1} (r_{kk}{}^{k-1})^* r_{kj}{}^{k-1} + r_{ij}{}^{k-1}$

Let us build the table when $k = 1$

$r_{ij}{}^k$	Computation for k=1	Reg. Exp.
$r_{11}{}^1$	$r_{ij}{}^k = r_{ik}{}^{k-1} (r_{kk}{}^{k-1})^* r_{kj}{}^{k-1} + r_{ij}{}^{k-1}$ $i = 1, j = 1, k = 1$ $r_{11}{}^1 = r_{11}{}^0 (r_{11}{}^0)^* (r_{11}{}^0) + r_{11}{}^0$ $= \varepsilon (\varepsilon)^*. \varepsilon + \varepsilon$ $r_{11}{}^1 = \varepsilon$	ε
$r_{12}{}^1$	$i = 1, j = 2, k = 1$ $r_{12}{}^1 = r_{11}{}^0 (r_{11}{}^0)^* (r_{12}{}^0) + r_{12}{}^0$ $r_{12}{}^1 = \varepsilon (\varepsilon)^* (0) + 0$ $= \varepsilon .0 + 0$ $r_{12}{}^1 = 0$	0
$r_{21}{}^1$	$i = 2, j = 1, k = 1$ $r_{21}{}^1 = r_{21}{}^0 (r_{11}{}^0)^* (r_{11}{}^0) + (r_{21}{}^0)$ $= \emptyset (\varepsilon)^* \varepsilon + \emptyset$ $= \emptyset + \emptyset \quad \therefore \quad \emptyset \varepsilon = \emptyset$ $r_{21}{}^1 = \emptyset$	$\emptyset$
$r_{22}{}^1$	$i = 2, j = 2, k = 1$ $r_{22}{}^1 = r_{21}{}^0 (r_{11}{}^0)^* (r_{12}{}^0) + r_{22}{}^0$ $= \emptyset.(\varepsilon)^* (0) + \varepsilon$ $= \emptyset + \varepsilon$ $r_{22}{}^1 = \varepsilon$	ε

Now let us compute for final state, which denotes the regular expression.

r_{12}^2 will be computed, because there are total 2 states and final state is q_2 whose start state is q_1. $r_{12}^2 = (r_{12}^1) \, (r_{22}^1)^* (r_{22}^1) + (r_{12}^1)$

$$= 0(\varepsilon)^*(\varepsilon) + 0$$

$$= 0 + 0$$

$r_{12}^2 = 0$ which is a final regular expression.

This is a language which accepts only a single **0**.

Example 3.15

Construct regular expression from given finite automata.

Solution:

As we have seen in previous example, we are following the important formula as

$$r_{ij}^k = r_{ik}^{k-1} \, (r_{kk}^{k-1})^* \, (r_{kj}^{k-1}) + r_{ij}^{k-1}$$

Let us compute when $k = 0$

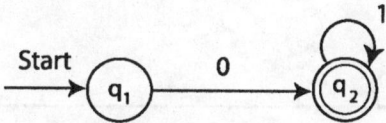

r_{ij}^k	regular expression
r_{11}^0	ε
r_{12}^0	0
r_{21}^0	$\varnothing$
r_{22}^0	$1 + \varepsilon$

Now we will calculate for k = 1

$r_{ij}^{\ k}$	**Computation for k=1**	**Reg. Exp.**
$r_{11}^{\ 1}$	$r_{ij}^{\ k} = r_{ik}^{\ k-1} (r_{kk}^{\ k-1})^* r_{kj}^{\ k-1} + r_{ij}^{\ k-1}$ $i = 1, j = 1, k = 1$ $r_{11}^{\ 1} = r_{11}^{\ 0} (r_{11}^{\ 0})^* (r_{11}^{\ 0}) + r_{11}^{\ 0}$ $\quad = \varepsilon\ (\varepsilon)^* (\varepsilon) +\ \varepsilon$ $r_{11}^{\ 1} =\ \varepsilon$	ε
$r_{12}^{\ 1}$	$i = 1, j = 2, k = 1$ $r_{12}^{\ 1} = r_{11}^{\ 0} (r_{11}^{\ 0})^* (r_{12}^{\ 0}) + r_{12}^{\ 0}$ $\quad = \varepsilon\ .0 + 0$ $\quad = 0 + 0$ $r_{12}^{\ 1} = 0$	0
$r_{21}^{\ 1}$	$i = 2, j = 1, k = 1$ $r_{21}^{\ 1} = r_{21}^{\ 0} (r_{11}^{\ 0})^* (r_{11}^{\ 0}) + (r_{21}^{\ 0})$ $\quad = \emptyset\ (\varepsilon)^*\ \varepsilon + \emptyset$ $r_{21}^{\ 1} = \emptyset$	$\emptyset$
$r_{22}^{\ 1}$	$i = 2, j = 2, k = 1$ $r_{22}^{\ 1} = r_{21}^{\ 0} (r_{11}^{\ 0})^* (r_{12}^{\ 0}) + r_{22}^{\ 0}$ $\quad = \emptyset.(\varepsilon)^* (0)\ + (1 + \varepsilon)$ $\quad = \emptyset + \varepsilon$ $r_{22}^{\ 1} =\ 1 + \varepsilon$	ε

Now for calculating regular expression we should compute for the path from start state to final state. That is from q_1 to q_2.

Considering $i = 1, j = 2, k = 2$

$$\therefore \quad r_{12}^{\ 2} = r_{12}^{\ 1} (r_{22}^{\ 1})^* r_{22}^{\ 1} + r_{12}^{\ 1}$$
$$= (0.1^*(1 + \varepsilon)) + 0$$
$$r_{12}^{\ 2} = 01^* + 0$$

This is a final regular expression. This is a language beginning with **0** and followed by any number of **1**'s or a single **0**.

3.4.1 ARDEN'S METHOD FOR CONVERTING DFA TO REGULAR EXPRESSION

As we have seen this theorem it is also used to write regular expression for a given deterministic finite automata or non-deterministic finite automata.

Let **P** and **Q** be regular expressions over Σ

If P does not contain ε, then the following equation in **R** i.e., $R = Q + R P$ has unique solution given by $R = Q P^*$

For a given DFA or NFA, regular expression is constructed as follows

Assume that transition graph does not have ε-moves and it has only one initial state say v_1. a_{ij} denotes a regular expression represents the set of labels of edges from v_i to v_j. Consequently, we can get the following set of equations, ε is added in equation of v_1 (initial state).

$$v_1 = v_1 a_{11} + v_2 a_{21} + \ldots + v_n a_{n1} + \varepsilon$$

$$v_2 = v_1 a_{12} + v_2 a_{22} + \ldots + v_n a_{n2}$$

$$v_m = v_1 a_{1m} + v_2 a_{2m +} \ldots + v_n a_{nm}$$

By repeatedly applying substitutions and Arden's lemma, we can get v_i (final state) in terms of a_{ij}'s. If there are more final states, then we have to take the union of all v_i's corresponding to final states.

Example 3.16

Construct RE from given DFA.

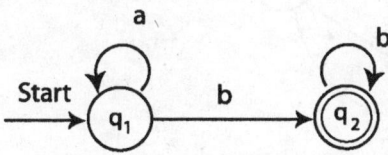

Solution: Let us write down the equations

$$q_1 = q_1 a + \varepsilon$$

Since q_1 is a start state, ε will be added and the input **a** is coming to q_1 from q_1 hence we write

State = input coming to it × source state of input

Similarly $\quad q_2 = q_1 b + q_2 b$

Let us simplify q_1 first

$$q_1 = q_1 a + \varepsilon$$

we can re-write it as $q_1 = \varepsilon + q_1 a$

which is similar to $R = Q + RP$ which further gets reduced to $R = QP^*$.

assuming $R = q_1, Q = \varepsilon, P = a$

we get $q_1 = \varepsilon\, a^*$

$q_1 = a^*$ (since $\varepsilon R = R$)

Substituting value of q_1 in q_2 we get

$$q_2 = q_1 b + q_2 b$$
$$q_2 = a^* b + q_2 b$$

We can compare this equation with $R = Q + R P$ assuming $R = q_2, Q = a^* b, P = b$

which gets reduced to $R = Q P^*$.

$\therefore$ $q_2 = a^* b\, b^*$

As $RR^* = R^+$

$q_2 = a^*\, b^+$

From the given DFA, if we want to find out the regular expression, we normally calculate the equation for final state.

Since in the given DFA q_2 is a final state and $q_2 = a^* b^+$.

$\therefore$ RE $= a^* b^+$ corresponding to the given DFA

Example 3.17

Find out the regular expression from given DFA.

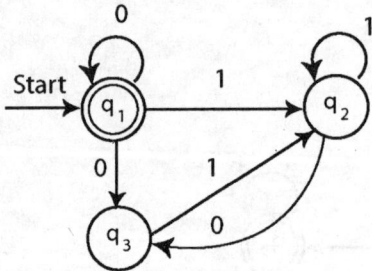

Solution: Let us solve the DFA by writing the regular expression, for each state.

$$q_1 = q_1\, 0 + q_3\, 0 + \varepsilon$$
$$q_2 = q_2 1 + q_3 1 + q_1 1$$
$$q_3 = q_2 0 + q_1 0$$

For getting the R.E for the DFA, we have to solve q_1 because q_1 is the only final state

$$q_2 = q_2 1 + q_2 01 + q_1 1$$
$$q_2 = q_2 (1 + 01) + q_1 1$$

we will compare R = Q + RP with above equation, so R = q_2, Q = $q_1 1$, P = (1 + 01)
which ultimately gets reduced to QP*.

$$q_2 = q_1 1 (1+01)^*$$

Substituting this value in **q_1**

$$q_1 = q_1 0 + q_3 0 + \varepsilon$$
$$= q_1 0 + q_2 00 + \varepsilon$$
$$= q_1 0 + q_1 (1(1 + 01)^*) 00 + \varepsilon$$
$$q_1 = q_1 (0 + 1(1 + 01)^* 00) + \varepsilon$$

again   R = Q + RP

where   R = q_1

            Q = ε

            P = 0 + 1(1 + 01)* 00

Hence $q_1 = \varepsilon [0 + p (1 + 01)^* 00]^*$

There fore R E $q_1 = [0 + 1(1 + 01)^* 00]^*$ since $\varepsilon R = R$

Example 3.18

Construct RE for the given DFA.

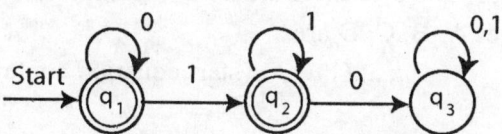

Solution:

Let us build the regular expression for each state.

$$q_1 = q_1 0 + \varepsilon$$
$$q_2 = q_1 1 + q_2 1$$
$$q_3 = q_2 0 + q_3 (0 + 1)$$

Since final states are **q_1** and **q_2**, we are interested in solving **q_1** and **q_2** only.

Let us see **q_1** first

$$q_1 = \varepsilon + q_1 0$$

which is   R = Q + RP equivalent so we can write

$$q_1 = \varepsilon (0)^*$$
$$q_1 = 0^* \qquad \text{(since } \varepsilon R = R)$$

Substituting this value into **q_2**, we will get

$$q_2 = 0^* 1 + q_2 1$$

$q_2 = 0^*1(1)^*$ (since $R = Q + RP \Rightarrow QP^*$)

Here we can observe that q_3 is dead state.

The regular expression is given by

$r = q_1 + q_2$
$\quad = 0^* + 0^*11^*$
$r = 0^* + 0^*1^+$ since $11^* = 1^+$
$RE = 0^* + 0^*1^+$

Example 3.19

Find the regular expression for the following DFA

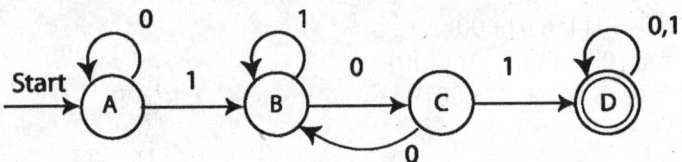

Solution:

There is only one initial state i.e., **A** and also there are no ε – moves.

∴   The equations for the states A, B, C, D are

$A = A0 + \varepsilon$ ....(1)   (start equation contains ε)

$B = A1 + B1 + C0$ ....(2)

$C = B0$ ....(3)

$D = C1 + D(0 + 1)$ ....(4)

Since **D** is the only final state, solve these equations for **D**

Consider equations (2) and (3), we have

$B = A1 + B1 + B00$

$B = A1 + B(1 + 00)$

It is of the form $R = Q + RP$

∴ $B = A1(1 + 00)^*$ .... (5)

(By Arden's theorem, if $R = Q + RP$ then $R = QP^*$)

Consider equation (4)

$D = C1 + D(0 + 1)$

$D = B01 + D(0 + 1)$ (since $C = B0$)

∴ By Arden's theorem, $D = B01(0 + 1)^*$

From (5), we have $D = A1(1 + 00)^* 01(0 + 1)^*$ .... (6)

From (1) $A = \varepsilon + A0$, by Arden's theorem

$A = \varepsilon 0^* \Rightarrow A = 0^*$

Substituting **A** in (6), D = 0*1 (1 + 00)* 01 (0 + 1)*

As **D** is the only final state, the regular expression is 0*1 (1 + 00)* 01 (0 + 1)*

Example 3.20
Construct regular expression corresponding to the state diagram given below

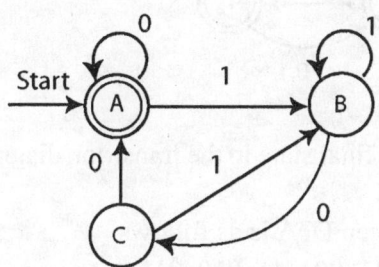

Solution:
If the initial state is equal to the final state in the transition diagram, then we have to include ε in the language

Therefore the language for the given DFA is as follows

∴　　L = {ε, 0, 100, 0100, 1100, ...}

From DFA there is only one initial state and there are no ε– moves.

The equations to A, B, C are

$$A = A0 + C0 + ε \qquad\qquad ....(1)$$
$$B = A1 + B1 + C1 \qquad\qquad ....(2)$$
$$C = B0 \qquad\qquad ....(3)$$

Solve the equations (1), (2) and (3) for **A**, since **A** is the only final state by applying substitutions and Arden's lemma.

Substitute　　C = B0 in (2), we get

$$B = A1 + B1 + B01$$
$$B = A1 + B (1 + 01),$$

By applying Arden's theorem (R = Q + RP ⇒ R = QP*) we get

$$B = A1 (1 + 01)^*$$

Consider　A = A0 + C0 + ε

$$A = A0 + B00 + ε$$
$$= A0 + A1 (1 + 01)^*00 + ε$$

∴　　A = ε + A (0 + 1 (1 + 01)* 00)

By Arden's theorem,　A = ε (0 + 1(1 + 01)*00)*

∴　A = (0 + 1 (1 + 01)*00)*

As **A** is the only final state, the R.E corresponding to the given diagram is
(0 + 1 (1 + 01)* 00)*

Example 3.21

Find regular expression for the following DFA

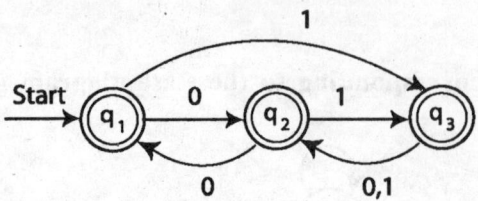

Solution: If the initial state is equal to the final state in the transition diagram, then we have to include ε in the language.

Therefore the language for the given DFA is as follows

$$\therefore \quad L = \{\varepsilon, 0, 1, 01, 10, 11, 001, 11, 010, 011 \ldots\}$$

There is only initial state and there are no null or $\varepsilon-$ moves

The equations corresponding to the states q_1, q_2, q_3 are as follows.

$$q_1 = q_2 0 + \varepsilon \qquad\qquad\qquad \ldots\ldots\ldots(1)$$
$$q_2 = q_1 0 + q_3 (0 + 1) \qquad\qquad \ldots\ldots\ldots (2)$$
$$q_3 = q_1 1 + q_2 1 \qquad\qquad\qquad \ldots\ldots\ldots (3)$$

Here q_1, q_2, q_3 are final states, we have to solve for q_1, q_2, q_3.

$$\therefore \quad R.E = q_1 + q_2 + q_3.$$

Consider equations (3) and (1), we have

$$q_3 = q_1 1 + q_2 1$$
$$= (q_2 0 + \varepsilon) 1 + q_2 1$$
$$\therefore \quad q_3 = q_2 (01 + 1) + 1 \qquad\qquad \ldots\ldots\ldots (4)$$

Substitute $\mathbf{q_3}$ and $\mathbf{q_1}$ in $\mathbf{q_2}$, we get

$$q_2 = (q_2 0 + \varepsilon) 0 + (q_2 (01 + 1) + 1) (0 + 1)$$
$$= q_2 00 + 0 + q_2 (01 + 1) + 1) (0 + 1) + 1(0 + 1)$$
$$\therefore \quad q_2 = 0 + 1(0 + 1) + q_2 (00 + (01 + 1) (0 + 1))$$

It is of the form $R = Q + RP$

By Arden's theorem if $\mathbf{R = Q + RP}$ then $\mathbf{R = QP^*}$

$$\therefore \quad q_2 = (0 + 1(0 + 1)) (00 + (01 + 1) (0 + 1))^* \qquad\qquad \ldots\ldots (5)$$

Substitute $\mathbf{q_2}$ in $\mathbf{q_1}$, we get

$$q_1 = q_2 0 + \varepsilon$$
$$\therefore \quad q_1 = (0 + 1(0 + 1)) (00 + (01 + 1) (0 + 1))^* 0 + \varepsilon \qquad \ldots\ldots\ldots(6)$$

Substitute $\mathbf{q_2}$ in $\mathbf{q_3}$ i.e., in equation (4), we get

$$q_3 = q_2 (01 + 1) + 1$$
$$\therefore \quad q_3 = (0 + 1 (0 + 1)) (00 + (01 + 1) (0 + 1))^* (01 + 1) + 1 \qquad \ldots\ldots\ldots (7)$$

Since the final states are q_1, q_2 and q_3, the regular expression $= q_1 + q_2 + q_3$.

$\therefore \quad q_1 + q_2 + q_3 = (0 + 1 (0 + 1)) (00 + (01 + 1) (0 + 1))^*0 + \varepsilon + (0 + 1 (0 + 1))$

$\quad (00 + (01 + 1) (0 + 1)) + (0 + 1 (0 + 1)) (00 + (01 + 1) (0 + 1))^* (01 + 1) + 1$

By taking common factor $(0 + 1 (0 + 1)) (00 + (01 + 1) (0 + 1))^*$

we get $\varepsilon + 1 + [(0 + 1 (0 + 1)) (00 + (01 + 1) (0 + 1))^*] [0 + \varepsilon + 01 + 1]$

$\therefore$ The regular expression for the given DFA is as follows.

$\quad \text{R.E} = \varepsilon + 1 + [(0 + 1 (0 + 1)) (00 + (01 + 1) (0 + 1))^*] [0 + \varepsilon + 01 + 1]$

$\quad\quad\quad = \varepsilon \mid 1 \mid [(0 + 1 (0 + 1)) (00 + (01 + 1) (0 + 1))^*] [0 \mid \varepsilon \mid 01 \mid 1]$

The language **L** from R.E is $L = \{\varepsilon, 1, 0, 10, 00, 01, 001, 000, 0001, 100, 101, 1001, \ldots\}$

The language **L** from R.E = The language **L** from DFA.

3.5 PUMPING LEMMA FOR REGULAR SETS

Theorem: This is a basic and important theorem used for checking whether a given string is accepted by regular expression or not.

Any language for which it is possible to design the finite automata is definitely the regular language.

Let $M = (Q, \Sigma, \delta, q_0, F)$ be a finite state machine with **n** states.

Let **L** be a regular language accepted by **M**.

Then a string $z \in L$ such that $\mid z \mid \geq n$ can be written as z = uvw satisfying the following conditions

$\quad\quad\quad\quad$ 1) $v \neq \varepsilon$

$\quad\quad\quad\quad$ 2) $\mid uv \mid \leq n$

$\quad\quad\quad\quad$ 3) The string $uv^iw \in L$ for all $i \geq 0$

The theorem states that given a string **z** belonging to **L**, we always find a substring v near the beginning of **z** that can be pumped repeated as many times as we like, keeps the resulting string in the language.

Proof: Suppose **L** is a regular language accepted by DFA, $M = (Q, \Sigma, \delta, q_0, F)$ with **n** states. Consider an input string $z = a_1 a_2 a_3 \ldots a_m$ where $m \geq n$ and a_i is an input symbol in Σ

The mapping function is defined as

$\quad\quad \delta (q_0, a_1\ a_2 \ldots a_m) = q_m$

Then we can write the string **z** as,

$\quad\quad z = uvw \quad\quad$ where

$\quad\quad u = a_1\ a_2 \ldots a_i$

$\quad\quad v = a_{i+1}\ a_{i+2} \ldots a_j$

$\quad\quad w = a_{j+1}\ a_{j+2} \ldots a_m$

This can be represented as shown below.

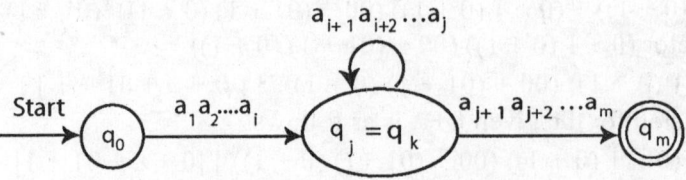

We observe that if **i= 0**, **u** may be empty.

If **j = n = m**, **w** may be empty, but **v** cannot be empty since i < j strictly.

Consider the automation on receiving of input **uv^{i}w** for i ≥ 0.

1) Case 1: If i = 0

      In this case, FA goes from initial state q_0 to q_j on input **u**.

Since $q_j = q_k$, it goes to accepting state q_m from q_k on input **w**.

      Hence it accepts the string **"uw"**

2) Case 2: If i > 0

      In this case, FA goes from initial state q_0 to q_j on input **u**, repeat many a times to q_j itself on input **v** and the goes to q_m on input **w**.

      Hence it accepts string **"uv^{i}w"** for all i ≥ 0

      ∴   uv^{i}w ∈ L

      Hence the theorem is proved.

Applications of pumping lemma:

1. Pumping lemma is used to prove that certain languages are not regular.

2. It is useful to determine whether the language accepted by a finite automaton is finite or infinite.

Example 3.22 **Prove that the language L = {a^p | p is prime} is not regular**

Solution: Let **L** be regular and **n** be the number of states in the FA accepting **L**.

      Let z = a^p.

      |z| = p and |z| > n

By pumping lemma, z can be represented by z = uvw with |uv| ≤ n, |v| ≥ 1, u, v, w are simply strings of **a**'s.

      Let us take v = a^m for some m ≥ 1,

      Let i = p + 1

we have z = uvw ⟹ |z| = |uvw| = p

Prove that |uv^{i}w| ≠ p   i.e., uv^{i}w ∉ L

$$|uv^iw| = |uvv^{i-1}w| = |uvw| + |v^{i-1}|$$
$$= |uvw| + |a^{m(i-1)}| \quad \text{(since } v = a^m\text{)}$$
$$= p + m\,(i\text{-}1)$$

$$= p + mp \qquad (\text{since } i = p+1)$$
$$= p\,(1 + m)$$

$|uv^i w| = p\,(1+m) \neq p$, which is not prime

$\therefore \quad uv^i w \notin L$, which is a contradiction

For example: $L = \{a, aa, aaa, aaaaa, \dots\}$

Let $z = aaa$

Applying pumping lemma

$uv^2 w = a\,(a)^2\,a = aaaa$, which is not in **L**.

$\therefore \quad uv^2 w \notin L$, which is a contradiction.

$\therefore \quad L = \{a^p \mid p \text{ is prime}\}$ is not regular.

Example **3.23** **Find whether the following is a regular set or not**
$$L = \{a^n b^n \mid n \geq 1\}$$

Solution: Given $L = \{a^n b^n \mid n \geq 1\}$

Assume **L** is regular and **n** is the number of states in FA accepting **L**

Let, $z = a^n b^n$

$$|z| = n + n = 2n > n$$

$\therefore \quad |z| > n$

By pumping lemma

$z = uvw$ such that $|uv| \leq n$ and $|v| \neq 0$.

We need to find **i** such that $uv^i w \notin L$ to get contradiction.

There are three cases for the string **v**

Case 1: **v** consists of only **a's** i.e., $v = a^k$ where $k \geq 1$

Let $i = 0$

$uv^i w = a^n b^n$

then $\qquad uw = a^{n-k} b^n$

since $k \geq 1$, $n-k \neq n$

$\therefore \quad uw \notin L$, which is a contradiction

For example: $L = \{ab, aabb, aaabbb, aaaabbbb, \dots\}$

Consider $\qquad z = aaabbb$. Here $n=3$

$\therefore$ Here, $\qquad v = aa = a^2$. (**v** consisting of only **a's**)

$\therefore \qquad\qquad k = 2 \geq 1$

$\therefore \qquad\qquad uw = abbb$

$$= a^{3-2}bbb = a^{n-k}b^n$$

$uw = a^1 b^3 \notin L$, which is a contradiction

Case 2: **v** consists of only **b's** , i.e., $v = b^l$ where $l \geq 1$

Let $i = 0$

$uv^i w = a^n b^n$

then $uw = a^n b^{n-l}$

since $l \geq 1$, $n \neq n-l$

∴ $uw \notin L$, which is a contradiction

For example: $L = \{ab, aabb, aaabbb, aaaabbbb, ...\}$

Consider $z = aaa\ bb\ b$

∴ $z = uvw$, $n = 3$

Here $v = bb = b^2$ (v consisting of only **b**'s)

∴ $v = b^l$, $l = 2 \geq 1$

∴ $uw = aaab$

∴ $uw = a^3b^l = a^nb^{n-l}$

∴ $uw = a^3b^l \notin L$, which is a contradiction

Case 3: **v** consists of both **a**'s and **b**'s i.e., $v = a^kb^l$ where $k, l \geq 1$

Let $i = 2$

$uv^iw = a^nb^n$

∴ $uv^iw = a^{n-k}a^kb^lb^{n-l}$

Then $uv^2w = a^na^kb^lb^n = a^{n+k}b^{l+n}$

∴ $uv^2w \notin L$, since $n+k \neq l+n$

∴ $uv^2w \notin L$, a contradiction.

For example: consider $aaabbb$, $n=3$

$v = a^2b^l$ i.e., a^kb^l where $k=1$, $l=1$

Consider uv^2w i.e., $aaa\ aab\ bbb = a^3\ a^2b^l\ b^3$

∴ $uv^2w = a^na^kb^lb^n$

∴ $uv^2w = a^5b^4 \notin L$

∴   In all the three cases, $uv^2w \notin L$ **L** is not regular

For example: consider the string $z = aaa\ abb\ bb$

Take $v = abbb = a^1b^3$

where $k=1$, $l = 3$

∴ $uv^2w = aaaaabbbbbbb$

$= aaaaabbbbbbb$

$= aaaa\ abbb\ bbbb$

$uv^2w = a^na^kb^lb^n$

$uv^2w \notin L$

∴ $L = \{a^n\ b^n \mid n \geq 1 \mid \}$ is not regular.

Example 3.24 **Show that the set $\{a^{i^2} \mid i \geq 1\}$ is not regular state and explain the theorem used**

Solution: Assume $L = \{a^{i^2} \mid i \geq 1\}$ consists of set of strings which are perfect squares.

Assume that **L** is regular.

$\therefore$ L = {a, aaaa, aaaaaaaaa, ...}

Consider the strings in the language **L**

a is a string of length **1**, which is a perfect square.

"aaaa" is a string of length **4**, which is a perfect square

"aaaaaaaaa" is a string of length **9**, which is a perfect square and so on

$\therefore$ L is a language consists of set of strings which are perfect squares.

Let $z = a^{i^2}$, then $|z| = n^2 > n$

$\therefore$ $|z| > n$

By pumping lemma

Let $z = uvw \in L$

$|z| = |uvw|$

$\therefore$ $n^2 = |u| + |v| + |w|$ .... (1)

Consider uv^2w

$|uv^2w| = |u| + 2|v| + |w|$ .... (2)

From (1) and (2) we know that $|u| + |v| + |w| < |u| + 2|v| + |w|$

i.e., $n^2 < |uv^2w|$ ....(3)

As $|uv| \le n$, we have $|v| \le n$ and $|v| \ge 1$

Therefore $|uv| \le n^2 + n$

We have $|v| \le n$ .... (4)

$\therefore$ $|uv^2w| = |u| + 2|v| + |w|$

$= |u| + |v| + |w| + |v|$

We have $|u| + |v| + |w| = n^2$ and $|v| \le n$.

$\therefore$ $|uv^2w| \le n^2 + n$. .... (5)

From (3) and (5), we have

$n^2 < |uv^2w| \le n^2 + n$ .... (6)

$\therefore$ $n^2 < |uv^2w| < n^2 + n + n + 1$ (since $n^2 + n < n^2 + n + n + 1$)

$\therefore$ $n^2 < |uv^2w| < n^2 + 2n + 1$

$\therefore$ $n^2 < |uv^2w| < (n+1)^2$

Hence $|uv^2w|$ strictly lies between n^2 and $(n+1)^2$, And there is no perfect square between n^2 and $(n+1)^2$

Therefore $uv^2w \notin L$, which is a contradiction.

$\therefore$ **L** is not regular.

For example: L = {a, aaaa, aaaaaaaaa, ...}

Let **L** be a regular language.

Take any string from the defined language and write it in **uvw** form such that $|v| \le |uv| \le n$ and $|v| \ge 1$ and apply pumping lemma theorem

Let z = aaaa where u = a, v = aa, w = a

uv^2w = aaaaaa, which is not a perfect square

$\therefore$ $uv^2w \notin L$, contradiction

$\therefore$ By pumping lemma if $uvw \in L$ then $uv^iw \in L$, for all $i \ge 0$

L is not regular

3.6 APPLICATIONS OF REGULAR EXPRESSIONS

The regular expressions can be modeled by finite automata. As we have solved many examples and experienced those regular expressions are effective representations of languages. The smaller unit of regular expressions can express the given language over certain input set.

1. Text editors: Text editors are some programs which are used for processing the text. For example, **UNIX** text editor uses the regular expression for substituting the strings, such as S/**bbb***/b/

Gives that substitute a single blank for the first string of two or more blanks found in a given line.

In the **UNIX** text editors any regular expression is converted to an NFA with ε-transitions, this NFA is then simulated directly.

2. Lexical analyzers: Compiler uses this program of lexical analyzer in the process of compilation. The task of lexical analyzer is to scan the input program and separate out the **tokens**.

For example: Identifier is a category of token in the source language and it can be identified by regular expression as

(letter) (letter + digit)*

If anything in the source language matches with this regular expression, then it is recognized as identifier. The letter is nothing but a set $\{A, B, \ldots Z, a, b, \ldots z\}^*$ and digit is $\{0, 1, \ldots 9\}^*$. The regular expression is effective way for identifying tokens from a language.

3.7 CLOSURE PROPERTIES OF REGULAR SETS

Applying operations on regular languages preserves the regular set. That is applying an operation such as union on two regular sets then the resulting set is also regular. These are called closure properties of regular sets.

The regular sets are closed under

1) Union, concatenation and kleene closure.
2) Complementation.
3) Intersection.
4) Substitution.
5) Homomorphisms.
6) Inverse homomorphisms.
7) Quotient with arbitrary sets.

1) Union, concatenation and kleene closure:

The regular sets are closed under union, concatenation and kleene closure. That is if L_1 and L_2 are two regular languages, then $L_1 \cup L_2$, $L_1 L_2$, L_1^* are also regular.

Proof: If L_1 and L_2 are regular sets then there exists regular expressions r_1 and r_2 denoting L_1 and L_2 respectively i.e., $L_1 = L(r_1)$ and $L_2 = L(r_2)$

By definition, $r_1 + r_2$ is a regular expression denoting the language $L_1 \cup L_2$
$$\text{i.e., } L_1 \cup L_2 = L(r_1 + r_2).$$
Therefore $L_1 \cup L_2$ is also regular
Similarly the concatenation and kleene closure of regular sets is also regular.

2) Complementation

The regular sets are closed under complementation.
i.e., if L is a regular set such that $L \subseteq \Sigma^*$, then $\Sigma^* - L$ is regular

Proof: We need to prove that if **L** is a regular set then its complement is also regular.

Let **L** be accepted by a DFA, $M = (Q, \Sigma, \delta, q_0, F)$ and $L \subseteq \Sigma^*$.
To accept $\overline{L}$ i.e., $(\Sigma^* - L)$, the machine must complement the final states of **M**. So the DFA accepting $\overline{L}$ is given by $M^1 = (Q, \Sigma, \delta, q_0, \{Q-F\})$. That is in M^1 all the non final states of **M** become as final states. This means the set of strings accepted by **M** are rejected by M^1 and vice –versa
$$\therefore \quad \overline{L} \text{ accepted by } M^1 \text{ is a regular set.}$$

3) Intersection

The class of regular sets is closed under intersection. i.e., if L_1 and L_2 are regular sets then $L_1 \cap L_2$ is also regular set.
Proof: We know that $\qquad L_1 \cap L_2 = \overline{\overline{L_1} \cup \overline{L_2}}$
Here the over bar denotes complementation of ($\overline{L_1} \cup \overline{L_2}$) over alphabets of L_1 and L_2
By applying the closure property of regular sets under complementation $\overline{L_1}$ and $\overline{L_2}$ are regular.
So their union $\overline{L_1} \cup \overline{L_2}$ is also regular, once again from the complementation property of regular sets $\overline{\overline{L_1} \cup \overline{L_2}}$ is regular.
$$\therefore \quad L_1 \cap L_2 = \text{ is regular.}$$

SOLVED PROBLEMS

Problem 1: What is a regular expression and define the same formerly? Explain with example?

Solution:

Regular sets: A regular set is a set of strings, for which there exists some finite automaton accepting that set i.e., if **R** is regular set then R = L(M) for some finite automaton **M**. Similarly if **M** is a finite automaton then **L(M)** is always a regular set.

Regular expressions: Regular expression is a notation to specify a regular set, hence for every regular expression there exist a finite automaton accepting the language specified by the regular expression. Similarly for every finite automata **M** there exists a regular expression notation specifying **L(M)**. Regular expressions are useful for representing certain sets of strings in an algebraic fashion.

A formal recursive definition of a regular expression over Σ as follows

1) Any terminal symbol (i.e., an element of Σ), ε are regular expressions when we view **a** in Σ as a regular expression, we denote it by **a**.

2) The union of two regular expressions R_1 and R_2, written as $R_1 + R_2$ is also a regular expression.

3) The concatenation of two regular expressions R_1 and R_2, written as R_1R_2 is also a regular expression.

4) The iteration (or closure) of a regular expression **R**, written as R^*, is also a regular expression.

5) If **R** is a regular expression, then (R) is also a regular expression.

6) The regular expressions over Σ are precisely those obtained recursively by the application of the rules 1– 5 once or several times

Note: Schematic representation of an DFA accepting **a**

Regular expression	Regular set	Finite automata
$\emptyset$	{}	q_o q_f
ε	{ε}	q_o
every **a** in Σ is a regular expression	{a}	$q_o \rightarrow q_f$
$r_1 + r_2$ or $r_1 \mid r_2$ is a regular expression	$R_1 \cup R_2$ (where R_1 and R_2 are regular sets corresponding to r_1 and r_2 respectively).	q_o, ε, $M(r_1)$, $M(r_2)$, ε, q_f
$r_1 r_2$ is a regular expression	R_1 / R_2 (where R_1 and R_2 are regular sets corresponding to r_1 and r_2 respectively)	$q_o \xrightarrow{\varepsilon} M(r_1) \xrightarrow{\varepsilon} M(r_2) \xrightarrow{\varepsilon} q_f$
r^* is a regular expression	R^* (where R is a regular set corresponding to **r**) {ε, a, aa, aaa....}	$q_0 \xrightarrow{\varepsilon} M(r) \xrightarrow{\varepsilon} q_f$, with ε loops
a+	{a, aa, aaa ....}	$q_0 \xrightarrow{\varepsilon} q_1 \xrightarrow{a} q_2 \xrightarrow{\varepsilon} q_f$, with ε loop

Problem 2: Find a regular expression to the language of all strings of {0, 1}*
        containing exactly two 0's

Solution:

The regular expression for the strings containing exactly two 0's is $1^*01^*01^*$
    This regular expression describes the words as 00, 001, 100, 010, 10101, 11001,
        101110, 10011....
            To make the word 001, first 1^* and second 1^* become ε and last 1^* becomes **1**.

Problem 3: What are regular expressions? Give rules for defining regular expressions
        over a given alphabet 'Σ'

Solution:

Regular expression: A regular expression is a notation to represent certain sets of strings
in an algebraic fashion. This notation involves a combination of strings of symbols from
some alphabet Σ, the symbols for null string ε, the star operator * and plus operator +.

Identity rules: An identity for an operator is a value such that when the operator is applied
to the identity and some other value, the result is other value. There are three laws for
regular expressions involving these concepts. They are

        (i) $\emptyset + L = L + \emptyset = L$

        (ii) $\varepsilon L = L \varepsilon = L$

        (iii) $\emptyset L = L \emptyset = \emptyset$

These three laws are powerful tools in simplification

Problem 4: Find the regular expression over alphabet {0,1}* *for*
        a) The language of strings containing odd length
        b) The language of strings of length less than or equal to 6
        c) The language of all strings containing atleast '1'

Solution:

    a)   In the given language, the string should have odd number of 1's
        so one possible regular expression is $(0 + 11)^*1$
        and another possible regular expression is $1(0 + 11)^*$
        $\therefore$ Required regular expression is $r = (0 + 11)^*1 + 1(0 + 11)^*$

    b)   The language of strings of length less than or equal to 6
        R.E for string of length one is 0+1
        R.E for string of length two is (0+1) (0+1)
similarly we can write the regular expressions for all strings of lengths upto 6

$\therefore$ The required regular expression is

R.E $= (0 + 1) + (0 + 1)(0 + 1) + (0 + 1)(0 + 1)(0 + 1) +$
$(0 + 1)(0 + 1)(0 + 1)(0 + 1) + (0 + 1)(0 + 1)(0 + 1)(0 + 1)(0 + 1) +$
$(0 + 1)(0 + 1)(0 + 1)(0 + 1)(0 + 1)(0 + 1) + \varepsilon$

c) The language of all string containing atleast **1**
In the given language, the strings should have atleast one **1**
so one possible R.E is $(0+1)^*1$ and
Another possible R.E is $1(0+1)^*$
$\therefore$ Required R.E is $r = (0+1)^*1+1(0+1)^*$.

Problem 5: Write regular expressions for each of the following languages over an alphabet {0, 1}
a) The set of all strings not containing "111"
b) The set of all strings in which every pair of adjacent 0's appears before any pair of adjacent 1's

Solution:

a) Strings should not contain "111",i.e. strings can have 0, 10, 110
so regular expression is $(0+10+110)^*$ and strings can have $(0+10+110)^*1$ and
$(0+10+110)^*11$
$\therefore$ The regular expression for the given set of strings
$(0+10+110)^*+(0+10+110)^*1+(0+10+110)^*11=(0+10+110)^*(\varepsilon +1+11)$
b) Strings should have "0011" for every pair of "00"
$\therefore$ All the strings containing "0011" are $(0011+1)^*$
$\therefore$ The required regular expression is $(01 + (0011+1)1^*) (0011+1)1^*$

Problem 6: Give the word description for the following regular expressions
a) $110^*(0+1)$
b) $1(0+1)^*101$

Solution:

a) $110^*(0+1)$
The NFA for the above regular expression is

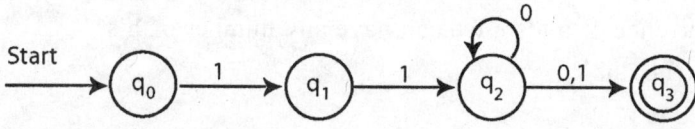

The language for the R.E $110^*(0+1)$ is L $= \{110, 111, 11000, , 1101, 11001, 110001......\}$
0^* means an empty string or only zeros

110* means two 1's are kept before the substring of 0*

110*(0+1) means 110* followed by either **0** or **1**

∴ It is a language that consists of string beginning with "11" and ending with either **0 or 1** and in between we can have any number of **0**'s or empty string(ε)

b) 1(0+1)*101

The language L = {1101, 1001101, 101101, 100101, 110101,......}

The NFA for the above regular expression is

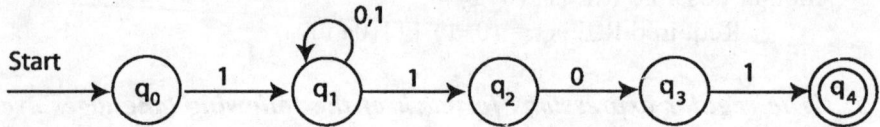

It is a language consisting of the strings beginning with **1** ending with **101** and in between we can have any combination of **0**'s and **1**'s

Problem 7: Construct a regular expression representing the following sets.
The set of all strings over {a, b} in which there are atleast two occurrences of 'b' between any two occurrences of 'a'

Solution:

Let ω be in the given set **L**

If ω has two **b**'s successively then ω = (bb)

If ω has atleast two occurrences of **b** means there should exist compulsory one pair of **b**. So ω = $(bb)^+$

If ω has only one **a** before the occurrence of pair of **b**'s, then

ω = a $(bb)^+$

If ω has one **a** after the occurrence of pair of **b**'s, then

ω = a $(bb)^+$a

Before the first **a**, we can have any number of **b**'s

ω = b^*a $(bb)^*$a

After the second occurrence of **a** also, we can have any number of **b**'s

∴ ω = b^*a (bb)*ab*

The language generated is

L = {abba, abbab, babbabb, bbabbab, babbbbabb, babbbbbbab,..}

Problem 8: Consider the following regular expression and construct the finite
       automaton *a)* *a + b*   *b)* *(a + b)**
              c) *a(a + b)**   *d)* *a(a + b)*b*

Solution:

a) Finite automata for the regular expression **a + b** is

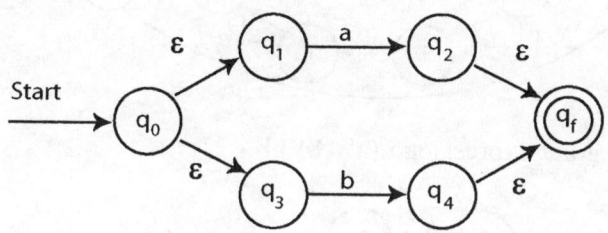

b) Finite automata for the regular expression **(a + b)*** is

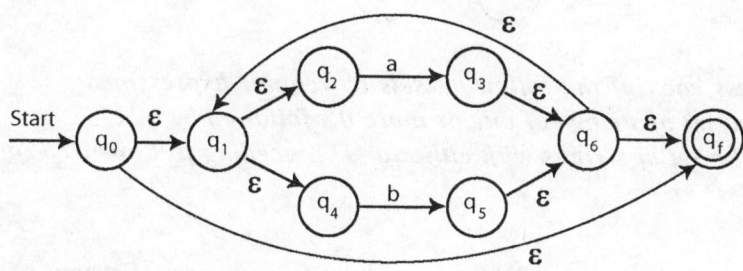

c) Finite automata for the regular expression **a(a + b)*** is

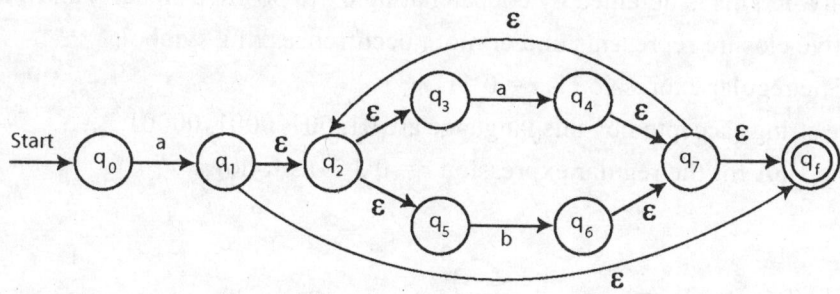

d) Finite automata for the regular expression **a(a + b)˙b** is

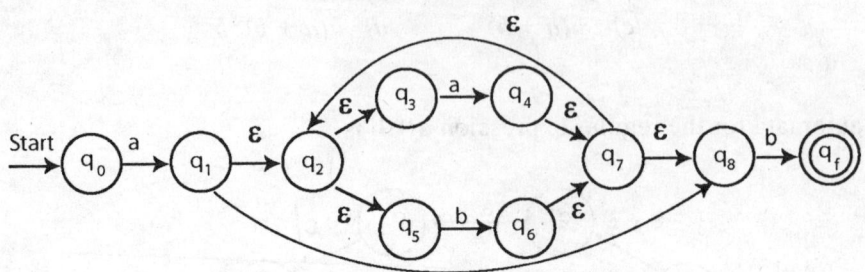

e) Finite automata for the regular expression **a(a + b)˙bb** is

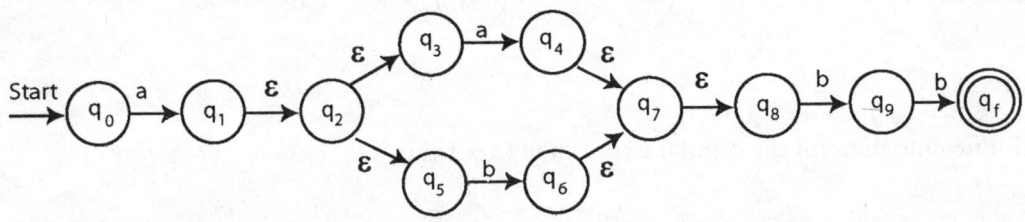

Problem 9: Express each of the following sets by regular expressions
        (i) The set of strings of one or more 0's followed by a '1'
        (ii) The set of strings with either a '1' preceding a '0' or a '0' preceding
        a '1'

Solution:

(i) The set of strings of one or more **0**'s followed by a **1** i.e., the set of strings of the language must start with atleast one **0**' followed by any number of **0**'s and end with a **1**. This is obtained by concatenating **0⁺** (a positive closure) and a **1**. The positive closure represents one or more occurrences of a symbol.

∴ The regular expression is r = 0^+1

The strings accepted by this language are 01, 001, 0001, 00001,........

The DFA for the regular expression r = 0^+1 is as follows

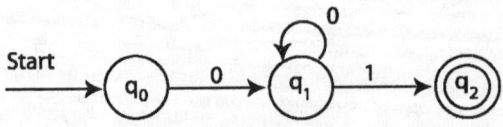

ii) The set of strings with either a **1** preceding a **0** or a **0** preceding a **1**

∴ The set of strings accepted by this language are ε, 10, 1010, 101010, 01, 0101, 010101, ...

∴ The regular expression is $(10)^* + (01)^*$.

The NFA with ε−moves for the R.E is as follows

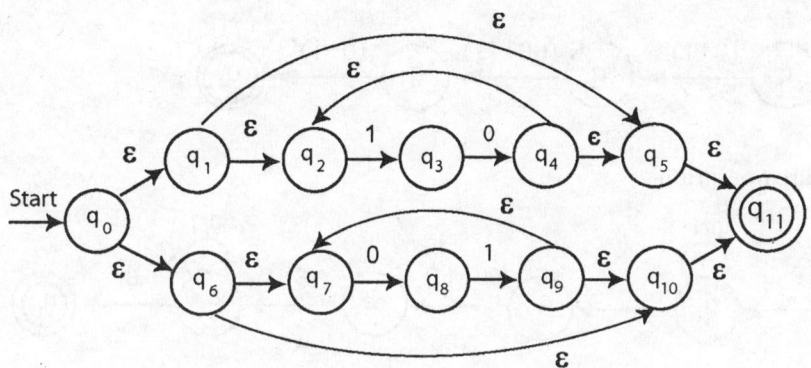

The DFA for the above NFA is as following

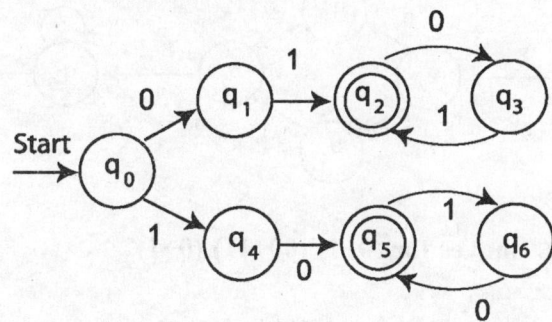

Problem 10: Construct finite automaton to accept the regular expression
$$(0+1)^*(00+11)(0+1)^*$$

Solution:

Given regular expression $r = (0+1)^*(00+11)(0+1)^*$

Step 1: Construct a transition graph equivalent to given regular expression using NFA with ε−moves

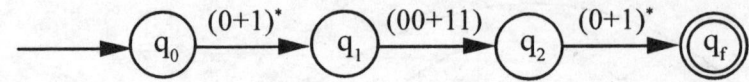

a) Eliminate concatenation operation

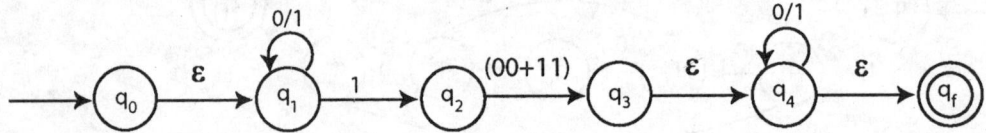

(b) Eliminate closure operation

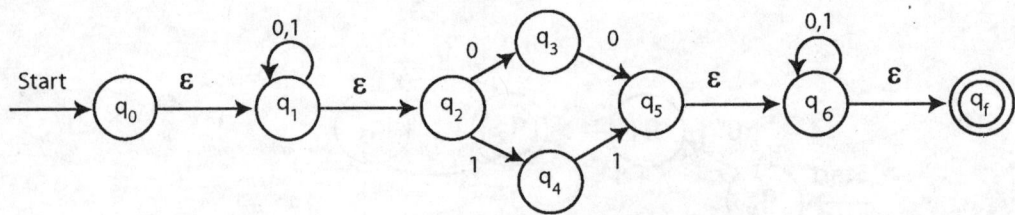

(c) Eliminate concatenation and union operations

NFA with ε−moves for (0+1)*(00+11) (0+1)*

Step 2: Convert NFA with ε−moves to NFA without ε−moves

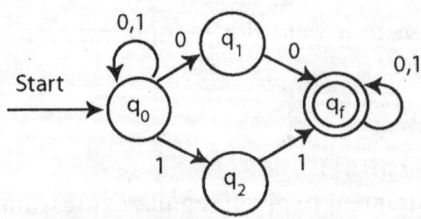

NFA without ε−moves for r= (0+1)*(00+11) (0+1)*

Step 3: Transition table for NFA

Q \ Σ	0	1
→ q_0	$\{q_0, q_1\}$	$\{q_0, q_2\}$
q_1	q_f	-
q_2	-	q_f
ⓠq_f	q_f	q_f

Step 4: Transition table for DFA

Q \ Σ	0	1
→$[q_0]$	$[q_0, q_1]$	$[q_0, q_2]$
$[q_0, q_1]$	$[q_0, q_1, q_f]$	$[q_0, q_2]$
$[q_0, q_2]$	$[q_0, q_1]$	$[q_0, q_2, q_f]$
$[q_0, q_1, q_f]$	$[q_0, q_1, q_f]$	$[q_0, q_2, q_f]$
$[q_0, q_2, q_f]$	$[q_0, q_1, q_f]$	$[q_0, q_2, q_f]$

Here we observe that $[q_0, q_1, q_f]$ and $[q_0, q_2, q_f]$ have identical transitions on inputs **0** and **1**. So we can reduce the number of states of DFA.

Replace all the occurrences of $[q_0, q_2, q_f]$ by $[q_0, q_1, q_f]$.

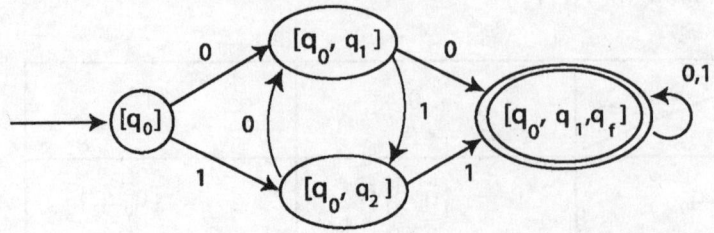

DFA for the R.E. r = (0+1)*(00+11) (0+1)*

*Problem 11: Construct NFA for the regular expression   (a|b)***aab*

Solution:

Let r = (a|b)*abb
r = r₁r₂ where r₁ = (a|b)* ,   r₂ = abb.
Consider r₁ = (a|b)*

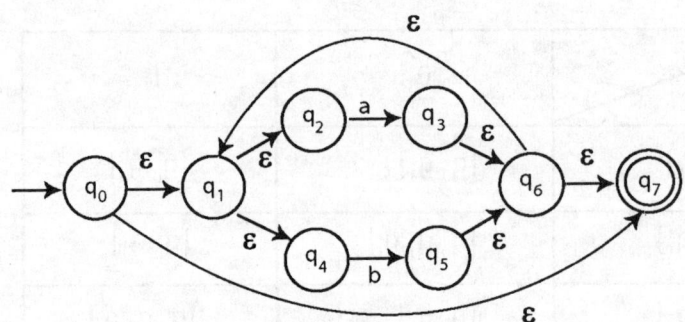

Consider r₂ = abb
Therefore, r = r₁r₂ i.e., by concatenating **r₁** and **r₂**

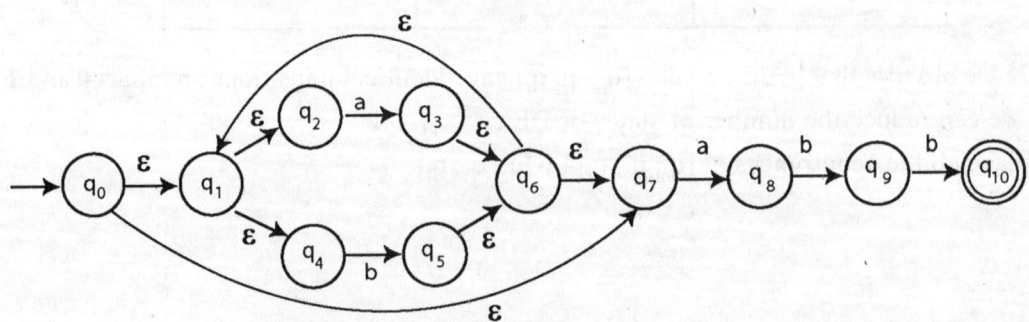

NFA without ε −moves

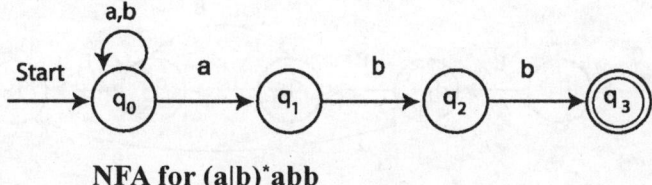

NFA for (a|b)*abb

Problem 12: Construct NFA for the R.E 01 + 10**

Solution:

Given regular expression $r = 01^* + 10^*$
Let $r = r_1 + r_2$ where $r_1 = 01^*$ and $r_2 = 10^*$
NFA for $r_1 = 01^*$
r_1 is obtained by concatenating **0** and **1***

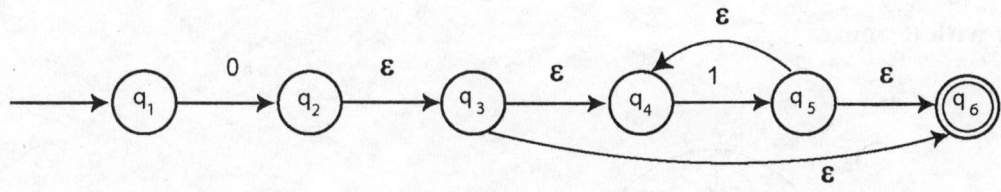

NFA for $r_2 = 10^*$
r_2 is obtained by concatenating **1** and **0***

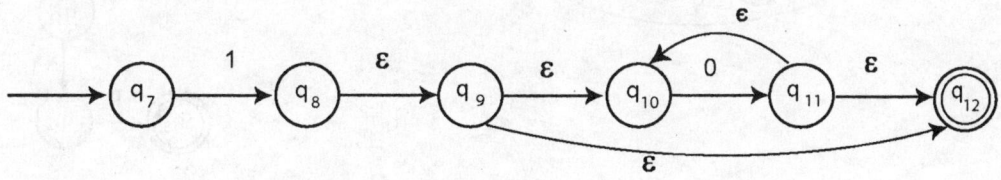

$\therefore r = r_1 + r_2$
$\therefore r = 01^* + 10^*$
r is obtained by taking the union of **01*** and **10***

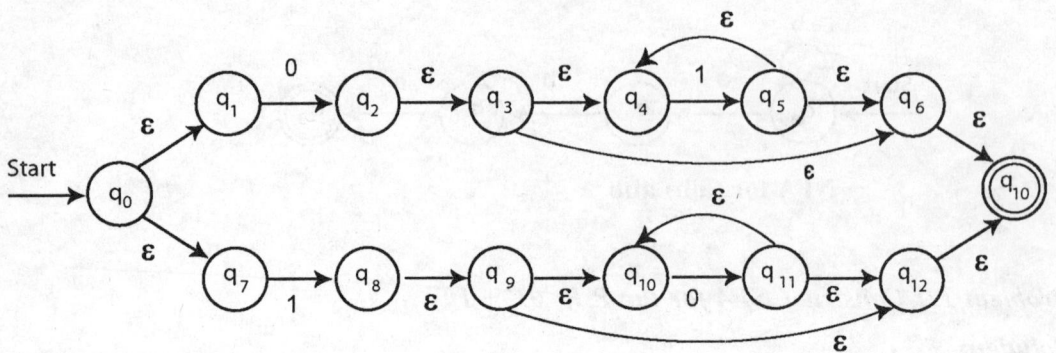

NFA to the R.E $r = 01^* + 10^*$

Problem 13: Construct DFA for the following R.E $(0|1)^*(00|11)110$

Solution:

NFA with ε−moves

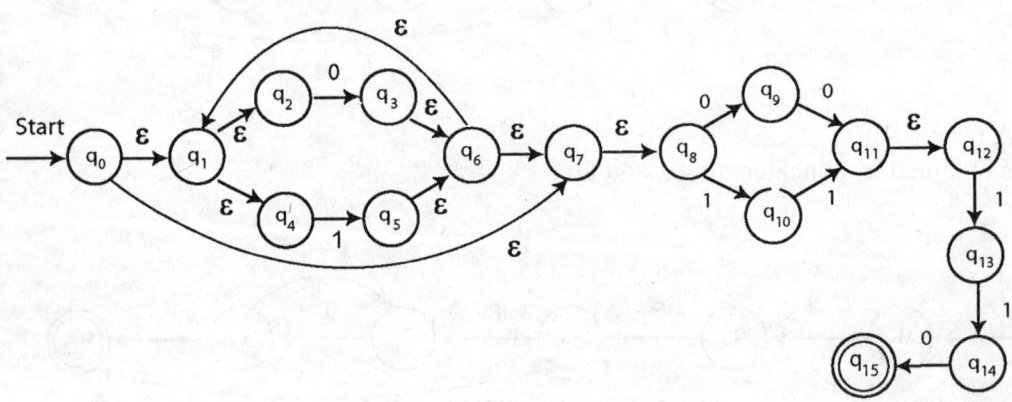

NFA with   ε−moves

NFA without ε –moves

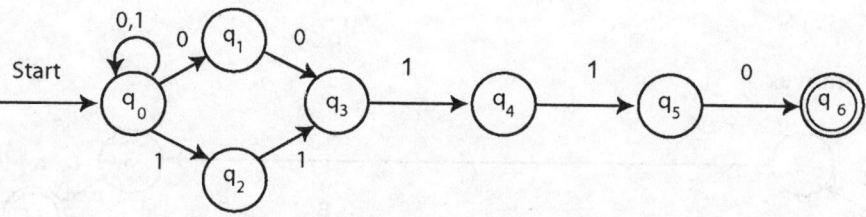

Transition table for DFA

Q \ Σ	0	1
→ $[q_0]$	$[q_0, q_1]$	$[q_0, q_2]$
$[q_0, q_1]$	$[q_0, q_1, q_3]$	$[q_0, q_2]$
$[q_0, q_2]$	$[q_0, q_1]$	$[q_0, q_2, q_3]$
$[q_0, q_1, q_3]$	$[q_0, q_1, q_3]$	$[q_0, q_2, q_4]$
$[q_0, q_2, q_3]$	$[q_0, q_1]$	$[q_0, q_2, q_3, q_4]$
$[q_0, q_2, q_4]$	$[q_0, q_1]$	$[q_0, q_2, q_3, q_5]$
$[q_0, q_2, q_3, q_4]$	$[q_0, q_1]$	$[q_0, q_2, q_3, q_4, q_5]$
$[q_0, q_2, q_3, q_5]$	$[q_0, q_1, q_6]$	$[q_0, q_2, q_3, q_4]$
$[q_0, q_1, q_6]$	$[q_0, q_1, q_3]$	$[q_0, q_2]$
$[q_0, q_2, q_3, q_4, q_5]$	$[q_0, q_1, q_6]$	$[q_0, q_2, q_3, q_4, q_5]$

Problem 14: *For the following R.E give the corresponding FSA (eliminate ε – transitions if any) R.E $0 + 1(01)^*$*

Solution:

NFA with ε –moves

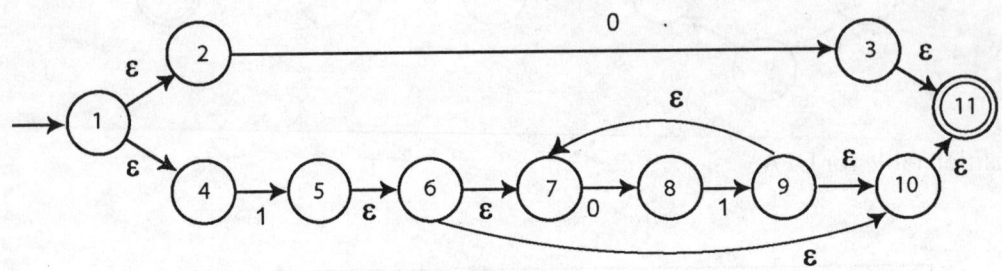

NFA for $0 + 1(01)^*$

Conversion to DFA (eliminating ε –moves)

Step 1: Make a state by writing the ε–closure of the initial state i.e., **1**

   ε–closure(1) = {1, 2, 4} = A

Step 2: Write ε–closures of **A** on **0** and **A** on **1**

   ε–closure(A, 0) = ε–closure(3) = {3, 11} = B

   ε–closure(A, 1) = ε–closure(5) = {5, 6, 7, 10, 11} = C

   ε–closure(B, 0) = ε–closure(Ø) = Ø

   ε–closure(B, 1) = ε–closure(Ø) = Ø

   ε–closure(C, 0) = ε–closure(8) = {8} = D

   ε–closure(C, 1) = ε–closure(Ø) = Ø

   ε–closure(D, 0) = ε–closure(Ø) = Ø

   ε–closure(D, 1) = ε–closure(9) = {7, 9, 10, 11} = E

   ε–closure(E, 0) = ε–closure(8) = {8} = D

   ε–closure(E, 1) = ε–closure(Ø) = Ø

Step 3: The final state NFA– ε is **11**, F = {11}

   The final states of DFA are **B, C, E**.

   ∴ F¹ = {B, C, E}

   The states of DFA are **A, B, C, D, E**.

Transition table for DFA

Q Σ	0	1
→ A	B	C
(B)	∅	∅
(C)	D	∅
D	∅	E
(E)	D	∅

Transition diagram for DFA.

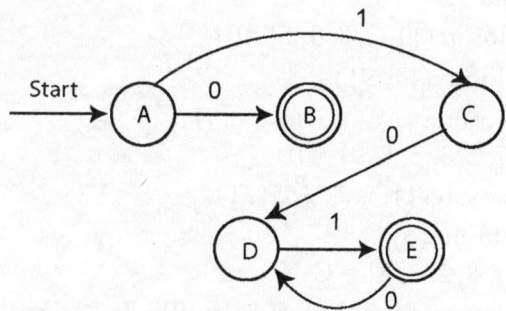

The language generated by the given regular expression is L= {0, 1, 101, 10101, 1010101, ...}

Since R.E = 0+1(01)*   i.e., 0|1(01)*

i.e., 0|1 {ε, 01, 0101, 010101, ...}

The above DFA accepts all the strings of the above language.

∴ The above DFA is the DFA corresponding to the given regular expression.

Problem 15: Convert the following regular expression to DFA

$$(0 + 1)^*01$$

Solution:

First we construct NFA– ε for (0|1)*01

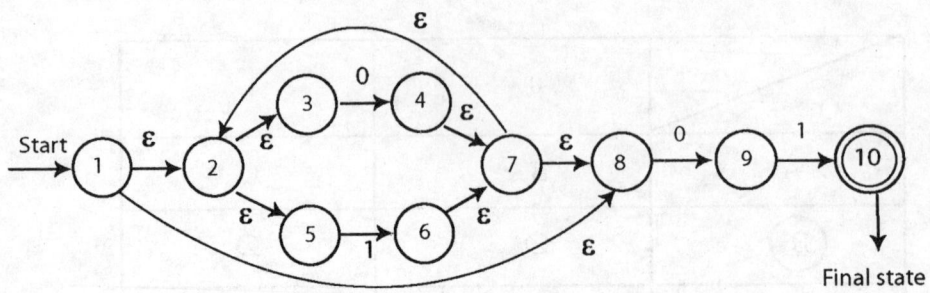

Final state

Conversion to DFA (eliminating ε- moves)

Step 1: First we make state for null-closure.

Take the initial state and write the ε–closure.

ε–closure(1) = {1, 2, 8, 3, 5} = A (since **1** is the initial state)

Step 2: Then we make states from **A** on input symbols **0** and **1**, and then for each new state

we make states on **0** and **1**.

$$\varepsilon\text{–closure}(A, 0) = \varepsilon\text{–closure}(\{1, 2, 8, 3, 5\}, 0)$$
$$= \varepsilon\text{–closure}(\{4, 9\})$$
$$= \varepsilon\text{–closure}(4) \cup \varepsilon\text{–closure}(9)$$
$$= \{4, 7, 2, 3, 5, 8, 9\} = B$$
$$\varepsilon\text{–closure}(A, 1) = \varepsilon\text{–closure}(\{1, 2, 8, 3, 5\}, 1)$$
$$= \varepsilon\text{–closure}(6)$$
$$= \{6, 7, 8, 2, 3, 5\} = C$$
$$\varepsilon\text{–closure}(B, 0) = \varepsilon\text{–closure}(\{4, 7, 2, 3, 5, 8, 9\}, 0)$$
$$= \varepsilon\text{–closure}(\{4, 9\}) = B$$
$$\varepsilon\text{–closure}(B, 1) = \varepsilon\text{closure}(\{4, 7, 2, 3, 5, 8, 9\}, 1)$$
$$= \varepsilon\text{–closure}(\{6, 10\}) = \{6, 7, 8, 2, 3, 5, 10\} = D$$
$$\varepsilon\text{–closure}(C, 0) = \varepsilon\text{–closure}(\{6, 7, 8, 2, 3, 5\}, 0) = \varepsilon\text{–closure}(\{4, 9\}) = B$$
$$\varepsilon\text{–closure}(C, 1) = \varepsilon\text{–closure}(\{6, 7, 8, 2, 3, 5\}, 1) = \varepsilon\text{–closure}(6) = C$$
$$\varepsilon\text{–closure}(D, 0) = \varepsilon\text{–closure}(\{6, 7, 8, 2, 3, 5, 10\}, 0) = \varepsilon\text{–closure}(4, 9) = B$$
$$\varepsilon\text{–closure}(D, 1) = \varepsilon\text{–closure}(6) = C$$

since (10) is the final state, therefore **D** is final state.

∴ The transition table for DFA is as follows

Q \ Σ	0	1
→ A	B	C
B	B	D
C	B	C
(D)	B	C

Here states **A** and **C** are same because both have same next states at the input symbols **0** and **1**, and both are non-final states.

So we can replace state **C** by **A**.

∴ The new transition table for DFA

Q \ Σ	0	1
→ A	B	A
B	B	D
(D)	B	A

Transition diagram for DFA

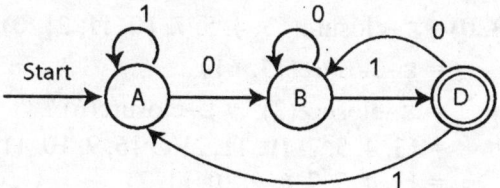

$L = \{01, 0001, 0101, 001, 101, 1101, 101001, \dots\}$

The R.E $= (0 + 1)^*01$ and the above DFA both accept the same language L.

Problem 16: Design a DFA for R = (0(0 + 1)*)⁺

Solution:

NFA– ε for $(0(0|1)^*)^+$

L = {0, 00, 01, 001, 011, 00101,...}

NFA– ε for $(0(0 + 1)^*)^+$ is as follows

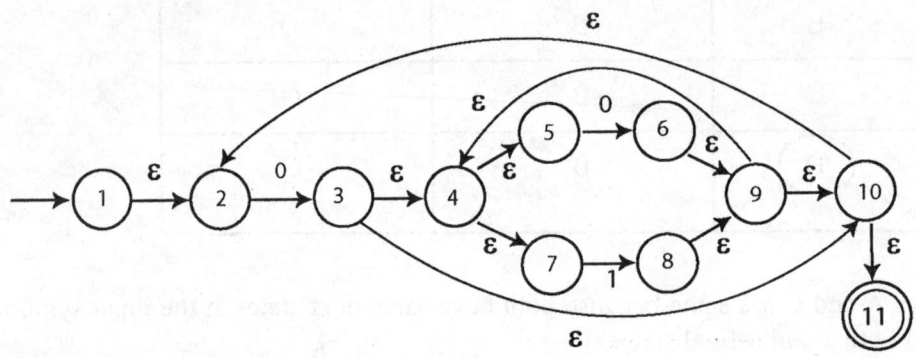

Conversion to DFA (eliminating ε-moves)

Step 1: First we make state for null-closure.

Take the initial state and write the ε–closure

Here the initial state is **1**

∴ ε–closure(1) = {1, 2} = A

Step 2: we make states from **A** on input symbols **0** and **1** and then for each new state we make states on **0** and **1**. We continue this process till there are no more new states.

$$ε–closure(A, 0) = ε–closure(\{1, 2\}, 0)$$
$$= ε–closure(3)$$
$$= \{3, 4, 5, 7, 10, 11, 2\} = B$$
$$ε–closure(A, 1) = ε–closure(\{1, 2\}, 1)$$
$$= ε–closure(∅)$$
$$= ∅$$
$$ε–closure(B, 0) = ε–closure(\{3, 4, 5, 7, 10, 11, 2\}, 0)$$
$$= ε–closure(\{3, 6\})$$
$$= ε–closure(3) ∪ ε–closure(6)$$
$$= \{3, 4, 5, 7, 10, 11, 2\} ∪ \{6, 9, 10, 11, 4, 5, 7\}$$
$$= \{3, 4, 5, 7, 6, 9, 10, 11, 2\}$$
$$= C$$
$$ε–closure(B, 1) = ε–closure(\{3, 4, 5, 7, 10, 11, 2\}, 1)$$
$$= ε–closure(8)$$

$$= \{8, 9, 10, 11, 4, 5, 7, 2\}$$
$$= D$$

$$\varepsilon\text{–closure}(C, 0) = \varepsilon\text{–closure}(\{2, 3, 4, 5, 6, 7, 9, 10, 11\}, 0)$$
$$= \varepsilon\text{–closure}(3, 6)$$
$$= C$$

$$\varepsilon\text{–closure}(C, 1) = \varepsilon\text{–closure}(\{2, 3, 4, 5, 6, 7, 9, 10, 11\}, 1)$$
$$= \varepsilon\text{–closure}(8)$$
$$= D$$

$$\varepsilon\text{–closure}(D, 0) = \varepsilon\text{closure}(\{2, 4, 5, 7, 8, 9, 10, 11\}, 0)$$
$$= \varepsilon\text{–closure}(3, 6)$$
$$= C$$

$$\varepsilon\text{–closure}(D, 1) = \varepsilon\text{–closure}(\{2, 4, 5, 7, 8, 9, 10, 11\}, 1)$$
$$= \varepsilon\text{–closure}(8)$$
$$= D$$

Transition table for DFA

Q \ Σ	0	1
→ A	B	∅
Ⓑ	C	D
Ⓒ	C	D
Ⓓ	C	D

As states **B, C, D** contains final state **11**, therefore all these states become final states. We can merge states **B, C, D,** because the next states are same for these states on **0** and **1** and all states **B, C, D** are final states B = C, B = D or C = B, D = B

Transition table for DFA (minimized)

Q \ Σ	0	1
→ A	B	∅
Ⓑ	B	B

Transition diagram

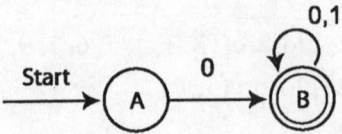

The above DFA accepts the language L= {0, 00, 01, 001, 011, 00101,....}
 This is the language L which is generated by the given R.E. and also accepted by the DFA.

Problem 17: Convert the R.E = (a|b)* into DFA

Solution:

The language L generated by the given regular expression is
L = { ε, a, b, ab, ba, abb, babb, ...}

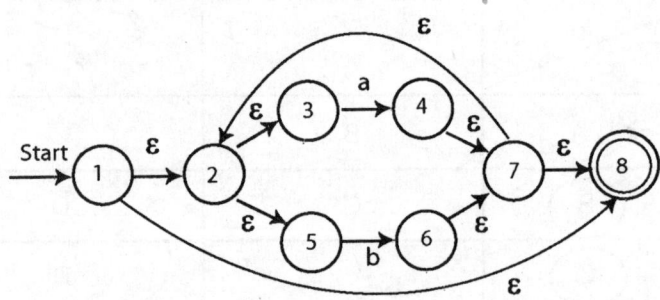

Step 1: Make a state by writing the ε-closure of the initial state.

$$A = \varepsilon\text{-closure}(1)$$
$$= \{1, 2, 3, 5, 8\}$$

Step 2: Make new states by writing the ε-closures of **A** on **a** and **A** on **b**

$$\therefore \varepsilon\text{-closure}(A, a) = \varepsilon\text{-closure}(4)$$
$$= \{4, 7, 2, 3, 5, 8\}$$
$$= B$$
$$\varepsilon\text{-closure}(A, b) = \varepsilon\text{-closure}(6)$$
$$= \{6, 7, 8, 2, 3, 5\}$$
$$= C$$
$$\varepsilon\text{-closure}(B, a) = \varepsilon\text{-closure}(4)$$
$$= B$$

ε–closure(B, b) = εclosure(6) = C

ε–closure(C, a) = B

ε–closure(C, b) = C

Since **A, B, C** contain the final state **8**, therefore the final states of DFA are **A, B, C**

The transition table for DFA

Q ∖ Σ	a	b
→(A)	B	C
(B)	B	C
(C)	B	C

Since **A, B, C** are final states, we can merge the states

$\therefore$ A = B = C

Transition table for DFA (minimized)

Q ∖ Σ	a	b
→(A)	B	C

Transition diagram for DFA

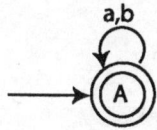

In the above DFA, we can observe initial state = final state, so we include ε in the language **L**.

L = {ε, a, b, aa, bba, abba, ...,}

$\therefore$ This language is same for both R.E. and DFA.

Problem 18: Construct FA for regular expression 0*1 + 10

Solution:

Given regular expression is r = 0*1 + 10

Step 1: Construct a transition graph equivalent to the regular expression using NFA with ε–moves. Let q_0 and q_f be the initial and final states respectively. So we get

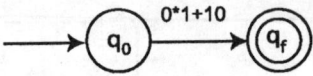

Transition graph for R.E.

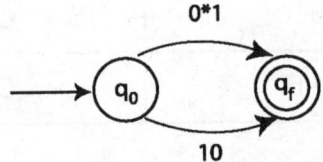

Eliminate + operation

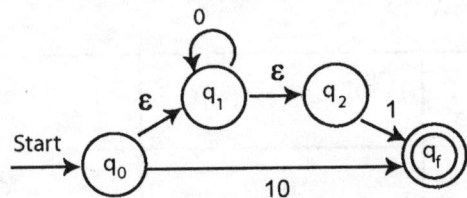

Eliminate *

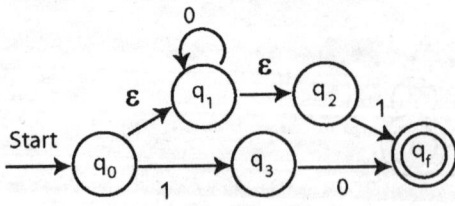

Eliminate concatenation.

The above diagram shows the NFA-ε for the R.E.

(or)

Given regular expression: $0^*1 + 10$

Let $r = r_1 + r_2$ where

             $r_1 = 0^*1$

             $r_2 = 10$

NFA with ε –moves for $r_1 = 0^*1$

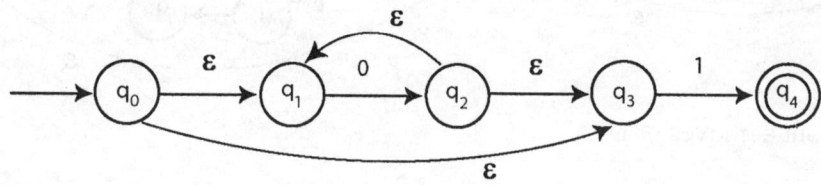

NFA with ε–moves for $r_2 = 10$

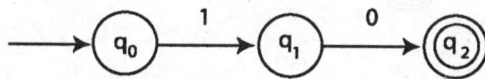

NFA with ε–moves for $r = r_1 + r_2$ i.e., $r = 0^*1 + 10$

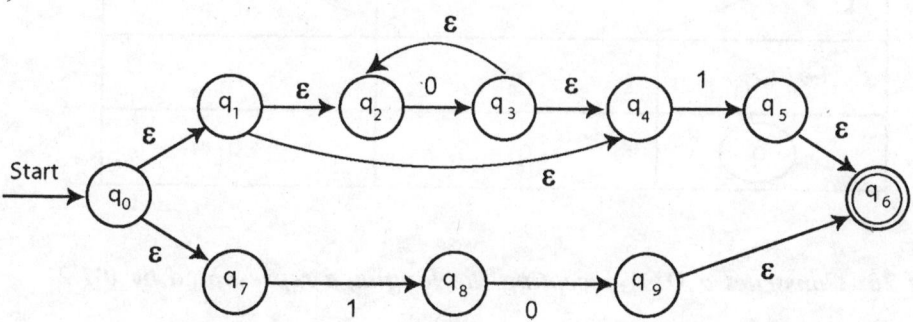

Problem 19: Construct a DFA for the regular set represented by $a^*b\,(a + b)^*$.

Solution:

NFA with ε–moves for $a^*b\,(a + b)^*$

   Let $r = a^*b\,(a + b)^*$

   Let $r = r_1 r_2 r_3$ where $r_1 = a^*$, $r_2 = b$ and $r_3 = (a + b)^*$

NFA– ε for **a*b (a + b)*** is as follows

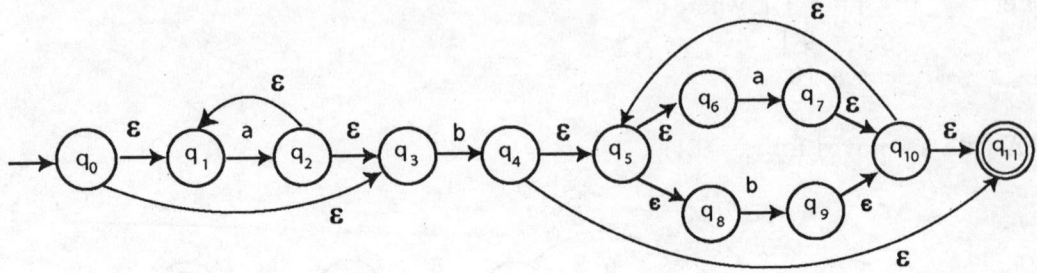

(ii) NFA without ε–moves or DFA

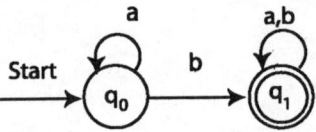

Transition table for DFA

Q $\diagdown$ Σ	a	b
→ q_0	q_0	q_1
⬭ q_1	q_1	q_1

Problem 20: Construct a DFA accepting the language represented by 0*1*2*

Solution:

The NFA–ε for $0^*1^*2^*$ is as follows.

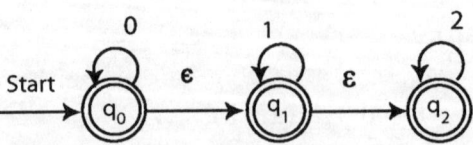

The transition table for NFA with ε- moves

Q \ Σ	0	1	2	ε
→ (q_0)	q_0	–	–	q_1
(q_1)	–	q_1	–	q_2
(q_2)	–	–	q_2	–

Conversion of NFA – ε to NFA

Step 1: Find ε–closure of all states

$$\varepsilon\text{–closure } (q_0) = \{q_0, q_1, q_2\}$$
$$\varepsilon\text{–closure } (q_1) = \{q_1, q_2\}$$
$$\varepsilon\text{–closure } (q_2) = \{q_2\}$$

Step 2: Calculation of $\hat{\delta}$ transitions

$$\hat{\delta} (q_0, \varepsilon) = \varepsilon\text{–closure}(q_0) = \{q_0, q_1, q_2\}$$
$$\hat{\delta} (q_0, 0) = \varepsilon\text{–closure}(\delta(\hat{\delta} (q_0, \varepsilon), 0)$$
$$= \varepsilon\text{–closure}(\delta(\{(q_0, q_1, q_2\}, 0))$$
$$= \varepsilon\text{–closure}(q_0) = \{q_0, q_1, q_2\}$$
$$\hat{\delta} (q_0, 1) = \varepsilon\text{–closure}(\delta(\hat{\delta} (q_0, \varepsilon), 1))$$
$$= \varepsilon\text{–closure}(\delta\{q_0, q_1, q_2\}, 1)$$
$$= \varepsilon\text{–closure}(q_1) = \{q_1, q_2\}$$
$$\hat{\delta} (q_0, 2) = \varepsilon\text{–closure}(\delta(\hat{\delta} \{q_1, q_2, q_3\}, 2))$$
$$= \varepsilon\text{–closure}(q_2) = \{q_2\}$$
$$\hat{\delta} (q_1, \varepsilon) = \varepsilon\text{–closure}(q_1) = \{q_1, q_2\}$$
$$\hat{\delta} (q_0, 0) = \varepsilon\text{–closure}(\delta(\hat{\delta} (q_1, \varepsilon), 0))$$
$$= \varepsilon\text{–closure}(\delta(\{q_1, q_2\}, 0)) = \varepsilon\text{–closure}(\varnothing) = \varnothing$$
$$\hat{\delta} (q_1, 1) = \varepsilon\text{–closure}(\delta(\{q_1, q_2\}, 1))$$
$$= \varepsilon\text{–closure}(q_1) = \{q_1, q_2\}$$
$$\hat{\delta} (q_1, 2) = \varepsilon\text{–closure}(\delta(\{q_1, q_2\}, 2))$$
$$= \varepsilon\text{–closure}(q_2)$$
$$= \{q_2\}$$
$$\hat{\delta} (q_2, \varepsilon) = \varepsilon\text{–closure}(q_2) = \{q_2\}$$
$$\hat{\delta} (q_2, 0) = \varepsilon\text{–closure}(\delta(q_0, 0))$$

$$= \varepsilon-closure(\emptyset)=\emptyset$$
$$\hat{\delta}\ (q_2,1) = \varepsilon-closure(\delta(q_2,1))$$
$$= \varepsilon-closure(\emptyset)=\emptyset$$
$$\hat{\delta}\ (q_2,2) = \varepsilon-closure(\delta(q_2,2))$$
$$= \varepsilon-closure(q_2)=\{q_2\}$$

Step 3: Final states of NFA wihtout ε- moves are the set of all sates whose ε- closure contains the final states of NFA -ε

$\therefore$ final staes of NFA are $\{q_0, q_1, q_2\}$

Transition table for NFA without ε- moves

Q \ Σ	0	1	2
→ q_0	$\{q_0, q_1, q_2\}$	$\{q_1, q_2\}$	$\{q_2\}$
q_1	$\emptyset$	$\{q_1, q_2\}$	$\{q_2\}$
q_2	$\emptyset$	$\emptyset$	$\{q_2\}$

Conversion of NFA to DFA

start from start state q_0, find the transitions from q_0 on 0,1,2.

Transition table for DFA

Q \ Σ	0	1	2
→ $[q_0]$	$[q_0, q_1, q_2]$	$[q_1, q_2]$	$[q_2]$
$[q_0, q_1, q_2]$	$[q_0, q_1, q_2]$	$[q_1, q_2]$	$[q_2]$
$[q_1, q_2]$	$\emptyset$	$[q_1, q_2\}$	$[q_2]$
$[q_2]$	$\emptyset$	$\emptyset$	$[q_2]$

Here we can merge states $[q_0]$ and $[q_0,q_1,q_2]$ because both have same transition states on input symbols 0,1 and 2 and both $[q_0]$ and $[q_0,q_1\ q_2]$ are final states.

∴ The transition table for DFA is as follows

Q \ Σ	0	1	2
→ $[q_0]$	$[q_0]$	$[q_1,q_2]$	$[q_2]$
$[q_1,q_2]$	∅	$[q_1,q_2]$	$[q_2]$
$[q_2]$	∅	∅	$[q_2]$

The transition diagram for the DFA is as follows.

Take $[q_0]$ =r, $[q_1,q_2]$ = s, $[q_2]$ =t.

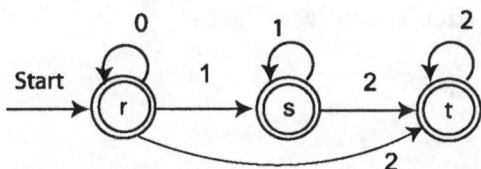

This DFA accepts strings of the form {0,1,2,01,12,02,012,001122,0122,001,112,022,...}

Note: we can merge two states by one, if both have same transition states at given input symbols and both will be either non−final states or both will be final states.

Problem 21: Consider the transition system what type of strings accepted by it. Prove that the strings recognized are (a + a (b + aa)*b)*a (b + aa)*a

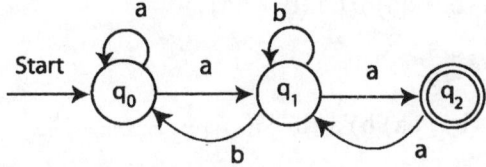

Solution:

The given transition system is NFA without ε−moves. We can directly apply the method i.e., NFA to regular expression, since the graph does not contain any ε−moves and there is only one initial state.

The three equations for q_0, q_1, q_2 can be written as follows.

$$q_0 = q_0a + q_1b + \varepsilon \qquad \qquad \qquad ......(1) \text{ -initial state}$$
$$q_1 = q_0a + q_1b + q_2a \qquad \qquad \qquad ......(2)$$
$$q_2 = q_1a \qquad \qquad \qquad \qquad ......(3)$$

since q_2 is the final state, solve the equations for q_2

$\therefore q_2 =$ the required regular expression.

To get q_2, apply the substitutions and Arden's lemma repeatedly.

Arden's lemma: let **P** and **Q** be regular expressions over . If P does not contain ε, then the following equation in **r** i.e., $R = Q + RP$ has unique solution given by $R = QP^*$. Consider equation (2) and (3)

$$q_2 = q_1a$$
$$q_1 = q_0a + q_1b + q_2a$$
$$q_1 = q_0a + q_1b + q_1aa$$
$$q_1 = q_0a + q1 \ (aa + b)$$

by applying **Arden's lemma**, we get

$$q_1 = q_0a \ (aa + b)^*$$
$$\therefore \ q_1 = q_0a \ (aa + b)^* \qquad \qquad \qquad ......(4)$$

Substitute q_1 in q_0, we get

$$q_0 = q_0a + q_1b + \varepsilon$$
$$q_0 = q_0a + q_0a \ (aa + b)^*b + \varepsilon$$
$$\therefore \ q_0 = \varepsilon + q_0 \ (a + a \ (aa + b)^*b)$$

By applying Arden's lemma, we get

$$q_0 = \varepsilon.q_0 \ (a + a \ (aa + b)^*b)$$
$$\therefore \ q_0 = \varepsilon.q_0 \ (a + a \ (b + aa)^*b) \qquad \qquad \qquad ......(5)$$

Substituting (5) in (4), we get

$$\therefore \ q_1 = q_0a \ (b + aa)^*$$
$$\therefore \ q_1 = (a + a \ (b + aa)^*b)^*a \ (b + aa)^* \qquad \qquad \qquad ......(6)$$

Substitute (6) i.e., q_1 in q_2 i.e., (3)

$$q_2 = q_1a$$
$$\therefore \ q_2 = (a + a \ (b + aa)^*b)^*a \ (b + aa)^*a$$

Since q_2 is the only final state, the set of strings recognized by the graph is given by

$$(a + a \ (b + aa)^*b)^*a \ (b + aa)^*a.$$

$\therefore$ R.E. for the given NFA is $(a + a \ (b + aa)^*b)^*a \ (b + aa)^*a$

Problem 22: Construct regular expression accepted by the following finite automaton

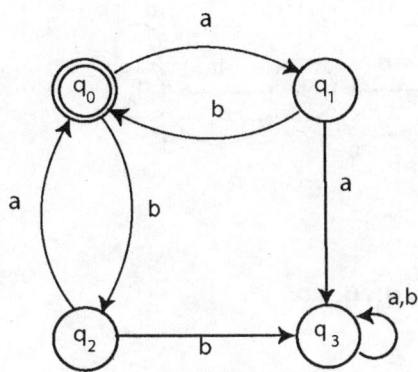

Solution:

If the initial state is equal to the final state in the transition diagram, then we have
to include ε in the language

Therefore the language for the given DFA is as follows

$L = \{\varepsilon, ab, ba, abba, baab,....\}$

Here q_3 is dead state

There is only initial state and there are no ε –moves.

The equations corresponding to the states q_0, q_1, q_2, q_3 are

$$q_0 = q_1 b + q_2 a + \varepsilon \qquad\qquad ......(1)$$
$$q_1 = q_0 a \qquad\qquad ......(2)$$
$$q_2 = q_0 b \qquad\qquad ......(3)$$
$$q_3 = q_1 a + q_2 + q_3 (a + b) \qquad\qquad ......(4)$$

Substituting $q_1 = q_0 a$ and $q_2 = q_0 b$ in equation (1) i.e., $q_0 = q_1 b + q_2 a + \varepsilon$

$$\therefore q_0 = q_0 ab + q_0 ba + \varepsilon$$
$$= q_0 (ab + ba) + \varepsilon$$
$$\therefore q_0 = \varepsilon + q_0 (ab + ba) \qquad\qquad ......(5)$$

It is of the form $R = Q + RP$ where $R = q_0$, $Q = \varepsilon$, $P = ab + ba$

By Arden's theorem, if $R = Q + RP$ then $R = QP^*$

By applying Arden's theorem to equation (5),

we get $q_0 = \varepsilon(ab + ba)^*$ (since $\varepsilon R = R$)

$\therefore$ As q_0 is the only final state, the regular expression corresponding to given DFA is

$$R = (ab + ba)^* \quad \therefore L = \{\varepsilon, ab, ba, abba, baab,...\}$$

Problem 23: Find the regular expression accepted by the following deterministic finite automaton.

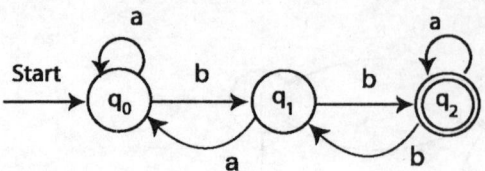

Solution:

The equations to the states q_0, q_1, q_2 are

$$q_0 = q_0 a + q_1 a + \varepsilon \qquad \qquad ......(1)$$
$$q_1 = q_0 b + q_2 b \qquad \qquad ......(2)$$
$$q_2 = q_1 b + q_2 a \qquad \qquad ......(3)$$

since q_2 is the final state, solving it will give the required expression.

so solve for q_2

consider $q_0 = q_0 a + q_1 a + \varepsilon$

$$q_0 = (q_1 a + \varepsilon) + q_0 a$$

By Arden's theorem, we get

$$q_0 = (q_1 a + \varepsilon) a^* \qquad \qquad ......(4)$$

substitute q_0 in q_1, we get

$$q_1 = q_0 b + q_2 b$$
$$q_1 = (q_1 a + \varepsilon) a^* b + q_2 b$$
$$q_1 = (a^* b + q_2 b) + q_1 a a^* b$$

By Arden's theorem, we get

$$q_1 = (a^* b + q_2 b)(a a^* b)^* \qquad \qquad ......(5)$$

substitute q_1 in q_2 i.e., in equation (3)

$$q_2 = q_1 b + q_2 a$$
$$= [(a^* b + q_2 b)(a a^* b)^*] b + q_2 a$$
$$\therefore q_2 = a^* b (a a^* b)^* b + q_2 (a + b(a a^* b)^* b)$$

By Arden's theorem, we get

$$q_2 = a^* b (a a^* b)^* b (a + b (a a^* b)^* b)^*$$

As q_2 is the only final state, the RE for the given DFA is

$$r = a^* b (a a^* b)^* b (a + b (a a^* b)^* b)^*$$

Problem 24: Find regular expression for the following automaton.

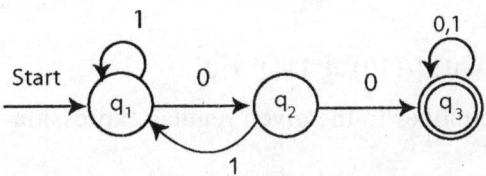

Solution:

Since there is only one initial state and there are no ε–moves, so Arden's theorem can be applied to the given DFA

The regular expressions for each state of the FA are

$$q_1 = q_1 1 + q_2 1 + \varepsilon \qquad \qquad ......(1)$$
$$q_2 = q_1 0 \qquad \qquad ......(2)$$
$$q_3 = q_2 0 + q_3 (0 + 1) \qquad \qquad ......(3)$$

since q_3 is the final state, we solve these equations for q_3.

substitute q_2 in q_1, we get

$$\begin{aligned} q_1 &= q_1 1 + q_2 1 + \varepsilon \\ &= q_1 1 + q_1 01 + \varepsilon \\ &= \varepsilon + q_1 (1 + 01) \end{aligned}$$

By Arden's theorem, $q_1 = \varepsilon(1 + 01)^*$

$$\therefore \ q_1 = (1 + 01)^* \qquad \qquad ......(4) \qquad (\text{since } \varepsilon R = R)$$

substitute q_2 in q_3, we get

$$\begin{aligned} q_3 &= q_2 0 + q_3 (0 + 1) \\ &= q_1 00 + q_3 (0 + 1) \qquad (\text{since } q_2 = q_1) \\ &= (1 + 01)^* 00 + q_3 (0 + 1) \qquad (\text{since } q_1 = (1 + 01)^*) \end{aligned}$$
$$\therefore \ q_3 = (1 + 01)^* 00 + q_3 (0 + 1)$$

By Arden's theorem, we get

$$q_3 = (1 + 01)^* 00(0 + 1)^*$$

As q_3 is the only final state for the given DFA, the R.E. for the given DFA is

$$r = (1 + 01)^* 00(0 + 1)^*$$

Problem 25: Construct FA equivalent to the following regular expression.
$$r = 01 \; [((10)^* + 111)^* + 0]^* \; 1$$

Solution:

Given regular expression is $r = 01 \; [((10)^* + 111)^* + 0]^* \; 1$

step 1: Construct NFA with ε-moves to the given regular expression

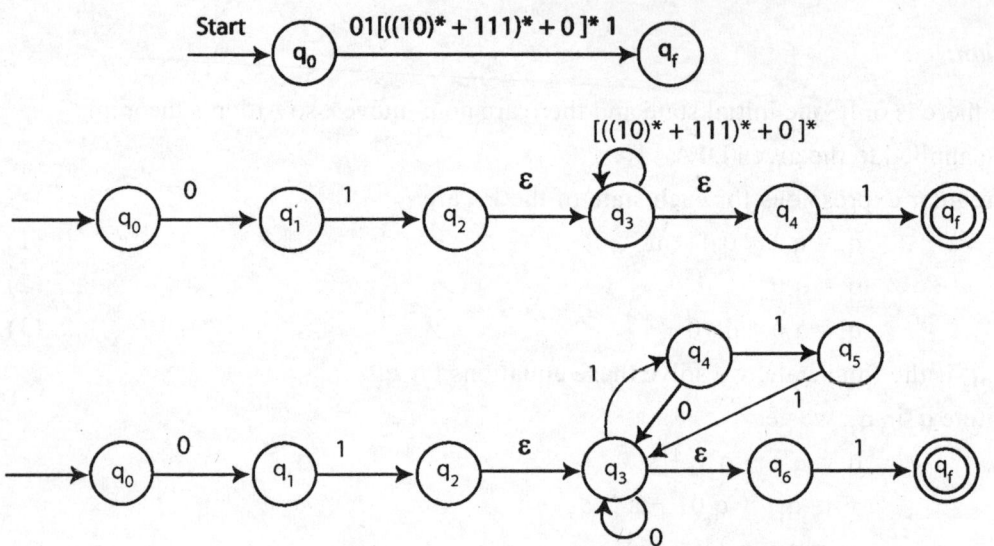

NFA with ε-moves

Step 2: NFA with ε-moves to NFA without ε-moves

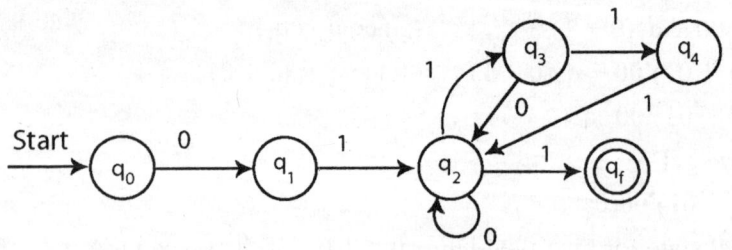

NFA without ε-moves for the given regular expression.

Step 3: convert NFA into equivalent DFA

Transition table for NFA

Q \ Σ	0	1
→ q_0	q_1	–
q_1	–	q_2
q_2	q_2	$\{q_3, q_f\}$
q_3	q_2	q_4
q_4	–	q_2
(q_f)	–	–

Transition table for DFA

Q \ Σ	0	1
→ $[q_0]$	$[q_1]$	–
$[q_1]$	–	$[q_2]$
$[q_2]$	$[q_2]$	$[q_3, q_f]$
$([q_3, q_f])$	$[q_2]$	$[q_4]$
$[q_4]$	–	$[q_2]$
$[q_f]$	–	–

Transition diagram for DFA

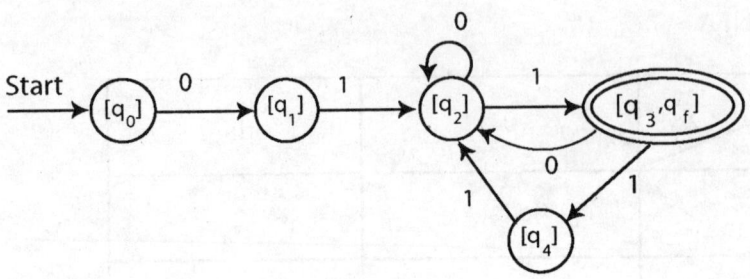

Problem 26:   *Find regular expressions for the following NFA's*

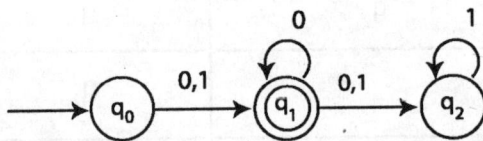

Solution:

The given transition graph does not contain any ε −moves and it has only one initial state.

The equations corresponding to q_0, q_1, q_2 are as follows.

$$q_0 = \varepsilon$$
$$q_1 = q_1 0 + q_0(0 + 1) \qquad \qquad \qquad ......(1)$$
$$q_2 = q_2 1 + q_1(0+1) \qquad \qquad \qquad ......(2)$$

Since q_1 is the only final state, therefore the regular expression is q_1.

consider equations (1) and (2)

$$q_0 = \varepsilon$$
$$q_1 = q_1 0 + \varepsilon(0+1) \qquad \qquad \qquad ......(1)$$
$$= (0+1) + q_1 0$$

Since q_1 is the only final state.

$\therefore$ RE $= 00^* + 10^*$

$L_1 = \{0, 1, 00, 10, 000, 1000, 0000, 10000, ...\}$

The set of strings recognized by the above finite automate is

$L_2 = \{0, 1, 00, 10, 000, 1000, ...\}$.

Problem 27: Find the regular expression for the following NFA

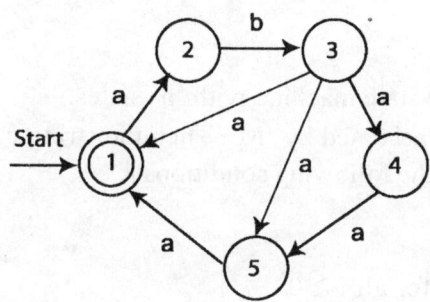

Solution:

The strings accepted by this finite automata are aba, abaa, abaaa.

Since I.S=F.S, include ε in L

∴ L = {ε , aba, abaa, abaaa, abaabaa,....}

The above transition graph does not contain any null or ε–moves, and it has only one initial state.

The equations corresponding to the states 1,2,3,4,5 are as follows.

1=3a+5a+ ε		......(1)
2=1a		......(2)
3=2b		......(3)
4=3a		......(4)
5=3a+4a		......(5)

Since '1' is the only final state, solve the equations for '1'.

Consider equations (2) and (3), we have

3=1ab   •   ......(6)   (since 2=1a)

Consider equations (4) and (5), we have

5=3a+3aa   (since 4=3a)

∴ 5=3(a+aa)

5=1ab(a+aa)=1aba+1abaa   ......(7)

substitute (6) and (7) in (1), we get

1=3a+5a+ ε

1=1aba+1abaa+1abaaa+ ε

∴ 1= ε+1(aba+abaa+abaaa).

∴ 1= ε (aba+abaa+abaaa)*

∴ 1=(aba+abaa+abaaa)*

∴ R.E=(aba+abaa+abaaa)*.

∴ 1={ε, aba,abaa,abaaa,abaaabaa...}

Problem 28: Give "pumping lemma"; explain its application with an example

Solution:

Pumping lemma:

Let $M=(Q,\Sigma,\delta,q_0,F)$ be a finite state machine with 'n' states.

Let 'L' be a regular language accepted by 'M'. Then the string $z \varepsilon L$ such that $|z| \geq n$ can be written as z=uvw satisfying the following conditions

1    $v \neq \varepsilon$ or $v \geq 1$

2    $|uv| \leq n$

3   The string $uv^iw \in L$ for all $i \geq 0$

The theorem states that given a string z belonging to L, we always find a substring v near the beginning of z that can be pumped or repeated as many times as we like, keeps the resulting string in the language.

Example: Let **L={a^{2n}|n≥1}** and prove that it is regular.

The language L consists of set of strings of length '**2n**' i.e., strings of even length.
Assume that L is regular and '**n**' be the number of states in the FA.

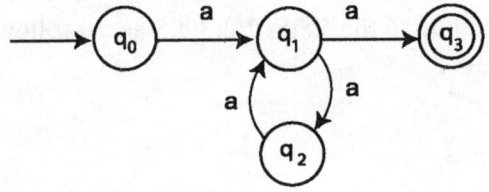

The corresponding finite automata is
$$L = \{ aa, aaaa, aaaaaa, \ldots \}$$
Select a string z such that $|uv| \leq n$ and $|v| \geq 0$

let z=uvw

$|z|=|uvw|=2n$

consider the string 'aaaa'

let z=aaaa which belongs to L i.e., $z \varepsilon L$

∴ z=aaaa   where u=a,   v=a^2,   w=a.

by pumping lemma,

If uvw ε L then prove that $uv^iw \varepsilon L$ for i≥0

let us consider uv^2w (i=2)

$$uv^2w = a(a^2)^2a$$
$$= aa^4a$$
$$= aaaaaa$$

∴ $|uv^2w|=|aaaaaa|=6$, which is even.

∴ $uv^2w \varepsilon L$

consider uv^3w $(i=3)$

$\quad uv^3w = a(a^2)^3a = aaaaaaaa$

$\therefore uv^3w$ is a string of length 8.

$\therefore uv^3w \ \varepsilon \ L$ which is even.

consider uv^0w $(i=0)$

$\quad uv^0w = uw = aa$

$\therefore uv^0w$ is a string of length 2, which is even.

$\therefore uv^0w \ \varepsilon \ L$

$\therefore uv^iw \ \varepsilon \ L$, for all $i \geq 0$

$\therefore$ The language $L=\{a^{2n} | n \geq 1\}$ is regular

Problem 29: *Show that $L=\{ww | w \in \{a,b\}^*\}$ is not regular. State and explain the theorem used.*

Solution:

Given $L=\{ww | w \ \varepsilon \ \{a,b\}^*\}$

Assume L is regular and n is the number of states in FA accepting L

$\qquad$ let $z = ww = a^n b \ a^n b$

$\qquad |z| = 2(n+1) > n \qquad \therefore |z| > n$

by pumping lemma

$\qquad |z| = uvw$ such that $|uv| \leq n$ and $|v| > 0$,

we need to find '**i**' such that $uv^iw \notin L$

There are two cases for string '**v**'

$\qquad L=\{abab, \ aabaab, aaabaaab, \dots\}$

Case 1: v does not contain **b**'s i.e., $v = a^k$, $k \geq 1$

consider the string aaa baaab

Here n = 3

$\qquad v = aa = a^2$

$\therefore k = 2$ (since $v = a^k$, $k \geq 1$)

let $v = a^k, k \geq 1$, then $uv^iw = a^n b a^n b$

let i = 0: $uw = abaaab = a^1 ba^3 b$

$\qquad = a^{n-k} ba^n b$

since $n-k \leq n$, $n-k \neq n$

$\quad uw \notin L$.

This is a contradiction.

Case 2: v consists of only one **b**

consider the string

z = aa aba aab

Here v=aba, it contains only one **b**

let i=0

$uv^iw=uw=aaaab$

∴ uw=aaaab

∴ uw has only one **b**

∴ uw∉L, since L consists of strings having even number of **a**'s and even number of **b**'s

It is not possible to take 'v' as two **b**'s

If so $|v| \geq n + 2$.

but we know that $|v| \leq |uv| \leq n$

∴ we got contradiction in both the cases

∴ L is not regular.

Problem 30: Show that $\{a^n b^{2n}|n>0\}$ is not a regular set

Solution:

Given $L=\{a^n b^{2n}|n>0\}$

Let '**n**' be the number of states in the FA accepting 'L'.

Let $z=a^n b^{2n}$

$$|z| = n+2n = 3n>n$$

$$\therefore |z|>n$$

∴ The length of the strings accepted by L is n+2n=3n, i.e., length is always a multiple of '3'.

∴ Set of strings accepted by language L is

L={abb,aabbbb,aaabbbbbb,...}

By pumping lemma, if z = uvw ∈ L then $uv^iw ∈ L$

Choose a string from the given language and apply pumping lemma principle.

(i) consider z = a b b   where u = a, v = b, w = b $|uv|{\leq}n; |v|{\geq}1$

Take i = 2

$uv^2w = a(b)^2b = abbb$, $|uv^2w| = 4$; which is not multiple of 3

∴ $uv^2w∉L$, which is a contradiction.

∴uvw ∈ L, but $uv^2w∉L$.

∴ By pumping lemma, we can say that it is not regular.

(ii) consider the string z=aa bbb b, v=bbb

∴ uv^2w=aabbbbbbb, which of length 9, i.e., multiple of 3

$\qquad\qquad$ =a^2b^7, but it is not in the form of a^nb^{2n}

∴ $uv^2w \notin L$

∴ L is not regular.

Problem 31: Prove that $L=\{ww^R|w \in (a|b)^*\}$ is not regular.

Solution:

Given $L=\{ww^R|w \in (a|b)^*\}$

$\qquad$ Assume L is regular and n is the number of states in FA accepting L

$\qquad$ L=\{abba,ababbaba,abbbba,...\}

$\qquad$ Let z=uvw=ababbaba where u=a,v=b, w=abbaba

$\qquad$ We find 'i' such that $uv^iw \notin L$.

(i) let i=0 :

$\qquad\qquad$ uv^0w = uw = aabbaba.

The string aabbaba $\neq ww^R$.

∴ $uv^0w \notin L$, which is a contradiction.

(ii) let i=2 :

$\qquad\qquad$ z=uv^2w=abbabbaba$\neq ww^R$.

∴ $uv^2w \notin L$, which is a contradiction

By pumping lemma if uvw $\in$ L then $uv^iw \in$ L, for i≥0.

Here uvw $\in$ L but $uv^0w \notin L$ and $uv^2w \notin L$

∴ $L=\{ww^R|w \in (a|b)\}$ is not regular.

Problem 32: Show that $\{0^n1^m|gcd(n,m)=1\}$ is not regular

Solution:

Let $L=\{0^n1^m|gcd(n,m)=1\}$ is regular and 'n' be the number of states of finite automata accepting '**L**'.

$\qquad\qquad$ Let z = 0^n1^m

$\qquad\qquad$ $|z|$ = n+m>n

$\qquad\qquad$ ∴ $|z| \geq$ n

L=\{01,011,0111,00111...\}

Select z=uvw such that |uv|≤n and |v|>0

Let z=00111 where u=0, v=01, w=11

Applying pumping lemma, for $z=uv^iw$ with i=2

$\qquad$ uv^2w = 0(01)^{2}11

$\qquad\qquad$ = 0010111

$= 0010111 \notin L$

In the given language, in all the strings, we have **0**'s followed by **1**'s

$\therefore$ $uv^2w \notin L$, which is a contradiction By pumping lemma, if $uvw \in L$ then $uv^iw \in L$, for all $i \geq 0$

$\therefore$ Here $uvw \in L$, but $uv^2w \notin L$.

$\therefore$ when i=2, we got contradiction.

$\therefore$ By pumping lemma, the given set
    $L = \{0^n 1^m | \gcd(n,m) = 1\}$ is not regular.

REVIEW QUESTIONS

1)   Construct a DFA for the regular expression $(a+b)^*(aa+bb)(a+b)^*$

2)   Construct non-deterministic finite automata equivalent to the following regular expression i.e., NFA with ε–moves to R.E $10+(0+11)0^*1$

3)   Construct a finite automaton for $a(a+b)^*bb$

4)   Construct NFA for $10(0+1)^*(10+01)^*$.

5)   Construct NFA for $(a|b)^*abb$

6)   Construct NFA and DFA for the following R.E $(0|1)^*(00|11)110$

7)   Give the corresponding DFA (eliminate ε–transitions if any) for R.E $(11)^*(00)^*101$.

8)   Construct NFA for the R.E (i) $(0+1)^*(01+110)$ (ii) 01^*+10^*

9)   Find the regular expression accepted by the following automaton.

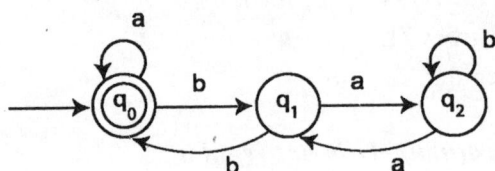

10)   Consider the FA given, construct a regular expression that is accepted by it

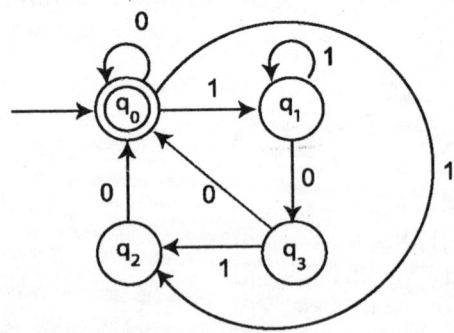

11) Construct a DFA for the R.E $10+(0+11)0^*1$ and optimize the states.

12) Give the regular expression accepted by the following finite automata.

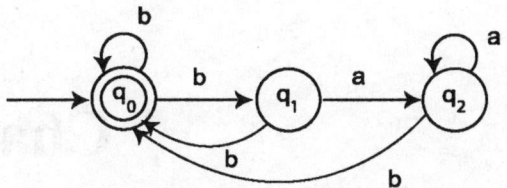

13) Construct the NFA for the R.E $r=0^*1((0+1)0^*1)^*(\varepsilon +(0+1)(00)^*)+0(00)^*$

14) Construct FA equivalent to the following regular expression.

$R=01[((10)^*+111)^*+0]^*1$

Chapter 4

Grammar Formalism

4.1 INTRODUCTION

A regular language can be depicted either as

1)   Regular Expressions

2)   Finite Automata orm

3)   Grammar

The effective way to represent regular language is in the form of a grammar. A grammar is nothing but a finite set of rules defined for a language. A new set of rules can be derived in order to accept the desired string.

In the current chapter, we shall be discussing about regular grammar, how a grammar is constructed and different representations of the grammar. Also the equivalence between regular grammar and FA is discussed. The basic concepts of context free grammar and derivations are discussed.

4.2 REGULAR GRAMMAR

4.2.1 DEFINITION

A regular grammar is defined as **G = (V, T, P, S)** where

V is a set of symbols called non terminals which are used to define the rules

T is set of symbols called terminals

P is a set of production rules and

S is the start symbol, where $S \in V$

4.2.2 RIGHT-LINEAR GRAMMAR AND LEFT-LINEAR GRAMMAR

If all the productions in a CFG are of the form A→wB or A→w, where A and B are variables and w is a (possibly empty) string of terminals, then such grammar is known as *right-linear grammar*.

If all the productions in a CFG are of the form A→Bw or A→w, where A and B are variables and w is a (possibly empty) string of terminals, then such grammar is known as *left-linear grammar*.

A right-linear or left-linear grammar is called as *regular grammar*.

Example 4.1

The right-linear grammar for the expression $0(10)^*$ is

$S \rightarrow 0A$

$A \rightarrow 10A \mid \varepsilon$

The other way of representing the given expression is

$S \rightarrow 01S \mid 0$

The left-linear grammar for the same expression is

$S \rightarrow S10 \mid 0$

NOTE: Reverse of a right-linear grammar is a left-linear grammar and vice-versa.

4.2.3 CONSTRUCTION OF REGULAR GRAMMAR FOR A GIVEN REGULAR EXPRESSION

A regular grammar can be constructed from a given regular expression by using the following method:

1) Construct NFA with ε-transitions for a given regular expression.
2) Convert the given NFA with ε-transitions to NFA without ε-transitions.
3) Construct an equivalent DFA for the above NFA.
4) For the given DFA, all the states become non terminals i.e., V and the transitions made from one state to another state on a given input symbol (from T) are represented as production rules.

Example 4.2

The equivalent production is $q_0 \rightarrow aq_1$

If q_1 is a final state, then we have another production $q_0 \rightarrow a$

Example 4.3

Construct a regular grammar for the regular expression a*b(a+b)*.
Let us construct DFA for the given regular expression.

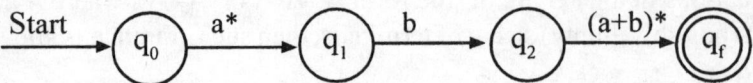

Now, convert it into NFA with ε -transition

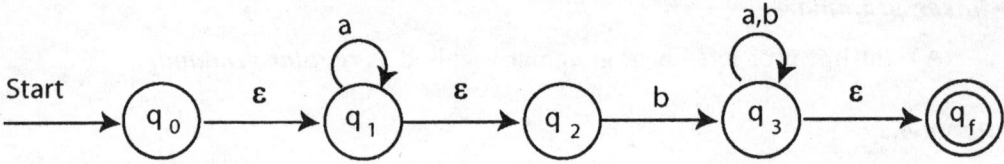

Now by eliminating ε-moves, the automata is

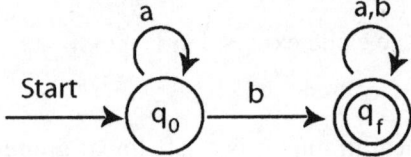

The regular grammar is given as
$$G = (V, T, P, S)$$
where $V = \{q_0, q_f\}$
$T = \{a, b\}$
$S = q_0$
$P = \{ q_0 \rightarrow a\, q_0 ,$
$q_0 \rightarrow b\, q_f ,$
$q_0 \rightarrow b,$
$q_f \rightarrow a\, q_f ,$
$q_f \rightarrow b\, q_f ,$
$q_f \rightarrow a ,$
$q_f \rightarrow b \}$

Thus G is the required regular grammar.
we can observe that, the above is a right-linear grammar.

Example 4.4

Construct right-linear and left linear grammar for the following regular expression.

$$(0+1)^* \ 00 \ (0+1)^* \quad \text{or} \quad (0 \mid 1)^* \ 00 \ (0 \mid 1)^*$$

Solution:

Given regular expression is $r = (0+1)^* \ 00 \ (0+1)^*$

The FA for r is as follows:

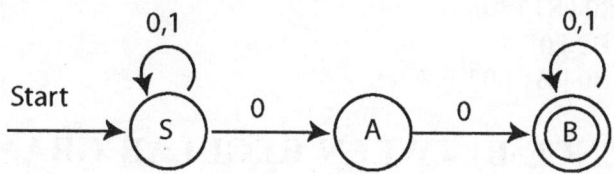

Right-linear grammar: Here initial state is not equal to final state, so apply case I. To obtain right-linear grammar, we apply the following steps.

step-1:

If there is a transition $\delta(q_i, a) = q_j$ in FA,

then include the production $q_i \rightarrow aq_j$ in P.

step-2:

If there is a transition $\delta(q_i, a) = q_f$, where q_f is final state in FA,

then include the production $q_i \rightarrow a$ in P.

step-3:

An initial state of FA is the start symbol in G.

Using step -1 We have the following productions

$\delta(S, 0) = S \Rightarrow S \rightarrow 0S$

$\delta(S, 1) = S \Rightarrow S \rightarrow 1S$

$\delta(S, 0) = A \Rightarrow S \rightarrow 0A$

$\delta(A, 0) = B \Rightarrow A \rightarrow 0B$

$\delta(B, 0) = B \Rightarrow B \rightarrow 0B$

$\delta(B, 1) = B \Rightarrow B \rightarrow 1B$

Using step- 2 We have the following productions

Since B is the final state of FA, the productions are as follows

$\delta(A, 0) = B \Rightarrow A \rightarrow 0$

$\delta(B, 0) = B \Rightarrow B \rightarrow 0$

$\delta(B, 1) = B \Rightarrow B \rightarrow 1$

Using step -3 The start symbol of G is S, where S is the start state of FA.

∴ The right-linear grammar is

$S \rightarrow 0S \mid 1S \mid 0A$

$A \rightarrow 0B \mid 0$

$B \rightarrow 0B \mid 1B \mid 0 \mid 1$

Left-linear grammar: To construct left-linear grammar, reverse the given regular expression, we get $r = (0+1)^* 00 (0+1)^*$, which is same as the given regular expression. So we get the same right-linear grammar.

To get the left-linear grammar, interchange the positions of terminals and non-terminals on R.H.S of productions of right-linear grammar.

$\therefore$ **Left-linear grammar is**

$S \rightarrow S0 \mid S1 \mid A0$

$A \rightarrow B0 \mid 0$

$B \rightarrow B0 \mid B1 \mid 0 \mid 1$

4.3 EQUIVALENCE BETWEEN REGULAR GRAMMAR AND FINITE AUTOMATA

4.3.1 CONSTRUCTION OF FA FROM A REGULAR GRAMMAR (RIGHT-LINEAR GRAMMAR)

Let $G = (V, T, P, S)$ be a regular grammar, we can construct FA

$$M = (Q, \Sigma, \delta, q_0, \{q_f\})$$

where

   i. States(Q) correspond to variables(V)

  ii. Initial state (q_0) correspond to start symbol(S)

 iii. Transitions in M correspond to productions in P

If there is a production of the form $A_i \rightarrow aA_j$, the automata is represented as follows

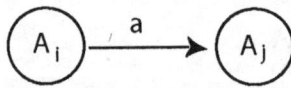

If there is a production of the form $A_i \rightarrow a$, the corresponding transition terminates at a new state. This is the unique final state. In this case the automata is represented as follows

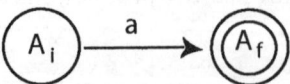

Then the transition function δ is defined as follows

  i.   Each production $A_i \rightarrow aA_j$ induces a transition function $\delta(A_i, a) = A_j$

  ii.   Each production $A_i \rightarrow a$ induces a transition function $\delta(A_i, a) = A_f$, where A_f is the final state.

Example 4.5

Construct finite automata recognizing the following regular grammar.

Solution :

$$S \rightarrow aS \mid bA \mid b$$
$$A \rightarrow aA \mid bS \mid a$$

The required finite automata will have three states S, A and an additional state say F, which is a final state.

For constructing FA for the given regular grammar, we apply the following rules:

1)   For each production $A_i \rightarrow aA_j$, we have $\delta(A_i, a) = A_j$, i.e., draw an edge labeled 'a' from state A_i to A_j.

2)   If production is $A_i \rightarrow a$, then $\delta(A_i, a) = A_f$ (new state) where A_f is a final state, i.e., we draw an edge labeled 'a' from A_i to A_f.

If we apply rule 1 for the productions

$$S \rightarrow aS \mid bA$$
$$A \rightarrow aA \mid bS$$

we have the following transitions

$$\delta(S, a) = S$$
$$\delta(S, b) = A$$
$$\delta(A, a) = A$$
$$\delta(A, b) = S$$

If we apply rule 2 for the productions

$$S \rightarrow b$$
$$A \rightarrow a$$

we have the following transitions

$$\delta(S, b) = F \quad \text{where F is the new final state}$$
$$\delta(A, a) = F$$

The resulting **FA** is as follows

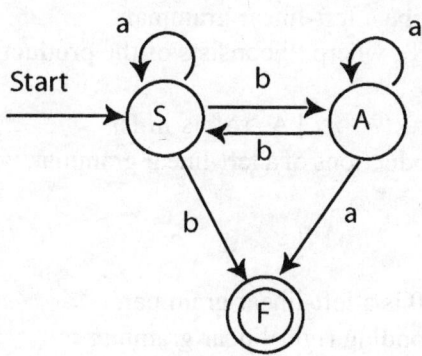

Example 4.6

Construct finite automata recognizing the following regular grammar

$$A_0 \rightarrow aA_1$$
$$A_1 \rightarrow bA_1 \mid bA_0 \mid a$$

Solution :

By applying procedure for constructing FA for the given regular grammar, we have the following. The finite automata for the given grammar will have three states A_0, A_1 and A_f is a new (final) state.

By applying rule 1 for the productions

$$A_0 \rightarrow aA_1$$
$$A_1 \rightarrow bA_1 \mid bA_0,\text{ we have the following transitions}$$
$$\delta(A_0, a) = A_1$$
$$\delta(A_1, b) = A_1$$
$$\delta(A_1, b) = A_0$$

By applying rule 2 for the production

$$A_1 \rightarrow a$$

we have the following transition, where F is the new final state

$$\delta(A_1, a) = F$$

The required FA is shown below

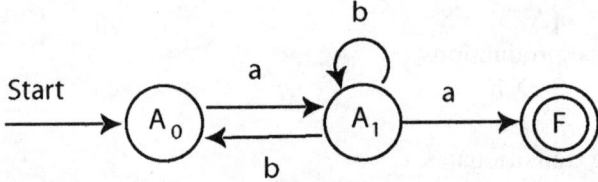

4.3.2 CONVERSION OF LEFT-LINEAR GRAMMAR TO RIGHT-LINEAR GRAMMAR

Let $G = (V, T, P, S)$ be a left-linear grammar.
Let $G^1 = (V, T, P^1, S)$, where P^1 consists of the productions of G with right side reversed, i.e.,

$$P^1 = \{A \rightarrow \alpha \mid A \rightarrow \alpha^R \text{ is in P}\}$$

If we reverse the productions of a left-linear grammar, we get a right-linear grammar and vice-versa.

Example 4.7

$S \rightarrow S10 \mid 0$ is a left-linear grammar.
The corresponding right-linear grammar is
$S \rightarrow 01S \mid 0$

4.3.3 CONVERSION OF A REGULAR GRAMMAR (RIGHT-LINEAR) TO NFA WITH ε

Suppose L = L (G) for some right-linear grammar G = (V, T, P, S). We construct NFA with ε M = (Q, T, δ, [S], {[ε]}) that simulates derivations in G.

Q consists of the symbols [α] such that α is S or a suffix of some right-hand side of a production P.

We define δ by:

1)   If **A** is a variable, then

$\delta([A], \varepsilon) = \{[\alpha] \mid A \rightarrow \alpha$ is a production$\}$

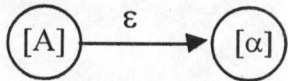

2)   If **a** is in T and α in $T^* \cup T^*V$, then

$\delta([a\alpha], a) = \{[\alpha]\}$

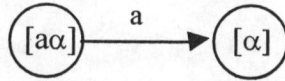

3)   [a] can be written as [aε

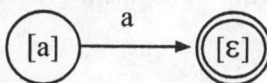

where [ε] is the final state

Example 4.8

Construct NFA with ε for the following right-linear grammar

S → 0A

A → 10A | ε

Solution:

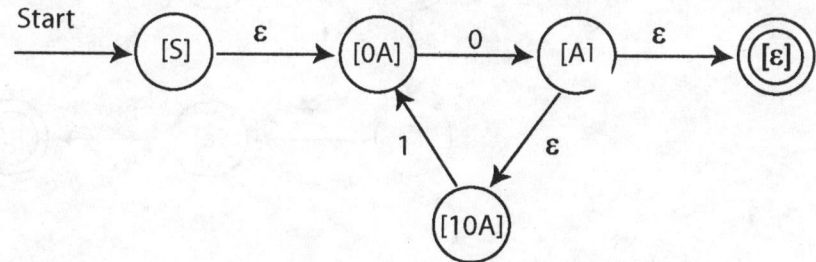

Example 4.9

Construct NFA with ε for the following right-linear grammar

$$S \to 01S \mid 0$$

Solution:

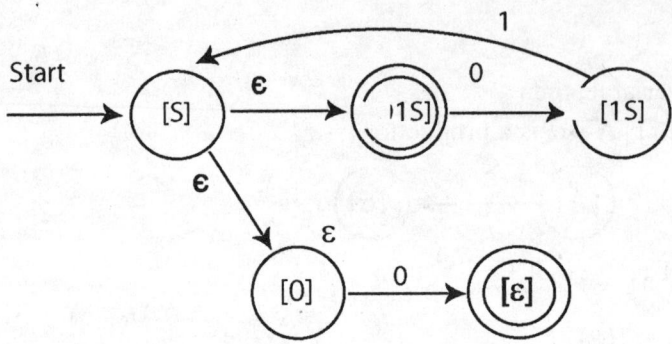

Example 4.10

Construct NFA with ε for the following right-linear grammar

$$S \to bB, \quad B \to bC$$
$$B \to a, \quad C \to a, \quad B \to b$$

Solution :

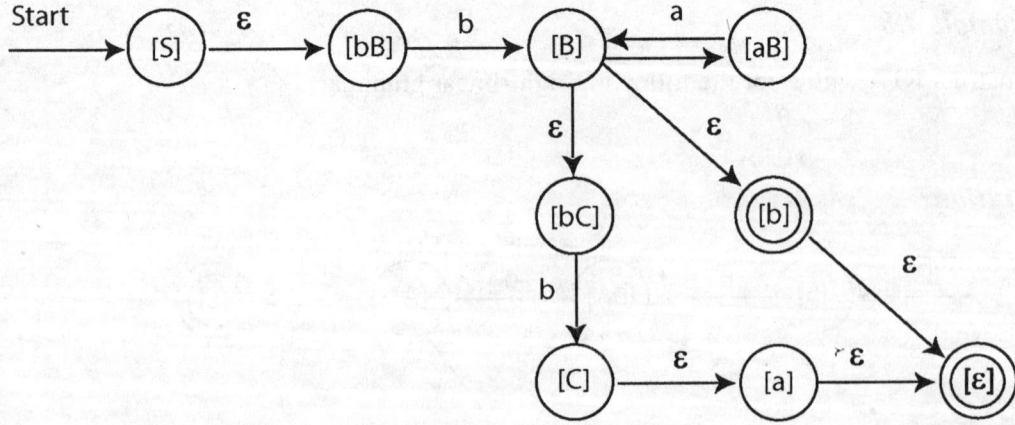

To construct NFA with ε transition for left-linear grammar generated for a given regular expression:

1) Convert the given grammar to right linear grammar by reversing the productions.
2) Construct NFA with ε-transitions for the converted right-linear grammar.
3) Reverse the edges of that NFA and exchange initial and final states, we get another NFA for the given regular expression.

Example 4.11

Construct NFA - ε for the regular expression 0(10)*

The left-linear grammar for the regular expression $0(10)^*$ is

$$S \rightarrow S10 \mid 0$$

Solution:

Step-I: By reversing the R.H.S of the production, we obtain right-linear grammar

$$S \rightarrow 01S \mid 0$$

Step-II: Construct NFA with ε for this grammar

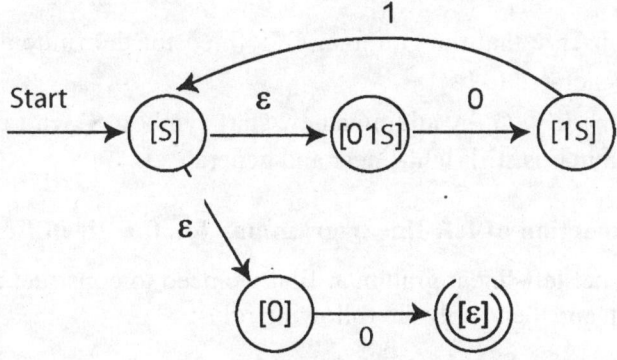

Step-III: Reverse the edges and exchange initial and final states

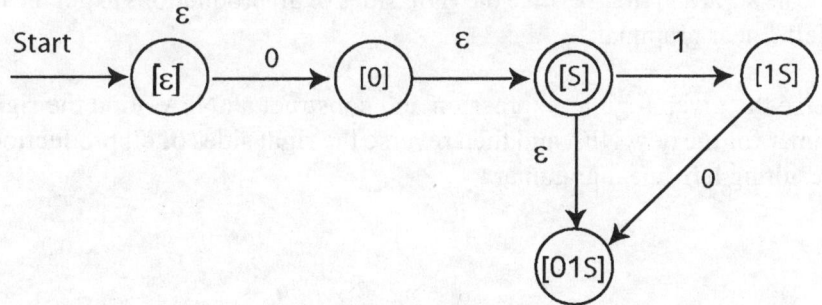

The given NFA accepts the regular expression $0(10)^*$

4.3.4 CONSTRUCTION OF REGULAR GRAMMAR FROM A GIVEN FA

a) Construction of right-linear grammar for the given finite automaton

Let the right-linear grammar be G = (V, T, P, S) where
$\quad$ V = set of all variables
$\quad$ T = set of all terminals
$\quad$ P = set of all productions
$\quad$ S = start symbol

To obtain the productions of the grammar i.e., the set P, we apply the following cases:

Case I: If the initial state $\neq$ final state i.e., $q_0 \notin F$, then apply the following steps

Step 1: If there is a transition $\delta(q_i, a) = q_j \notin F$, then include a production $q_i \to aq_j$ in P.

Step 2: If there is a transition $\delta(q_i, a) = q_j \in F$, then include the productions $q_i \to aq_j$ and $q_i \to a$ in P.

Step 3: An initial state of FA is the start symbol of 'G'.

Case II: If the initial state = final state i.e., $q_0 \in F$ then the language contains 'ε'.
$$L = \{\varepsilon, \dots..\}$$

We can observe that the grammar G defined for the finite automata given in the above example generates L-$\{\varepsilon\}$.

We may modify 'G' by adding a new start symbol 'S' with productions $S \to q_0 \mid \varepsilon$. The resulting grammar is still right-linear and generates L.

b) Construction of left-linear grammar for the given finite automaton

To construct left-linear grammar, first we need to construct an NFA for the given regular expression and then apply the following rules:

1) Reverse the directions of all the edges.
2) Make the initial state as final state and final state as initial state.
3) After construction of new finite automaton, find the right-linear grammar for the new FA and then reverse the right sides of all productions to get the resulting left-linear grammar.

(or)

Reverse the given regular expression and construct an NFA. Find the right-linear grammar for the new NFA and then reverse the right sides of all productions to get the resulting left-linear grammar.

Example 4.12

Find the regular grammar that represents the following DFA

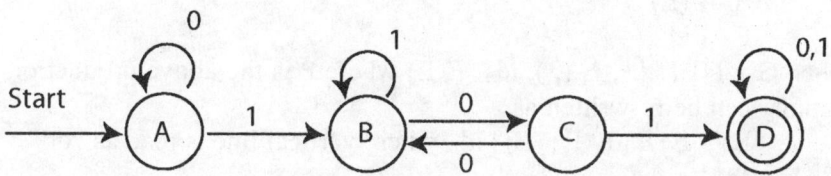

Solution :

Here initial state is not equal to final state, so apply case I only to the given DFA. The productions in P are as follows

$\delta(A, 0) = A \Rightarrow A \to 0A$

$\delta(A, 1) = B \Rightarrow A \to 1B$

$\delta(B, 0) = C \Rightarrow B \to 0C$

$\delta(B, 1) = B \Rightarrow B \to 1B$

$\delta(C, 0) = B \Rightarrow C \to 0B$

$\delta(C, 1) = D \Rightarrow C \to 1D$

$\qquad\qquad\qquad C \to 1 \qquad$ Since 'D' is final state

$\delta(D, 0) = D \Rightarrow D \to 0D$

$\qquad\qquad\qquad D \to 0 \qquad$ Since 'D' is final state

$\delta(D, 1) = D \Rightarrow D \to 1D$

$\qquad\qquad\qquad D \to 1 \qquad$ Since 'D' is final state

$\therefore$ The regular grammar (i.e., right-linear grammar) is G = (V, T, P, S) where V = {A, B, C, D}, T = {0, 1}, S=A and P is given by

$\qquad A \to 0A \mid 1B$

$\qquad B \to 0C \mid 1B$

$\qquad C \to 0B \mid 1D \mid 1$

$\qquad D \to 0D \mid 1D \mid 0 \mid 1$

4.4 CONTEXT FREE GRAMMAR (CFG)

4.4.1 DEFINITION

Let G = (V, T, P, S) be a context free grammar where

V denotes the set of variables (non terminals)

T denotes the set of terminals

P is the set of production rules of the form A→α where A∈V and α∈(V ∪ T)* and S is the starting variable.

NOTE: V and T are disjoint.

Example 4.13

$$E \rightarrow E+E$$
$$E \rightarrow E*E$$
$$E \rightarrow (E)$$
$$E \rightarrow id$$

Then $G = (\{E\}, \{+, *, (,), id\}, P, E)$ where P is the above production rules. The above grammar can be re-written as

$$E \rightarrow E+E \mid E*E \mid (E) \mid id, \text{ where vertical line is read as "or"}$$

The following conventions are used regarding grammars

1. The capital letters A, B, C, D, E and S denote variables. S is the start symbol unless otherwise stated.

2. The lower case letters a, b, c, d, e, digits and boldface strings are terminals.

3. The capital letters X, Y and Z denote symbols that may be either terminals or non terminals (variables).

4. The lower-case letters u, v, w, x, y and z denote string of terminals.

5. The lower-case Greek letters α, β and γ denote string of variables and terminals.

4.4.2 DERIVATIONS AND LANGUAGES

The language generated by a grammar $G = (V, T, P, S)$ is formally represented by a derivation.

Let us define two relations $\Rightarrow G$ and   G between strings in $(V \cup T)^*$,

If $A \rightarrow \beta$ is a production of P and α and γ are any strings in $(V \cup T)^*$, then $\alpha A \gamma \Rightarrow \alpha \beta \gamma$. We say that the production $A \rightarrow \beta$ is applied to the string αA to obtain $\alpha \beta \gamma$ or that $\alpha A \gamma$ directly derives $\alpha \beta \gamma$ in grammar G.

Suppose that $\alpha_1, \alpha_2, \ldots\ldots \alpha_m$ are strings in $(V \cup T)^*$, $m \geq 1$ and

$$\alpha_1 \Rightarrow \alpha_2$$
$$\alpha_2 \Rightarrow \alpha_3$$
$$\alpha_{m-1} \Rightarrow \alpha_m$$

Then we say $\alpha_1 \Rightarrow \alpha_m$ or α_1 derives α_m in grammar G.

The *language generated* by G (denoted as L(G)) is $\{w \mid w$ is in T^* and $S \Rightarrow w\}$ i.e., the string is in L(G) if:

1) The string consists of terminals only

2) The string can be derived from S

We call L as context free language (CFL), if it is L(G) for some CFG G.

A string of terminals and variables α is called a *sentential form* if $S \Rightarrow \alpha$.

Example 4.14

Consider a grammar G = (V, T, P, S), where V = {S}, T = {a, b} and P = {S→aSb, S→ab}. Write the language generate by the grammar G.

Solution:

By applying first production n-1 times and then the second production we have

$$S \Rightarrow aSb$$
$$\Rightarrow aaSbb$$
$$\Rightarrow aaaSbbb$$
$$\Rightarrow a^4Sb^4$$
$$\Rightarrow a^{n-1}Sb^{n-1}$$
$$\Rightarrow a^{n-1}abb^{n-1} \quad \text{i.e., } a^nb^n$$

Therefore, $S \Rightarrow a^nb^n$

Hence, $L(G) = \{ a^nb^n \mid n > = 1 \}$

4.5 DERIVATION TREES

It is useful to display derivations as trees. Derivation trees (or parse trees) are useful in applications such as compilation of programming languages.

4.5.1 DEFINITION

Let $G = (V, T, P, S)$ be CFG. A tree is a derivation tree (or parse tree) for G if:

1) Every vertex has a label which is a symbol of $V \cup T \cup \{\varepsilon\}$.
2) The label of the root is S.
3) If a vertex is interior and has label A, then A must be in V (i.e., set of non-terminals).
4) If an interior vertex n is labeled A, and the sons of n are labeled X_1, X_2, ...X_n from the left, then $A \rightarrow X_1, X_2 \ldots \ldots X_n$ must be production in P.
5) If vertex n has label ε, then n is a leaf and is the only son of its father.

Example 4.15

Consider the following CFG

$$E \rightarrow E+E \mid E*E \mid (E) \mid id$$

Derive a string id*id+id and give the derivation tree.

Solution :

Let $E \rightarrow E+E$ .... (1)
$E \rightarrow E*E$ ....(2)
$E \rightarrow (E)$.................(3)
$E \rightarrow id$ ....(4)

The derivation is

$$E \Rightarrow E*E \qquad \text{(Using 2)}$$
$$\Rightarrow id*E \qquad \text{(Using 4)}$$
$$\Rightarrow id*E+E \qquad \text{(Using 1)}$$

$\Rightarrow$ id*id+E   (Using 4)

$\Rightarrow$ id*id+id   (Using 4)

Hence the required string is obtained from the grammar.

The derivation tree is

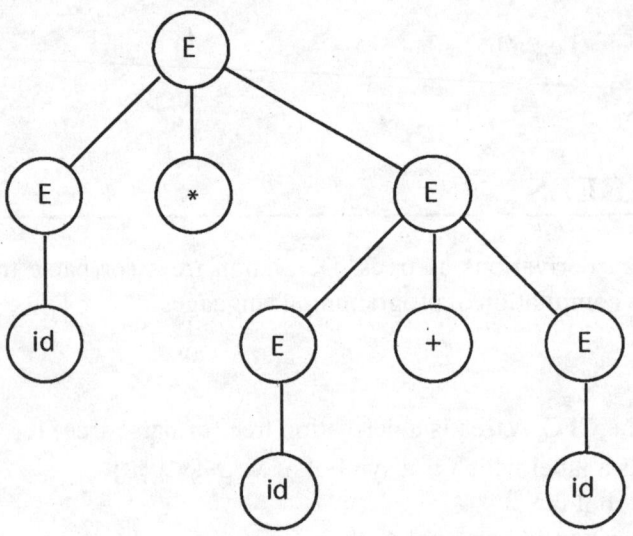

If we concatenate the leaf nodes from left to right, we get the desired string i.e., id*id+id

4.5.2 LEFT-MOST AND RIGHT-MOST DERIVATION OF STRINGS

Left-Most Derivation (LMD):

A derivation S$\Rightarrow$w is called left-most derivation, if we apply the production rules only to the left-most non-terminal at each step.

Example 4.16

Consider the CFG

S$\rightarrow$ aAS | a

A$\rightarrow$SbA | SS | ba

Use left-most derivation to derive a string aabbaa

Solution :

The left-most derivation for the given grammar is

S $\Rightarrow$ aAS   [S$\rightarrow$aAS]

$\Rightarrow$ aSbAS	[A→SbA]
$\Rightarrow$ aabAS	[S→a]
$\Rightarrow$ aabbaS	[A→ba]
$\Rightarrow$ aabbaa	[S→aAS]

Therefore, S $\Rightarrow$ aabbaa

The derivation tree is

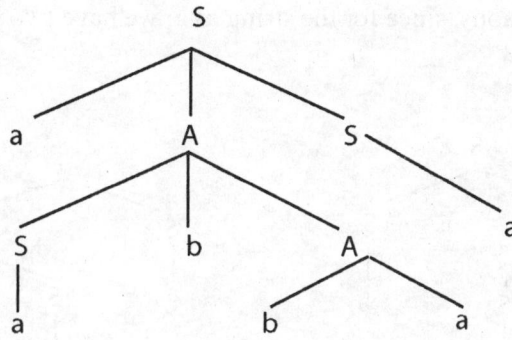

By concatenating leaf nodes we get "aabbaa", which is the desired string.

Right-Most Derivation (RMD):

A derivation S⇒w is called right-most derivation, if we apply the production rules only to the right-most non-terminal at each step.

For the above grammar, to derive **aabbaa** using right-most derivation is

S $\Rightarrow$ aAS	[S → aAS]
$\Rightarrow$ aAa	[S → a]
$\Rightarrow$ aSbAa	[A → SbA]
$\Rightarrow$ aSbbaa	[A → ba]
$\Rightarrow$ aabbaa	[S → a]

Ambiguous Grammar:

The terminal string **w** belonging to L(G) is **ambiguous,** if there exists more than one LMD or more than one RMD for the same string, which is generated by G.

Example 4.17

Consider the grammar and prove it is ambiguous.

S→SS | a | b

Solution:

Let us derive string from this grammar.

(i)   S ⇒ SS   (ii)   S ⇒ SS
       S ⇒ SSS   S ⇒ aS
       S ⇒ aSS   S ⇒ aSS
       S ⇒ abS   S ⇒ abS
       S ⇒ aba   S ⇒ aba

The above two derivations derive the same string, using left-most derivations in different ways.

∴ The grammar is ambiguous, since for the string **aba**, we have two left-most derivations.
Derivation trees for aba

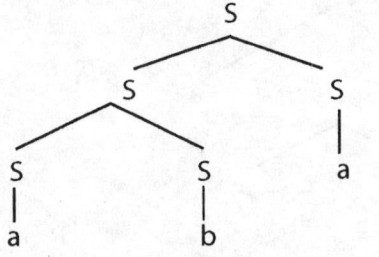 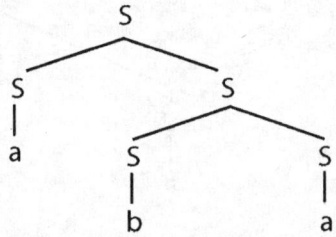

Since there are two derivation trees for the string **aba**, the given grammar is ambiguous.

Example 4.18

Show that the grammar is ambiguous
  S → a | abSb | aAb
  A → bS | aAAb

Solution:

Using left-most derivation

i)   S ⇒ aAb   (S → aAb)
       ⇒ abSb   (A → bS)
       ⇒ ababSbb   (S → abSb)
       ⇒ abababb   (S → a)
    ∴ S ⇒ abababb

ii)   S ⇒ abSb   (S → abSb)
       ⇒ abaAbb   (S → aAb)
       ⇒ ababSbb   (A → bS)
       ⇒ abababb   (S → a)
    ∴ S ⇒ abababb

The above two derivations derive the same string **'abababb'** using left-most derivations in different ways.

Now construct derivation trees for the string **'abababb'**.

Derivation trees for 'abababb'

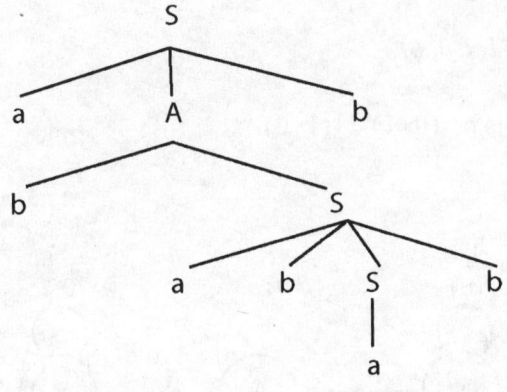

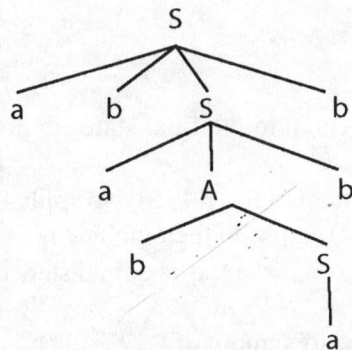

Since two distinct derivation trees are possible for the string **'abababb'**, the given grammar is ambiguous.

Solved Problems

Problem 1:** **Construct right-linear and left-linear grammars for the following regular expression.

$$(0 + 1)^* 11(1 + 0)^*$$

Solution:

Given regular expression is $r = (0+1)^* 11(1+0)^*$

The FA for **r** is as follows:

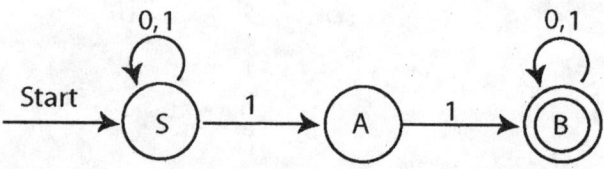

Right-linear grammar:

Here the initial state is not equal to the final state, so no need to include ε in the language.

To obtain right-linear grammar $G = (V, T, P, S)$, we apply the following steps.

Step-1: If there is a transition $\delta(q_i, a) = q_j \notin F$, then include $q_i \rightarrow a \, q_j$ in P.

Step-2: If there is a transition $\delta(q_i, a) = q_f$, where q_f is a final state then include a production $q_i \rightarrow a$ in P.

Step-3: An initial state of FA is the start symbol of G.

Using Step-1, we have the following productions

$$\delta(S, 0) = S \quad \Rightarrow \quad S \rightarrow 0S$$
$$\delta(S, 1) = S \quad \Rightarrow \quad S \rightarrow 1S$$
$$\delta(S, 1) = A \quad \Rightarrow \quad S \rightarrow 1A$$
$$\delta(A, 1) = B \quad \Rightarrow \quad A \rightarrow 1B$$
$$\delta(B, 0) = B \quad \Rightarrow \quad B \rightarrow 0B$$
$$\delta(B, 1) = B \quad \Rightarrow \quad B \rightarrow 1B$$

Using Step-2, we have the following productions, in the FA, B is the final state

$$\delta(A, 1) = B \quad \Rightarrow \quad A \rightarrow 1$$

$$\delta(B, 0) = B \quad \Rightarrow \quad B \to 0$$
$$\delta(B, 1) = B \quad \Rightarrow \quad B \to 1$$

Using Step-3, we have the start symbol of the right-linear grammar G is the start state of FA i.e., start symbol of G=S

∴ The right-linear grammar G = (V, T, P, S), where V={S, A, B}, S=S, T={0, 1}

P is the set of all productions and are given below.

S → 0S | 1S | 1A

A → 1B | 1

B → 0B | 1B | 0 | 1

Since initial state is not equal to final sate, no need to include ε in the language.

Left-linear grammar:

To construct the left-linear grammar, reverse the given regular expression, we get r= (1+0)* 11 (0+1)*, which is same as the given regular expression, so we get the same right-linear grammar.

To get the left-linear grammar, reverse the R.H.S. of the productions of right-linear grammar. Therefore, the left-linear grammar is

S → S0 | S1 | A1

A → B1 | 1

B → B0 | B1 | 0 | 1

Problem 2: Construct right-linear and left-linear grammars for the following regular expression.

$$0^* \ (1(0+1))^*$$

Solution :

Given regular expression is 0* (1(0+1))*, construct finite automata for the regular expression

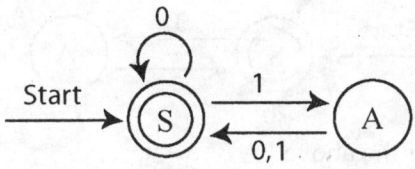

Right-linear grammar:

Here initial state is equal to final state, so include ε in the language, and also add S $\rightarrow$ ε in the productions.

To obtain a right –linear grammar, G= (V, T, P, S), apply the following steps:

Step-1: If there is a transition $\delta(q_i, a) = q_j$ in FA, then include $q_i \rightarrow aq_j$ in P.
Step-2: If there is a transition $\delta(q_i, a) = q_f$, a final state or accepting state in FA, then include a production $q_i \rightarrow a$ in P.
Step-3: An initial state of FA is the start symbol of G.

Using step-1:

$$\delta(S, 0) = S \Rightarrow S \rightarrow 0S \mid \varepsilon$$
$$\delta(S, 1) = A \Rightarrow S \rightarrow 1A$$
$$\delta(A, 0) = S \Rightarrow A \rightarrow 0S$$
$$\delta(A, 1) = S \Rightarrow A \rightarrow 1S$$

Using step-2: (S is final state)

$$\delta(A, 0) = S \Rightarrow A \rightarrow 0$$
$$\delta(A, 1) = S \Rightarrow A \rightarrow 1$$
$$\delta(S, 0) = S \Rightarrow S \rightarrow 0$$

Using step-3:

Since the start state and final state are same i.e. S, the right-linear grammar G = (V, T, P,S), where V = { S, A}, T = {0,1}, S = S, and the P is the of productions given as
$$S \rightarrow 0S \mid 1A \mid 0 \mid \varepsilon$$
$$A \rightarrow 0S \mid 1S \mid 0 \mid 1$$

Left-linear grammar :

To construct left-linear grammar, reverse the given regular expression i.e., $((0+1)1)^* 0^*$
The new FA is shown below

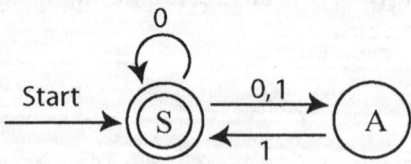

The right-linear grammar for the above FA is
$$\delta(S, 0) = S \Rightarrow S \rightarrow 0S \mid \varepsilon$$

$\delta (S, 0) = A \Rightarrow S \rightarrow 0A$

$\delta (S, 1) = A \Rightarrow S \rightarrow 1A$

$\delta (A, 1) = S \Rightarrow A \rightarrow 1S$

$\delta (A, 1) = S \Rightarrow A \rightarrow 1$

$\delta (S, 0) = S \Rightarrow S \rightarrow 0$

$\therefore$ The right-linear grammar for the reversed regular expression is

$S \rightarrow 0S \mid 0A \mid 1A \mid 0|\varepsilon$

$A \rightarrow 1S \mid 1$

To get the left-linear grammar, reverse the R.H.S of the above productions

$S \rightarrow S0$

$S \rightarrow A0$

$S \rightarrow A1$

$A \rightarrow S1$

$A \rightarrow 1$

$S \rightarrow 0$

$S \rightarrow \varepsilon$

$\therefore$ The left-linear grammar is $\quad S \rightarrow S0 \mid A0 \mid A1 \mid 0 \mid \varepsilon$

$A \rightarrow S1 \mid 1$

Note: If $0^* = \varepsilon$, $(1(0+1))^* = \varepsilon$ then $0^*(1(0+1))^* = \varepsilon$ i.e., initial state should be final state.

Problem 3: *For the following languages, give the corresponding grammar and finite state machines.*

$$((0^*1^*)\ 10)^*$$

Solution:

The NFA with ε - moves for the given regular expression is as follows

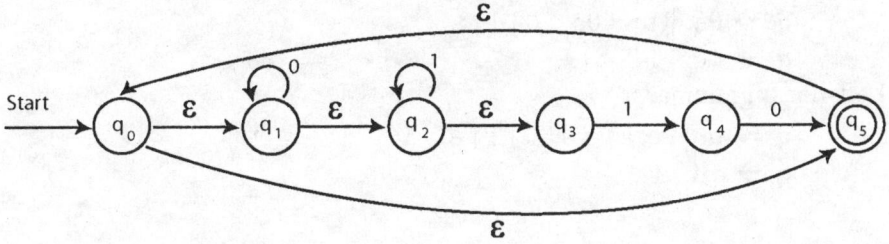

The NFA without ε-moves is as follows:

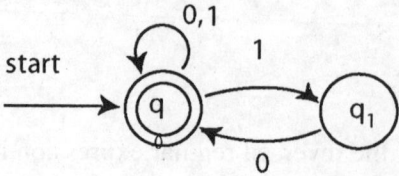

Right-linear grammar:

Since start state and final state are same, add ε to the language

$$q_0 \rightarrow 0q_0 \mid 1q_0 \mid 1q_1 \mid 0 \mid 1 \mid \varepsilon$$
$$q_1 \rightarrow 0q_0 \mid 0$$

The above grammar is right-linear, because all productions are of the form
$A \rightarrow wB$ or $A \rightarrow w$.

Left-linear grammar:

To construct the left-linear grammar, first we need to construct a FA as follows:
1)   Reverse the directions of all edges.
2)   Make initial state as final state and final state as initial state.

The new FA is shown below:

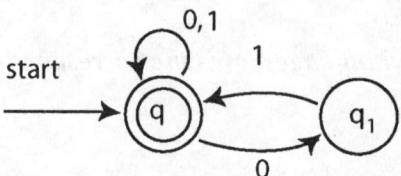

Here initial and final states are same; to get the left-linear grammar, first find the
right-linear grammar for the new FA and interchange the positions of terminals and non-
terminals in the R.H.S productions of the right-linear grammar.

Right-linear grammar for the new FA is

$$q_0 \rightarrow 0q_0 \mid 1q_0 \mid 0q_1 \mid 0 \mid 1 \mid \varepsilon$$
$$q_1 \rightarrow 1q_0 \mid 1$$

Left-linear grammar is

$$q_0 \rightarrow q_0 0 \mid q_0 1 \mid q_1 0 \mid 0 \mid 1 \mid \varepsilon$$
$$q_1 \rightarrow q_0 1 \mid 1$$

The above grammar is left-linear, because each production is of the form A → Bw or A → w

Here in the FA q_0, q_1 are considered as variables.

Problem 4: *Generate regular grammar for the following expression*
$$(0^* + 1^*)$$

Solution :

NFA with ε- moves for the given regular expression is shown below.

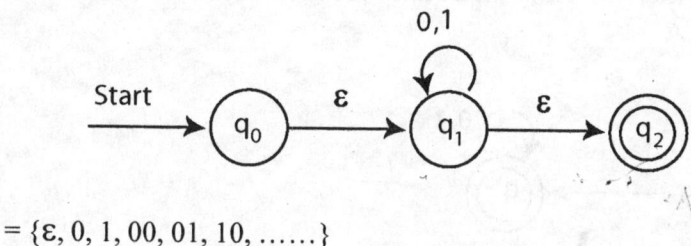

L = {ε, 0, 1, 00, 01, 10, ...}

NFA without ε- moves is

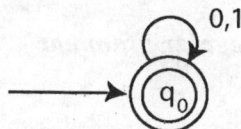

Transition table for DFA

Q \ Σ	0	1
→ q_0	q_0	q_0

Right-linear grammar:

$$\delta(q_0, 0) = q_0 \Rightarrow q_0 \to 0q_0$$
$$q_0 \to 0$$
$$q_0 \to \varepsilon \quad \text{since initial state and final state are same}$$
$$\delta(q_0, 1) = q_0 \Rightarrow q_0 \to 1q_0$$
$$q_0 \to 1$$

$\therefore$ Right-linear grammar is $q_0 \to 0q_0 \mid 1q_0 \mid 0 \mid 1 \mid \varepsilon$

Left-linear grammar:

Here in the above FA, the initial and final states are same and if we reverse the edges of FA, we get the same FA

$\therefore$ New FA is

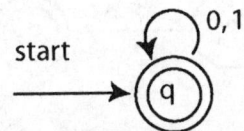

$\therefore$ Right-linear grammar $\Rightarrow q_0 \to 0q_0 \mid 1q_0 \mid 0 \mid 1 \mid \varepsilon$

$\therefore$ Left-linear grammar $\Rightarrow q_0 \to q_0 0 \mid q_0 1 \mid 0 \mid 1 \mid \varepsilon$

Here q_0 is considered as variable.

Problem 5: Construct a DFA for the following regular grammar

$$S \to Aa$$
$$A \to Sb \mid Ab \mid \varepsilon$$

Solution :

Given regular grammar is

$$S \to Aa$$
$$A \to Sb \mid Ab \mid \varepsilon$$

All the above productions are in the form of either $A \to Bw$ or $A \to w$
where $A, B \in V$, $w \in T^*$

$\therefore$ The given grammar is left-linear grammar, to construct DFA, we need to first convert left-linear grammar to right-linear grammar.

To achieve this, we just reverse the positions of terminals and non-terminals on R.H.S of each production.

$\therefore$ The right-linear grammar is given by

$$S \to aA$$
$$A \to bS \mid bA \mid \varepsilon, \text{ where A is the final state}$$

The transition functions are defined as follows

$$S \rightarrow aA \Rightarrow \delta(S, a) = A$$
$$A \rightarrow bS \Rightarrow \delta(A, b) = S$$
$$A \rightarrow bA \Rightarrow \delta(A, b) = A$$
$$A \rightarrow \varepsilon \Rightarrow \delta(A, \varepsilon) = A, \text{ where A is the final state}$$

The FA for the right-linear grammar is as follows

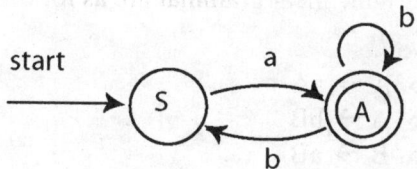

To construct DFA for the left-linear grammar, apply the following steps

    1. Reverse the direction of all edges.

    2. Make initial state as final state and final state as initial state.

    ∴ The DFA for the given grammar is shown below

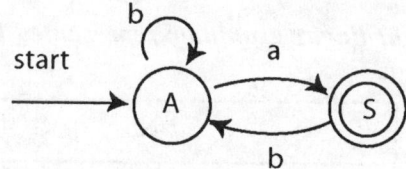

Problem 6: *Give the right-linear grammar for the following DFA*

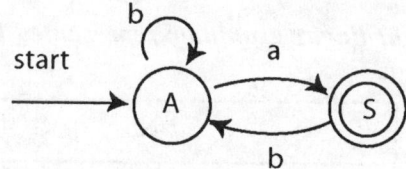

Solution :

Right-linear grammar: A grammar $G = (V, T, P, S)$ is said to be right-linear, if all productions are of the form $A \rightarrow wB$ or $A \rightarrow w$ where $A, B \in V$ and w is a string of terminals i.e., $w \in T^*$

Given DFA is

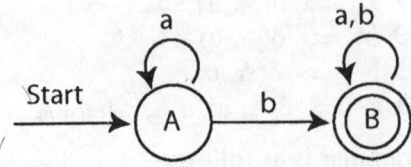

The productions for the right-linear grammar are as follows:

Using step-1: We have

$$\delta(A, a) = A \Rightarrow A \rightarrow aA$$
$$\delta(A, b) = B \Rightarrow A \rightarrow bB$$
$$\delta(B, a) = B \Rightarrow B \rightarrow aB$$
$$\delta(B, b) = B \Rightarrow B \rightarrow bB$$

Using step-2: We have

$$\delta(A, b) = B \Rightarrow A \rightarrow b$$
$$\delta(B, a) = B \Rightarrow B \rightarrow a$$
$$\delta(B, b) = B \Rightarrow B \rightarrow b$$

∴ The regular grammar is A → aA | bB | b

B → aB | bB | a | b

Problem 7: Obtain the right-linear grammar represented by the following DFA

PS	NS	
	0	1
→ A	A	B
B	C	B
Ⓒ	B	D
D	D	D

Solution :

The transition diagram is as follows

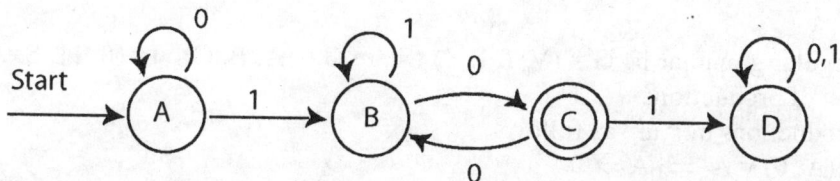

The productions in P are as follows

$$\delta(A, 0) = A \Rightarrow A \rightarrow 0A$$
$$\delta(A, 1) = B \Rightarrow A \rightarrow 1B$$
$$\delta(B, 0) = C \Rightarrow B \rightarrow 0C \text{ and } B \rightarrow 0, \text{ since C is the final state}$$
$$\delta(B, 1) = B \Rightarrow B \rightarrow 1B$$
$$\delta(C, 0) = B \Rightarrow C \rightarrow 0B$$
$$\delta(C, 1) = D \Rightarrow C \rightarrow 1D$$
$$\delta(D, 0) = D \Rightarrow D \rightarrow 0D$$
$$\delta(D, 1) = D \Rightarrow D \rightarrow 1D$$

$\therefore$ The right-linear grammar is

$$A \rightarrow 0A \mid 1B$$
$$B \rightarrow 0C \mid 1B \mid 0$$
$$C \rightarrow 0B \mid 1D$$
$$D \rightarrow 0D \mid 1D$$

In the above grammar, 'D' is not deriving any terminal string.

$\therefore$ D is useless symbol.

$\therefore$ Eliminate 'D' and its productions

Eliminate the useless productions C $\rightarrow$ 1D, D $\rightarrow$ 0D, D $\rightarrow$ 1D from the grammar.

The simplified right-linear grammar is

$$A \rightarrow 0A \mid 1B$$
$$B \rightarrow 0C \mid 1B \mid 0$$
$$C \rightarrow 0B$$

The above grammar is right-linear, because all productions are of the form A $\rightarrow$ wB
or   A $\rightarrow$ w

Problem 8: *For the following finite state automata, construct an equivalent grammar*

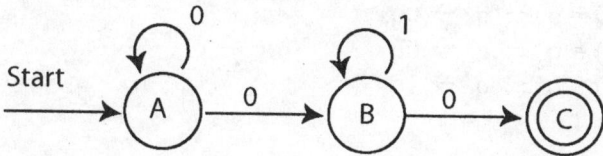

Solution :

Let the regular grammar be G = (V, T, P, S) where G ={A, B, C}, T={0, 1}, S=A,
P is the set of productions

∴ The productions in P are as follows

$$\delta(A, 0) = A \implies A \rightarrow 0A$$
$$\delta(A, 0) = B \implies A \rightarrow 0B$$
$$\delta(B, 1) = B \implies B \rightarrow 1B$$
$$\delta(B, 0) = C \implies B \rightarrow 0C$$
$$B \rightarrow 0, \text{ since 'C' is the final state}$$

∴ The regular grammar (i.e., right-linear) is

$$A \rightarrow 0A \mid 0B$$
$$B \rightarrow 1B \mid 0C \mid 0$$

Since there is no production on 'C', we can delete the useless production B → 0C from the
grammar.

∴ The regular grammar is

$$A \rightarrow 0A \mid 0B$$
$$B \rightarrow 1B \mid 0$$

Problem 9: Find the regular grammar for the following automata

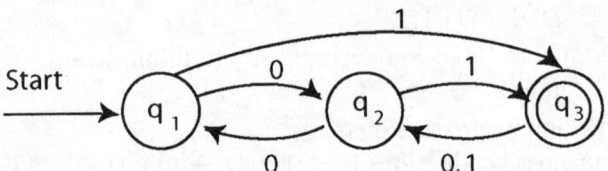

Solution:

Let the regular grammar be G = (V, T, P, S)

The productions in P are as follows

$$\delta(q_1, 0) = q_2 \implies q_1 \rightarrow 0q_2$$
$$\delta(q_1, 1) = q_3 \implies q_1 \rightarrow 1q_3$$
$$q_1 \rightarrow 1, \quad \text{since '}q_3\text{' is final state}$$
$$\delta(q_2, 1) = q_3 \implies q_2 \rightarrow 1q_3$$
$$q_2 \rightarrow 1, \quad \text{since '}q_3\text{' is final state}$$
$$\delta(q_2, 0) = q_1 \implies q_2 \rightarrow 0q_1$$
$$\delta(q_3, 0) = q_2 \implies q_3 \rightarrow 0q_2$$
$$\delta(q_3, 1) = q_2 \implies q_3 \rightarrow 1q_2$$

∴ The regular grammar (i.e., right-linear grammar) for the given FA is as follows
$G = (V, T, P, S)$ where $V = \{q_1, q_2, q_3\}$, $T = \{0, 1\}$, P is the set of productions.

$$q_1 \to 0q_2 \mid 1q_3 \mid 1 \qquad\qquad q_1 \to \varepsilon$$
$$q_2 \to 1q_3 \mid 0q_1 \mid 1$$
$$q_3 \to 0q_2 \mid 1q_2$$

here q_1, q_2, q_3 are variables.

Note: In the above productions, all productions are in the form of $A \to wB$ or $A \to w$. In every grammar upper case letters are variables. In FA, the states are called variables.

Problem 10: Find the regular grammar for the following automata

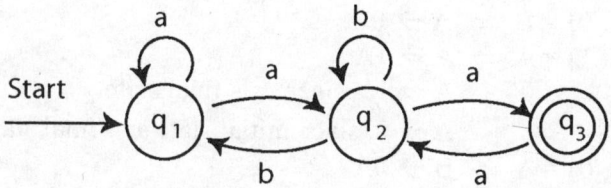

Solution :

Regular grammar: A grammar which is either right-linear or left-linear is a regular grammar.

The productions in P as follows

$$\delta(q_1, a) = q_1 \;\Rightarrow\; q_1 \to aq_1$$
$$\delta(q_1, a) = q_2 \;\Rightarrow\; q_1 \to aq_2$$
$$\delta(q_2, a) = q_3 \;\Rightarrow\; q_2 \to aq_3$$
$$\qquad\qquad\qquad\quad q_2 \to a, \quad \text{since } q_3 \text{ is the final state}$$
$$\delta(q_2, b) = q_2 \;\Rightarrow\; q_2 \to bq_2$$
$$\delta(q_2, b) = q_1 \;\Rightarrow\; q_2 \to bq_1$$
$$\delta(q_3, a) = q_2 \;\Rightarrow\; q_3 \to aq_2$$

∴ The regular grammar for the given FA is

$$q_1 \to aq_1 \mid aq_2$$
$$q_2 \to aq_3 \mid bq_2 \mid bq_1 \mid a$$
$$q_3 \to aq_2$$

Note: The states in the finite automata are considered as variables. Therefore, in the above FA q_1, q_2, q_3 are called variables.

Problem 11: *Give the regular grammar for the following finite state automata*

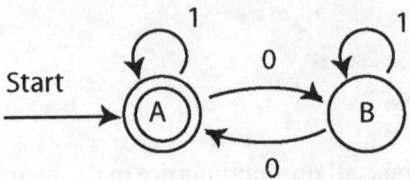

Solution :

A regular grammar is either a right-linear grammar or a left-linear grammar.
Let the regular grammar be right-linear grammar
The productions in P are as follows

$\delta(A, 0) = B \implies A \rightarrow 0B$

$\delta(A, 1) = A \implies A \rightarrow 1A$

$\qquad\qquad\qquad A \rightarrow 1$, since A is final state

$\qquad\qquad\qquad A \rightarrow \varepsilon$, since initial state and final state are same

$\delta(B, 0) = A \implies B \rightarrow 0A$

$\qquad\qquad\qquad B \rightarrow 0$, since A is final state

$\delta(B, 1) = B \implies B \rightarrow 1B$

The regular grammar G = (V, T, P, S) where

$\qquad\qquad$ V = {A, B} = set of variables or non-terminals

$\qquad\qquad$ T = {0, 1} = set of terminals

$\qquad\qquad$ P = set of productions and is given by

$\qquad\qquad$ P = { A $\rightarrow$ 0B | 1A | 1 | ε

$\qquad\qquad\qquad$ B $\rightarrow$ 0A | 1B | 0 }

$\qquad$ S = A = start symbol of the grammar

Problem 12: *Obtain a regular grammar to find the set of all strings not containing three consecutive 0's.*

Solution :

First we design a FA to accept strings with three consecutive 0's

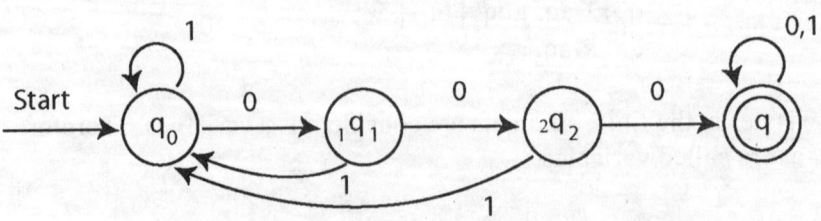

In the above DFA, q_3 is the final state.

Complementing the states, we get a FA to accept without three consecutive 0's

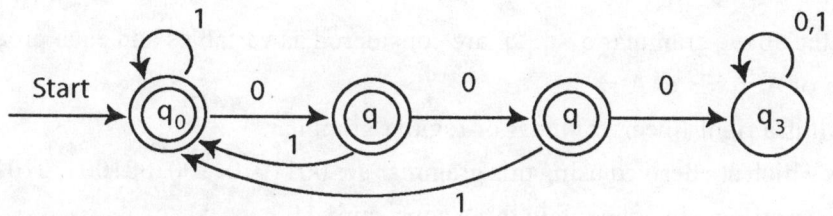

Right-linear grammar

$$\delta(q_0, 0) = q_1 \Rightarrow q_0 \rightarrow 0q_1 \text{ and } q_0 \rightarrow 0$$

$$\delta(q_0, 1) = q_0 \Rightarrow q_0 \rightarrow 1q_0 \text{ and } q_0 \rightarrow 1$$

$$\delta(q_1, 0) = q_2 \Rightarrow q_1 \rightarrow 0q_2 \text{ and } q_1 \rightarrow 0$$

$$\delta(q_1, 1) = q_0 \Rightarrow q_1 \rightarrow 1q_0 \text{ and } q_1 \rightarrow 1$$

$$\delta(q_2, 0) = q_3 \Rightarrow q_2 \rightarrow 0q_3$$

$$\delta(q_2, 1) = q_0 \Rightarrow q_2 \rightarrow 1q_0 \text{ and } q_2 \rightarrow 1$$

$$\delta(q_3, 0) = q_3 \Rightarrow q_3 \rightarrow 0q_3$$

$$\delta(q_3, 1) = q_3 \Rightarrow q_3 \rightarrow 1q_3$$

$\therefore$ The right-linear grammar or regular grammar is as follows

$$q_0 \rightarrow 0q_1 \mid 1q_0 \mid 0 \mid 1$$
$$q_1 \rightarrow 0q_2 \mid 1q_0 \mid 0 \mid 1$$
$$q_2 \rightarrow 0q_3 \mid 1q_0 \mid 1$$
$$q_3 \rightarrow 0q_3 \mid 1q$$

Since in the above productions, q_3 is not deriving any terminal string

$\therefore$ q_3 is useless symbol.

$\therefore$ Remove the productions $q_2 \rightarrow 0q_3$, $q_3 \rightarrow 0q_3$ and $q_3 \rightarrow 1q_3$ from the grammar

$\therefore$ The resultant regular grammar is

$$q_0 \rightarrow 0q_1 \mid 1q_0 \mid 0 \mid 1 \mid \varepsilon$$
$$q_1 \rightarrow 0q_2 \mid 1q_0 \mid 0 \mid 1$$
$$q_2 \rightarrow 1q_0 \mid 1$$

In the above grammar q_0, q_1, q_2 are considered as variables and each production is in the form of A $\rightarrow$ wB or A $\rightarrow$ w

$\therefore$ It is a right-linear grammar or regular grammar.

The strings which are derived using this grammar are 00110, 01100, 001001, 010100,.....

These strings do not contain three consecutive 0's.

Problem 13: *Construct the left-most and right-most derivations and parse trees for the following grammar*

$$S \rightarrow aB \mid bA$$
$$A \rightarrow aS \mid bAA \mid a$$
$$B \rightarrow bS \mid aBB \mid b \text{ which accepts the string "aaabbabbba".}$$

Solution:

Consider the string **"aaabbabbba"**

Left-most Derivation:

S $\Rightarrow$ aB	
$\Rightarrow$ aaBB	(since B $\rightarrow$ aBB)
$\Rightarrow$ aaaBBB	(since B $\rightarrow$ aBB)
$\Rightarrow$ aaabBB	(since B $\rightarrow$ b)
$\Rightarrow$ aaabbB	(since B $\rightarrow$ b)
$\Rightarrow$ aaabbaBB	(since B $\rightarrow$ aBB)
$\Rightarrow$ aaabbabB	(since B $\rightarrow$ b)
$\Rightarrow$ aaabbabbS	(since B $\rightarrow$ bS)
$\Rightarrow$ aaabbabbbA	(since S $\rightarrow$ bA)
$\Rightarrow$ aaabbabbba	(since A $\rightarrow$ a)

$\therefore$ S $\Rightarrow$ aaabbabbba

$\therefore$ 'S' derives "aaabbabbba" using left-most derivation.

Derivation tree or parse tree:

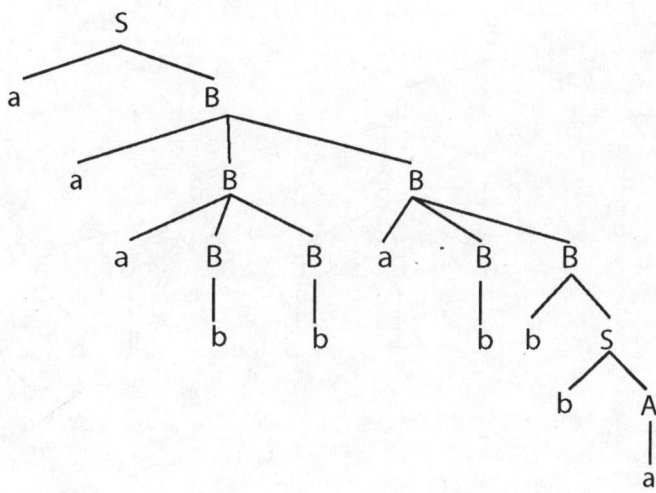

Right-most Derivation

$$S \Rightarrow aB$$

$\Rightarrow aaBB$	(since B → aBB)
$\Rightarrow aaBaBB$	(since B → aBB)
$\Rightarrow aaBaBbS$	(since B → bS)
$\Rightarrow aaBaBbbA$	(since S → bA)
$\Rightarrow aaBaBbba$	(since A → a)
$\Rightarrow aaBabbba$	(since B → b)
$\Rightarrow aaaBBabbba$	(since B → aBB)
$\Rightarrow aaaBbabbba$	(since B → b)
$\Rightarrow aaabbabbba$	(since B → b)

$\therefore S \Rightarrow$ aaabbabbba

$\therefore$ S derives "aaabbabbba" using right-most derivation.

Derivation tree or parse tree:

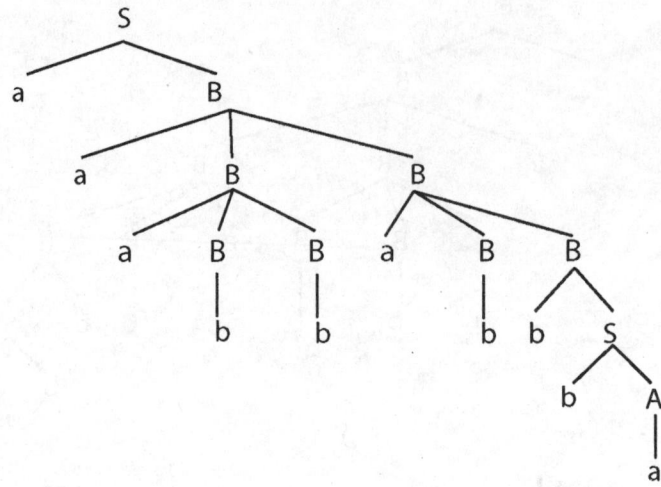

Problem 14: What is meant by ambiguous grammar? Test whether the grammar is ambiguous or not.

Solution:

Ambiguous grammar:

A grammar is said to be ambiguous, if it produces more than one parse tree (or derivation tree) for some string generated by it.

An ambiguous grammar contains productions of the form **S→SαS** where α is a string of terminals, appears twice on R.H.S of a production.

Consider the productions

$$S \to SS$$
$$S \to a$$
$$S \to b$$

Here, the grammar contains a production of the form **S→SαS.**

∴ The grammar is ambiguous.

Left-most derivations for the string "abba"

1) S ⇒ SS
    ⇒ SSS   (since S→SS)
    ⇒ SSSS   (since S→SS)

	⇒ aSSS	(since S→a)
	⇒ abSS	(since S→b)
	⇒ abbS	(since S→b)
	⇒ abba	(since S→a)

∴ S ⇒ **abba**

2)   S ⇒ SS

	⇒ aS	(since S→a)
	⇒ aSS	(since S→SS)
	⇒ abS	(since S→b)
	⇒ abSS	(since S→SS)
	⇒ abbS	(since S→b)
	⇒ abba	(since S→a)

∴ S ⇒ **abba**

Parse trees:

1)

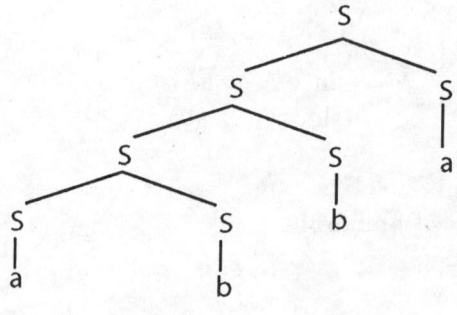

2)

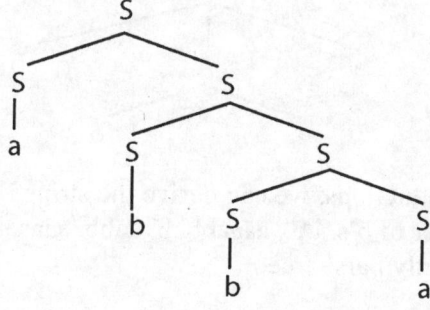

∴ For the string **"abba"** there are two distinct parse trees.

∴ The grammar is ambiguous.

Problem 15: *What is meant by ambiguous grammar? Test whether the grammar is ambiguous or not.*

$$S \rightarrow A \mid B$$
$$A \rightarrow aAb \mid ab$$
$$B \rightarrow abB \mid \varepsilon$$

Solution:

Given grammar is

$$S \rightarrow A \mid B$$
$$A \rightarrow aAb \mid ab$$
$$B \rightarrow abB \mid \varepsilon$$

A grammar (CFG) is said to be ambiguous, if for some L (G), there exist at least two distinct derivation trees.

(or)

Ambiguous grammar means, there exist two or more leftmost or rightmost derivations.

Derivation of the string: aaabbb

Leftmost derivation

$\quad$ S $\Rightarrow$ A(since S $\rightarrow$ A)

$\quad$ S $\Rightarrow$ aAb

$\quad\quad \Rightarrow$ aaAbb $\quad\quad$ (since A $\rightarrow$ aAb)

$\quad\quad \Rightarrow$ aaabbb $\quad\quad$ (since A $\rightarrow$ ab)

$\quad \therefore$ S $\Rightarrow$ aaabbb

Parse tree for w = aaabbb

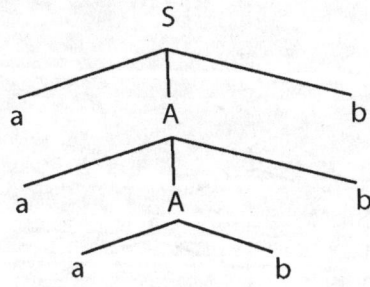

Using these productions, we can derive the strings of the form, any number of a's followed by any number of b's. i.e., aaaabbbb, aabb, aaaaabbbbb etc. for these type of strings we can construct only parse tree.

Derivation of the string: ababab

$\quad$ S $\Rightarrow$ B

$\Rightarrow$ abB	(since B $\rightarrow$ abB)
$\Rightarrow$ ababB	(since B $\rightarrow$ abB)
$\Rightarrow$ abababB	(since B $\rightarrow$ abB)
$\Rightarrow$ ababab	(since B $\rightarrow$ ε)

$\therefore$ S $\Rightarrow$ **ababab**

Parse tree for w=ababab

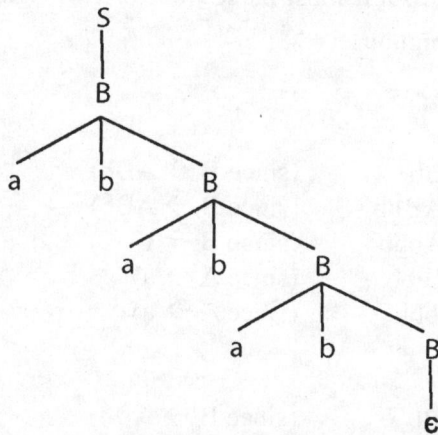

Using these productions, we can derive strings of the form ,any number of a's and b's alternately. Starting with 'a' and length of the string is greater than '2'.

For these type of strings, we can construct only one parse tree.

Construction of parse trees for the string w=ab.

Derivation of the string w=ab

    I)   S $\Rightarrow$ A $\Rightarrow$ ab

    II)   S $\Rightarrow$ B $\Rightarrow$ abB $\Rightarrow$ ab   derivaticn trees for these are

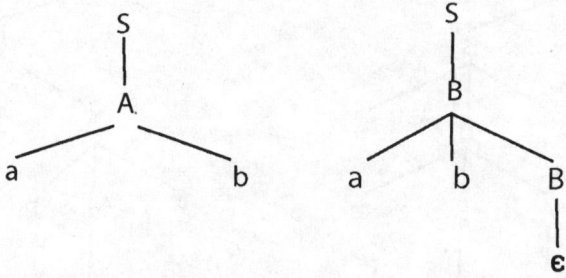

Two different derivation trees are possible for the string w=ab.

$\therefore$ The given grammar is ambiguous.

Problem 16: *Show that the grammar is ambiguous*

$$S \rightarrow aB \mid ab$$
$$A \rightarrow aAB \mid a$$
$$B \rightarrow ABb \mid b$$

Solution:

If the grammar has two left-most parse trees or two right-most parse trees, then the grammar is said to be ambiguous.

Using rightmost derivation:

$$S \Rightarrow aB$$
$$\Rightarrow aABb \qquad (\text{since } B \rightarrow ABb)$$
$$\Rightarrow aAABbb \qquad (\text{since } B \rightarrow ABb)$$
$$\Rightarrow aAAbbb \qquad (\text{since } B \rightarrow b)$$
$$\Rightarrow aAabbb \qquad (\text{since } A \rightarrow a)$$
$$\Rightarrow aaabbb \qquad (\text{since } A \rightarrow a)$$

$$S \Rightarrow aB$$
$$\Rightarrow aABb \qquad (\text{since } B \rightarrow ABb)$$
$$\Rightarrow aAbb \qquad (\text{since } B \rightarrow b)$$
$$\Rightarrow aaABbb \qquad (\text{since } A \rightarrow aAB)$$
$$\Rightarrow aaAbbb \qquad (\text{since } B \rightarrow b)$$
$$\Rightarrow aaabbb \qquad (\text{since } A \rightarrow a)$$

$\therefore$ For the string "aaabbb:, there are two rightmost derivations.

$\therefore$ The grammar is ambiguous.

Parse trees:

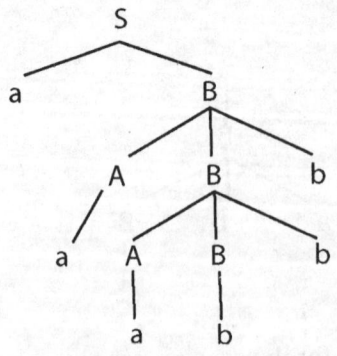 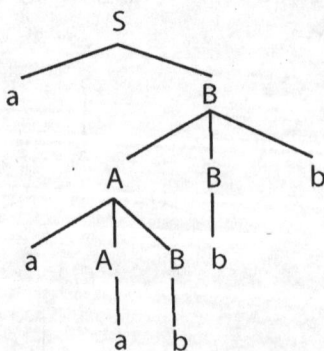

For the string "aaabbb" there are two distinct right parse trees. Therefore, the grammar is ambiguous.

REVIEW QUESTIONS

1. Define grammar what are the derivation, parse tree and sentential form.
2. Construct left-linear and right-linear grammar for the following languages. $(((01+10)^*11)^*00)^*$
3. For the following languages, give the corresponding grammars and finite state machines.
   a) $((0^* 1^*)10)^*$
   b) (0^*+1^*)

4. What is the left linear and right linear grammar?

5. Construct regular grammar G generating the regular set represented by $a^* b(a+b)^*$

6. What are linear grammars distinguish between right linear and left linear grammars and give examples of each type?

7. Construct right-linear and left- linear grammars for the following regular expression
   $S \rightarrow aS \mid bA \mid b$
   $A \rightarrow a A \mid bS \mid a$

8. For the regular expression $0(10)^*$ construct right-linear and left-linear grammars.

9. Obtain the regular grammar to accept the strings containing even number of zeroes.

10. Obtain the regular grammar for the following finite automata as shown below.

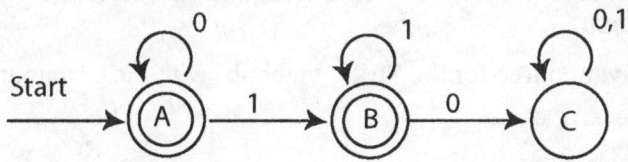

11. Obtain a regular grammar to the following finite automata.

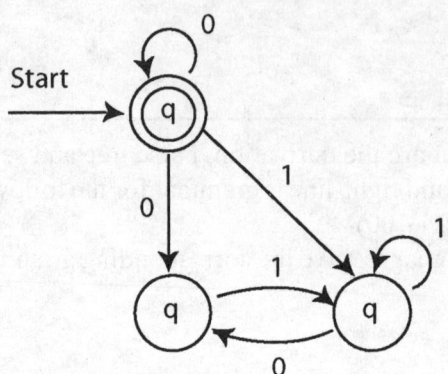

12.   Obtain a right-linear grammar for the language $L = \{a^n b^m \mid n>2, m>3\}$.

13.   Obtain a left-linear grammar for the following DFA.

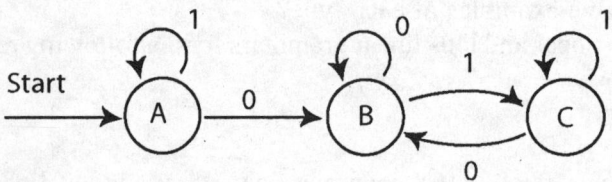

14.   Obtain the left and right -linear grammar for $(0^*(10+01)^*11)^*$.

15.   Given the grammar G as

       S → 0B | 1A
       A → 0 | 0S | 1AA
       B → 1 | 1S | 0BB

Give the left-most and right-most derivations and derivation trees to derive the string '00110101'

16.   Show a derivation tree for the String 'aabbbb' with the grammar

       S → AB| ε

       A → a B

       B → S b

17.   Find and also derive the

i)   Left-most derivation

ii)   Right-most derivation

iii) Derivation tree for the grammar G, which accepts the string "abababa" without using left-most or right-most derivation, which is in L(G), where G is S → SbS | a

18.   Prove that given grammar is ambiguous.

   a)   S → aSb | aaSb | ε

   b)   S → a | Sa | bSS | SSb | SbS

   c)   S → aSb | Sb | ε

   d)   S → aB | aaB

        A → a | Aa

        B → b

   e)   S → ABA

        A → aA | ε

        B → bB | ε

   f)   S → S | S

        S → a

   g)   S → a | aAb | abSb

        A → aAAb | bS

   h)   E → E+E | E*E | (E) | id

   i)   V={S, A}, T={a, b}, P={S→AA, A→AAA, A→a, A→bA, A→Ab}

Chapter 5

Context Free Grammar

5.1 INTRODUCTION

Context Free Grammar (CFG) is of great practical importance. The grammar is used

- ➤ In defining programming languages
- ➤ In string-processing applications
- ➤ For parsing the program by constructing syntax tree
- ➤ For translation of programming languages
- ➤ For describing arithmetic expression
- ➤ In computer linguistics
- ➤ In construction of compilers

The basic concepts of CFG have been discussed in the previous chapter. Also we have seen how the ambiguous grammar is derived from string derivations. In the present chapter, we shall discuss in detail the ambiguity in CFG's. Minimization of CFG and normal forms on the grammar are carried out in order to eliminate ambiguity. Pumping lemma is used to check whether the given grammar is context free or not. Context free languages and their properties are discussed.

5.2 AMBIGUITY IN CONTEXT FREE GRAMMAR

The CFG given by $G = (V, T, P, S)$ can be represented either by leftmost derivation or rightmost derivation. A derivation tree, which is also called as parse tree or syntax tree, can be drawn to arrive at a valid string.

A grammar is said to be ambiguous, if for a given string there exists more than one leftmost derivation or more than one rightmost derivation. An ambiguous grammar can have more than one parse tree or derivation tree.

Example 5.1 **Check whether the given grammar is ambiguous or not**

S → iCtS
S → iCtSeS
S → a
C → b

Solution: To check whether given grammar is ambiguous or not. Let us consider a string and draw derivation trees.

Let the string be **ibtibtibtaea**

The leftmost derivations for the string **ibtibtibtaea**

(i)   S ⇒ iCtSeS
⇒ ibtSeS
⇒ ibtiCtSeS
⇒ ibtibtSeS
⇒ ibtibtiCtSeS
⇒ ibtibtibtSeS
⇒ ibtibtibtaeS
⇒ ibtibtibtaea

(ii)   S ⇒ iCtS
⇒ ibtS
⇒ ibtiCtS
⇒ ibtibtS
⇒ ibtibtiCtSeS
⇒ ibtibtibtSeS
⇒ ibtibtibtaeS
⇒ ibtibtibtaea

Hence, the given string has two leftmost derivations.

Now draw the derivation tree for the string **ibtibtibtaea**

(i)

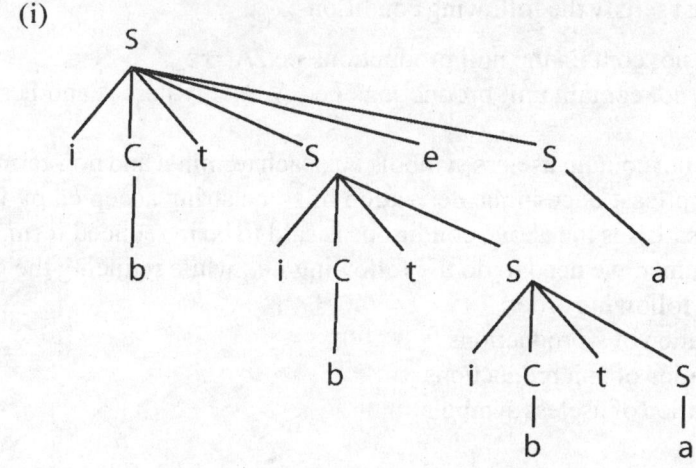

(ii)

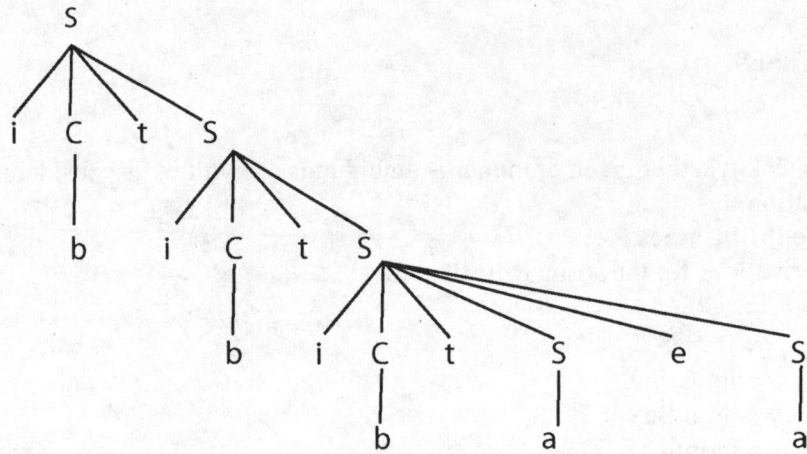

Hence the given grammar is ambiguous.

5.3 MINIMIZATION OF CONTEXT FREE GRAMMAR

Many of the programming languages can be represented by contextfree grammars. Hence the grammar should be effective and optimized. In some grammars, we may find terminals and non-terminals which are not of great significance in derivations. Such symbols can be identified and eliminated from the grammar, thus reducing the grammar.

Simplification of grammar is nothing but reducing the grammar by eliminating unnecessary symbols (terminals or non-terminals)

A reduced grammar must satisfy the following conditions:

- ➢ It does not contain the null productions i.e., A →ε
- ➢ It does not contain unit productions i.e., A → B, where A and B are non-terminals
- ➢ It does not contain useless symbols i.e., each terminal and non-terminal that appear atleast once in the derivation of some string accepted by L(G).

Any grammar that satisfies the above conditions is said to be in reduced form. In order to get reduced grammar, we need to do the following i.e., while reducing the grammar we must follow the following order:

1) Elimination of ε-productions
2) Elimination of unit productions
3) Elimination of useless symbols

5.3.1 ELIMINATION OF ε - PRODUCTIONS

A production A → ε is called a **null production or ε- production**. If a grammar contains ε- productions then these can be removed without changing the meaning of the grammar. We can eliminate these ε- productions by **replacement method**.

To eliminate the ε-productions, we need to apply the following steps:

a) Find the nullable variable set (V_n) for the given grammar G i.e., the set includes all the variables which are deriving ε- symbol.

For example: If S → AB, A →ε, B →ε then V_n ={S, A, B}

Here, S is also nullable variable because by substituting A →ε, B →ε it is deriving 'ε'

b) Include all the productions whose R.H.S does not have any nullable variables.

For example: The production D → b is not deriving any nullable variable, so it is directly included.

c) Now consider the productions whose R.H.S contains nullable variables and replace nullable variables with 'ε' in all possible combinations.

For example: 1) S → AB, A →ε, B →ε

S → AB gives S → AB | A | B

2) S → aA, A →ε

S → aA gives S → aA | a

By applying the above three steps to the given grammar, we get the resultant grammar. This resultant grammar does not conatin any null or ε-productions.

Example 5.2

Consider the grammar

S → aS | bS|ε **and eliminate ε- productions, if any.**

Solution:

The above grammar contains only one ε-production i.e.,

S → ε. It can be removed by placing

S → ε in the place of S.

When we place S → ε in S→ aS, we get S → a

When we place S → ε in S → bS, we get S → b

After eliminating null production, the grammar becomes S → aS | bS | a | b

Example 5.3

Remove the ε-productions from the following CFG by preserving meaning of it.

S → XYX

X → 0X | ε

Y → 1Y | ε

Solution:

Now, while removing ε-productions, we are deleting the rules X →ε and Y→ε. To preserve meaning of CFG, we are actually placing ε at right hand side where ever X and Y appear.

Let us take S → XYX

(i)   If first X at right hand side is ε then S → YX

(ii)   Similarly if last X at right hand side is ε then S → XY

(ii)   If Y = ε then   S → XX

(iii)   If X and Y are ε then S → X

(iv)   If both X's are replaced by ε then S → Y

$$\therefore S \to XY\,|\,YX\,|\,XX\,|\,X\,|\,Y$$

(v)   Now let us consider   X → 0X

If we place ε at right hand side for X then

$$X \to 0$$
$$X \to 0X\,|\,0$$

(vi)   Similarly,   Y → 1Y | 1

Collectively we can rewrite the CFG with removed ε-productions as

$$S \to XY\,|\,YX\,|\,XX\,|\,X\,|\,Y$$
$$X \to 0X\,|\,0$$
$$Y \to 1Y\,|\,1$$

Example 5.4

For the CFG given below eliminate ε-production if any

$$S \to aSa$$
$$S \to bSb$$
$$S \to ε$$

Solution:

According to the replacement procedure, we will place ε at S in the sentential form.

(i)   Put   S →   ε in S → aSa

Then S →   aa

(ii)   Put   S →   ε in S → bSb

Then S →   bb

Therefore the grammar after elimination of ε production is

$$S \to aSa\,|\,bSb\,|\,aa\,|\,bb$$

5.3.2 ELIMINATION OF UNIT PRODUCTIONS

A production in which one non-terminal derives another non-terminal i.e., A → B is called **Unit Production**.

We can eliminate these unit productions by **substitution method.**

Example 5.5
Consider the grammar

$$S \rightarrow AB$$
$$A \rightarrow a$$
$$B \rightarrow C$$
$$C \rightarrow D$$
$$D \rightarrow b \quad \textbf{eliminate the unit productions if any.}$$

Solution:

Here the productions $B \rightarrow C$ and $C \rightarrow D$ are unit productions

$C \rightarrow D$ can be written as $C \rightarrow b$
$B \rightarrow C$ can be written as $B \rightarrow b$

$\therefore$ The grammar becomes

$$S \rightarrow AB$$
$$A \rightarrow a$$
$$B \rightarrow b$$
$$C \rightarrow b$$
$$D \rightarrow b$$

Example 5.6
Eliminate the unit productions from the following grammar

$$S \rightarrow AB$$
$$A \rightarrow a$$
$$B \rightarrow C \mid b$$
$$C \rightarrow D$$
$$D \rightarrow E \mid bC$$
$$E \rightarrow d \mid Ab$$

Solution:

Here there are no ε-productions. Now let us try to eliminate unit productions.

As, $S \rightarrow AB$ and $A \rightarrow a$

There is only one rule with A giving terminal symbol. Hence there is no question of getting unit production with A.

Now consider, $B \rightarrow C$
$$C \rightarrow D$$
$$D \rightarrow E$$

It is clear that B, C and D are unit productions.

As $E \rightarrow d \mid Ab$, now replace the value of E in D.

Then, $D \rightarrow d \mid Ab \mid bC$

As $C \rightarrow D$ now replace D with the above

$$C \rightarrow d \mid Ab \mid bC$$

As $B \rightarrow C$ now replace C with the above

B → d | Ab | bC | b

Thus the grammar after removing unit productions is as follows

S → AB

A → a

B → d | Ab | bC | b

C → d | Ab | bC

D → d | Ab | bC

E → d | Ab

Now there is no path for D and E from the start state, so we can remove them by considering them as useless symbols. The optimized grammar will be

S → AB

A → a

B → d | Ab | bC | b

C → d | Ab | bC

5.3.3 ELIMINATION OF USELESS SYMBOLS

A variable is said to be useful, if and only if

➢   it can derive a terminal string and

➢   it can be reached from the start symbol.

Using dependency graph we can say whether a variable can be reached from the start symbol

For CFG's dependency graph will be useful to say whether a particular variable can be reached from start symbol or not.

A dependency graph has its vertices labeled with variables, with an edge between vertices A and B iff there is a production of form **A → xBy**

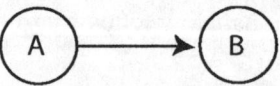

A variable is useful only if there is a path from the start symbol labeled 'S' to the vertex labeled with that variable.

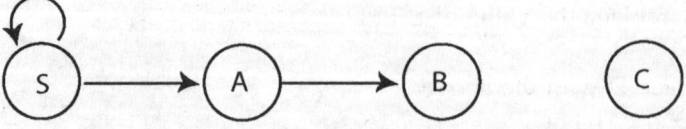

For example in the above figure, S, A, B are useful variables, C is useless variable.

If any one condition fails, it tends to be useless i.e., either it cannot generate a terminal string or it cannot be reached from the start symbol.

If both conditions are satisfied then it is useful variable. If a production involves any useless symbol, that production is called useless production.

(or)

A symbol of G is useful when it appears on R.H.S of a production and if it involves in sentential form. Otherwise it is said to be useless.

If there is a production of the form

$$S \rightarrow \alpha A\beta \qquad \text{where} \quad \alpha A\beta \overset{*}{\Rightarrow} w$$

Then 'A' is said to be useful symbol. Here α and β may be string of terminals and non-terminals.

Example 5.7
Consider the grammar

  S → AB | C
  A → a
  B → b **Here S, A, B, C are non-terminals.**

Identify useless variables.

Solution:

(i) Each variable should derive a terminal string. Here C is not deriving any terminal string since there are no productions for 'C'.

∴ 'C' is useless variable, we can remove it.

(ii) Dependency graph

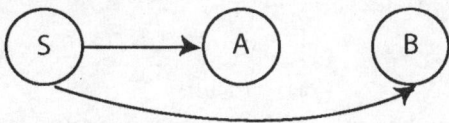

The variables A and B can be reached from the start symbol.

∴ The variables S, A, and B are useful.

∴ The reduced grammar is

  S → AB
  A → a
  B → b

Example 5.8

Consider the CFG G = {V, T, P, S}

> Where V = {S, A, B} T = {0, 1}
> P = {S → A 11 B | 11A
> S → 1B | 11
> A → 0B → BB}   **Remove useless symbols from the grammar.**

Solution:

(i) Now in the given CFG, if we try to derive any string, 'A' gives some terminal symbol as 0 but B does not gives any terminal string. By following the rules with B, we simply get ample number of B's and no significance string. Hence we can declare B as useless symbol and can remove the rules associated with it. Hence after removal of useless symbols we get,

> S → 11A | 11
> A → 0

(ii) Dependency graph

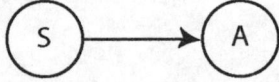

∴ S and A are useful

∴ The reduced grammar is

Example 5.9

Find CFG with no useless symbols equivalent to

> S → AB | CA   B → BC | AB
> A → a   C → aB | b

Solution:

> Consider, the rule
> S → AB   (1)
> S → CA   (2)

In the rule (2), C and A can be replaced by some terminating string. But in rule (1), B cannot be replaced by terminating string. It tends to form a never ending loop.

For example: S ⇒ AB ⇒ aAB ⇒ aaAB ⇒ aaaAB and so on.

Thus we come to know that B is a useless symbol. Removing B and its productions, the resultant grammar is as follows

> S → CA
> A → a
> C → b

5.4 NORMAL FORMS

The CFG can be simplified by removal of useless symbols, elimination of ε-productions and removal of unit productions.

By using the above, the grammar is minimized but not standardized. Since the R.H.S of the productions have no specific format. In order to standardize the given grammar, we normalize it. The most frequently used normal forms are:

      1)   Chomsky Normal Form (CNF)

      2)   Greibach Normal Form (GNF)

5.4.1 CHOMSKY NORMAL FORM (CNF)

A context free grammar G is said to be in Chomsky Normal Form (CNF) if all productions are of the form

      A → BC or A → a

where A, B, C are non-terminals and 'a' is a terminal. The CNF restricts the number of symbols on the right side of a production to be two, and the two symbols should be non-terminals or a single terminal.

For example, the grammar
      S → AB
      A → a
      B → b   is in CNF

But the grammar,   S → ABB
      A → a
      B → bb   is not in CNF, because the productions S → ABB and B → bb violets the rules of CNF

A reduced grammar i.e., a grammar without useless symbols, ε-productions and unit productions can be converted to CNF.

Steps to convert the given grammar into CNF

Step 1: First eliminate ε-productions, unit productions, useless productions.

Step 2: Write all productions in the form of A → BC or A → a

Example 5.10

Reduce the following grammar to Chomsky Normal Form
      S → 1A | 0B

$$A \to 1AA \mid 0S \mid 0$$
$$B \to 0BB \mid 1S \mid 1$$

Solution:

A reduced grammar i.e., a grammar without productions, unit ε-productions, useless variables can be converted to CNF.

Step 1: First eliminate productions, unit productions, useless productions.

Step 2: Write all productions in the form of A → BC or A → a

Step 1:

1) Elimination of ε- productions: The given grammar does not contain ε-productions. So no need to eliminate ε-productions.

2) Elimination of unit productions (A → B): The given grammar does not contain unit productions. So no need to eliminate unit productions.

3) Elimination of useless variables or productions:

➤ All the variables are generating terminal strings.
➤ For every variable there is a path from start symbol.

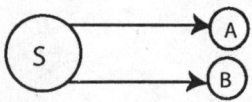

∴ All the variables are useful

∴ The given grammar does not contain ε-productions unit productions and useless productions i.e., the given grammar itself is the reduced grammar.

Step 2:
Reduction to Chomsky Normal Form:

Reduce all productions of the given grammar in the form of A → BC or A → a
Consider the grammar

$$S \to 1A \mid 0B$$
$$A \to 1AA \mid 0S \mid 0$$
$$B \to 0BB \mid 1S \mid 1$$

Step I: The only productions already in the proper form are A → 0 and B → 1

Step II: We my begin by replacing terminals on the right hand side of the above productions by variables except in the case of productions A → 0 and B →1
Replace the terminal symbols 0 and 1 by C and D and introduce new productions C → 0 and D → 1
The given grammar is reduced to

$$S \to DA \mid CB$$

$$A \rightarrow DAA \mid CS \mid 0$$
$$B \rightarrow CBB \mid DS \mid 1$$

Step 3:

Consider the productions which are not in the form of $A \rightarrow BC$ and reduce them in the form of $A \rightarrow BC$

Consider the productions

$$A \rightarrow DAA$$
$$B \rightarrow CBB$$

Let us introduce new productions

$$E \rightarrow AA$$
$$F \rightarrow BB$$

∴ $A \rightarrow DAA$ becomes $A \rightarrow DE$

$B \rightarrow CBB$ becomes $B \rightarrow CF$

∴   The productions for the grammar in Chomsky normal form are

$$S \rightarrow DA \mid CB$$
$$A \rightarrow DE \mid CS \mid 0$$
$$B \rightarrow CF \mid DS \mid 1$$
$$C \rightarrow 0$$
$$D \rightarrow 1$$
$$E \rightarrow AA$$
$$F \rightarrow BB$$

5.4.2 GREIBACH NORMAL FORM (GNF)

A context free grammar G is said to be in Greibach Normal Form (GNF), if all productions are of the form $A \rightarrow a\alpha$ where 'a' is a terminal and α is any number of non-terminals (possibly empty).

The GNF restricts the position of terminals and non-terminals where they appear. In Greibach normal form grammar, each production must start with a terminal, it should not start with a non-terminal.

Example 5.11

Consider the grammar

$$S \rightarrow bABC$$
$$A \rightarrow a$$
$$B \rightarrow bC$$
$$C \rightarrow c \text{ is in GNF}$$

But the grammar

$$S \rightarrow ABC$$
$$A \rightarrow a$$
$$B \rightarrow bC$$

C → C is not in GNF, because the production S → ABC violates the rule of GNF. Because on R.H.S, it starts with a non-terminal.

Importance of GNF: It is a useful grammatical form, has many theoretical and practical consequences. In this, restriction is not on the length of the right side of a production, but on the positions in which terminals and variables can appear.

If productions are observed, they are of the form A → ax, a ∈ T, x ∈ V* is similar to the productions of simple grammar. But GNF does not carry the restrictions that the pair (A, a) occurs at most once.

This feature gives GNF a generality not possessed by other grammars.

- Each use of a production introduces exactly one terminal into a sentential form i.e. a string of length 'n' has a derivation of exactly 'n' steps.
- If applied to PDA, results a PDA with no e-productions.
- GNF grammar is the one where every production body starts with a terminal.

Example 5.12

Convert the following grammar to Griebach Normal Form (GNF)

$G = (\{A_1, A_2, A_3\}, \{a, b\}, P, A_1)$

Where P consists of the following

$$A_1 → A_2 A_3 \qquad - (1)$$
$$A_2 → A_3 A_1 \mid b \qquad - (2)$$
$$A_3 → A_1 A_2 \mid a \qquad - (3)$$

Solution:

The given grammar does not contain null or unit or useless productions.

Step 1: Since the right-hand side of the productions for A_1 and A_2 start with non terminal of higher-numbered variables ($A_i → A_j \gamma$, $j \geq i$)

We begin with the production $A_3 → A_1 A_2$ and substitute the string $A_2 A_3$ for A_1 in (3)

Note that $A_1 → A_2 A_3$ is the only production with A_1 on the left.

The resulting set of productions is:

$$A_1 → A_2 A_3$$
$$A_2 → A_3 A_1 \mid b$$
$$A_3 → A_2 A_3 A_2 \mid a$$

Since the right-side of the production $A_3 → A_2 A_3 A_2$ begins with a lower-numbered variable, we substitute for the first occurrence of A_2 with both $A_3 A_1$ and b

$\therefore A_3 → A_2 A_3 A_2$ is replaced by

$$A_3 → A_3 A_1 A_3 A_2 \text{ and}$$

$$A_3 → b A_3 A_2$$

The new set is $A_1 \rightarrow A_2 A_3$

$$A_2 \rightarrow A_3 A_1 \mid b$$

$$A_3 \rightarrow A_3 A_1 A_3 A_2 \mid b A_3 A_2 \mid a$$

We now apply the following lemma to the production

$$A_3 \rightarrow A_3 A_1 A_3 A_2 \mid b A_3 A_2 \mid a$$

Lemma: Let $G = (V, T, P, S)$ be a CFG

Let $A \rightarrow A\alpha_1 \mid A\alpha_2 \mid A\alpha_3 \ldots\ldots\mid A\alpha_r$ be the set of A-productions for which A is the leftmost symbol on the right-hand side of the production. Let $A \rightarrow \beta_1 \mid \beta_2 \mid \ldots\ldots\mid \beta_s$ be the remaining A-productions.

Let $G_1 = (V \cup \{B\}, T, P_1, S)$ be the CFG formed by adding the variable B to V and replacing all the A-productions by the productions

1) $A \rightarrow \beta_i$ $\left.\begin{array}{l}\\ \\\end{array}\right\}$ $1 \le i \le s$

    $A \rightarrow \beta_i B$

2) $B \rightarrow \alpha_i$ $\left.\begin{array}{l}\\ \\\end{array}\right\}$ $1 \le i \le r$

    $B \rightarrow \alpha_i B$

    Then $L(G) = L(G_1)$

Now consider $A_3 \rightarrow \underbrace{A_3 A_1 A_3 A_2}_{\downarrow} \mid \underbrace{b A_3 A_2}_{\downarrow} \mid \underbrace{a}_{\downarrow}$

$$\qquad\qquad\qquad\qquad\quad \alpha_1 \qquad\quad \beta_1 \qquad \beta_2$$

Here $r = 1$, $s = 2$

Now replace all the A_3-productions by the production (1) and (2) and symbol B_3 is introduced

(1)	We have $s = 2$ $(1 \le i \le s)$		(2)	We have $r = 1 (1 \le i \le r)$

(1)   We have $s = 2$ $(1 \le i \le s)$ (2)   We have $r = 1(1 \le i \le r)$

     $\therefore 1 \le i \le 2 \Rightarrow i = 1$ and 2 $\therefore i = 1$

When $i = 1$ When $i = 2$ when $i = 1$

$A_3 \rightarrow \beta_1$ $A_3 \rightarrow \beta_2$ $B_3 \rightarrow \alpha_1$

$A_3 \rightarrow \beta_1 B_3$ $A_3 \rightarrow \beta_2 B_3$ $B_3 \rightarrow \alpha_1 B_3$

$\therefore$ From **(1)**, by substituting β_1 and β_2, we get $A_3 \rightarrow bA_3A_2B_3 \mid bA_3A_2 \mid aB_3 \mid a$

From **(2)**, by substituting α_1, we get $B_3 \rightarrow A_1A_3A_2 \mid A_1A_3A_2B_3$

$\therefore$ The resulting set is

$$A_1 \rightarrow A_2A_3$$
$$A_2 \rightarrow A_3A_1 \mid b$$
$$A_3 \rightarrow bA_3A_2B_3 \mid aB_3 \mid bA_3A_2 \mid a$$
$$B_3 \rightarrow A_1A_3A_2 \mid A_1A_3A_2B_3$$

Step 2:   Now all the productions with A_3 on the left have right-hand side that start with terminals.

These are used to replace 'A_3' in the production $A_2 \rightarrow A_3A_1$ and then the productions with 'A_2' on the left are used replace A_2 in the production $A_1 \rightarrow A_2A_3$
The result is the following

$A_3 \rightarrow bA_3A_2B_3$	$A_3 \rightarrow bA_3A_2$
$A_3 \rightarrow aB_3$	$A_3 \rightarrow a$
$A_2 \rightarrow bA_3A_2B_3A_1$	$A_2 \rightarrow bA_3A_2A_1$
$A_2 \rightarrow aB_3A_1$	$A_2 \rightarrow aA_1$
$A_2 \rightarrow b$	
$A_1 \rightarrow bA_3A_2B_3A_1A_3$	$A_1 \rightarrow bA_3A_2A_1A_3$
$A_1 \rightarrow aB_3A_1A_3$	$A_1 \rightarrow aA_1A_3$
$A_2 \rightarrow bA_3$	
$B_3 \rightarrow A_1A_3A_2$	$B_3 \rightarrow A_1A_3A_2B_3$

Step 3:   The two B_3-productions are converted to proper form, resulting in ten more productions.

Substitute the five productions of A_1 in both the B_3's.

$B_3 \rightarrow bA_3A_2B_3A_1A_3A_3A_2$	$B_3 \rightarrow bA_3A_2B_3A_1A_3A_3A_2B_3$
$B_3 \rightarrow aB_3A_1A_3A_3A_2$	$B_3 \rightarrow aB_3A_1A_3A_3A_2B_3$
$B_3 \rightarrow bA_3A_3A_2B_3$	$B_3 \rightarrow bA_3A_3A_2$
$B_3 \rightarrow bA_3A_2A_1A_3A_3A_2$	$B_3 \rightarrow bA_3A_2A_1A_3A_3A_2B_3$
$B_3 \rightarrow aA_1A_3A_3A_2$	$B_3 \rightarrow aA_1A_3A_3A_2B_3$

The final set of productions is

$A_3 \rightarrow bA_3A_2B_3$	$A_3 \rightarrow bA_3A_2$
$A_3 \rightarrow aB_3$	$A_3 \rightarrow a$
$A_2 \rightarrow bA_3A_2B_3A_1$	$A_2 \rightarrow bA_3A_2A_1$
$A_2 \rightarrow aB_3A_1$	$A_2 \rightarrow aA_1$

$A_2 \rightarrow b$

$A_1 \rightarrow bA_3A_2B_3A_1A_3$ $A_1 \rightarrow bA_3A_2A_1A_3$

$A_1 \rightarrow aB_3A_1A_3$ $A_1 \rightarrow aA_1A_3$

$A_1 \rightarrow bA_3$

$B_3 \rightarrow bA_3A_2B_3A_1A_3A_3A_2B_3$ $B_3 \rightarrow bA_3A_2B_3A_1A_3A_3A_2$

$B_3 \rightarrow aB_3A_1A_3A_3A_2B_3$ $B_3 \rightarrow aB_3A_1A_3A_3A_2$

$B_3 \rightarrow bA_3A_3A_2B_3$ $B_3 \rightarrow bA_3A_3A_2$

$B_3 \rightarrow bA_3A_2A_1A_3A_3A_2B_3$ $B_3 \rightarrow bA_{-3}A_2A_1A_3A_3A_2$

$B_{-3} \rightarrow aA_1A_3A_3A_2B_3$ $B_3 \rightarrow aA_1A_3A_3A_2$

which is in GNF

5.5 PUMPING LEMMA FOR CFLS

The pumping lemma for context-free languages is a way to generate many strings from a given string in a context free language. The lemma is useful for proving whether certain languages are context-free languages or not.

Lemma

The pumping lemma for CFL's states that a sufficiently given long string always contains two short substrings close together that can be repeated as many times a required.

Let 'L' be a context free language. Then there exists a constant 'n' which depends on 'L'.

If the string $z \in L$ and $|z| \geq n$, then according to lemma we can write z as z = uvwxy satisfying the following conditions:

1) $|vx| \geq 1$

2) $|vwx| \geq n$

3) $w\,v^i w\,x^i y \in L$ for all $i \geq 0$.

Proof: Let 'G' be a context-free grammar which is in Chomsky's Normal Form generating a language 'L' which is free from unit productions and null productions.

If there is a string 'z' in L (G), then there exists a parse tree for 'z' having a long path. We can show by induction on 'i' that if the parse tree of 'w' has length less than or equal to 'i', then the word length is less than or equal to 2^{i-1}

Basis (i=1)

Let the grammar G be $S \rightarrow a$

The word z that can be derived from G is z = a. The parse tree has length 1. According to the rule if parse tree has length **i**, then the word length is 2^{i-1} i.e., $|z| = 2^{1-1} = 1$ which is true.

Further G is in CNF. This language is regular that means there is a basis for induction.

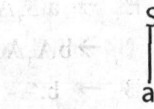

Parse tree for z = a

Induction (i>1)

Let z be a string that is derived by a grammar G and i > 1. In this case the parse tree has two children as shown below

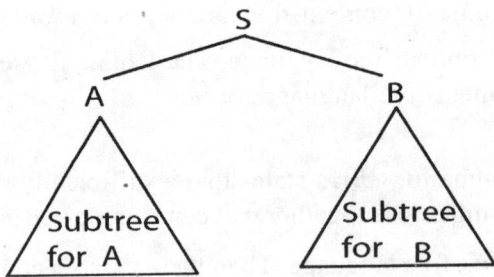

If all the possible paths in subtrees for A and B have length less than i-1, then both the subtrees generate strings of length 2^{i-2}

Thus the entire tree can generate words of length 2^{i-1}

Let there are 'k' variables in 'G' and $n=2^k$. If z ∈ L then $|z| \geq n$, then since $|z| > 2^{k-1}$, any parse tree constructed for z must have the path length atleast k+1. Such a path has atleast k+2 nodes where the last nodes are variables.

Thus we can pump a substring to z such that the new string to z such that the new string 'z¹' has path length i and $|z^1| \leq 2^{i-1}$.

Then G is called a context free grammar. Further G is in CNF, the grammar 'G' is

        S → AB
        A → AS
        B → AS
        S → a

$$A \rightarrow b$$

Assume a string in L (G), z = bbbaba.

The parse tree for **z = bbbaba** is shown below

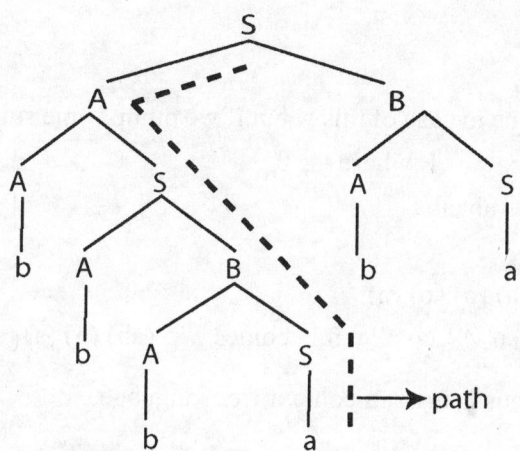

Let v=bb, w=a and x= ε

Then we know that

$$S \overset{*}{\Rightarrow} vSx \text{ and}$$
$$S \overset{*}{\Rightarrow} z$$

Where $| vwx| \leq 2^k = n$

i.e. the path length i=3 and $|z| \leq 2^{i-1}$ i.e. $3 < 2^2$

$$S \overset{*}{\Rightarrow} v^i w x^i \text{ for all } i \geq 0$$

So we can write z as follows z=uvwxy for some u and y which satisfies the condition

$i \leq |z| \leq 2^{i-1} \leq n$.

Hence the grammar is regular.

Example 5.13

Prove whether the given language L= {SS^T | S $\in$ (a, b)*} is context free or not.

Solution:

This is a language of even palindrome. In order to recognize a string in the language, machine needs to remember the first half so as to be able to compare it to the second. For the palindromes we use last-in, first-out principle and stack has the same property.

If we consider z = uvwxy be the string which belongs to L and L is a language of palindrome. Let us say L is a context free language.

According to lemma

1. $|vx| \geq 1$

2. $|vwx| \geq n$

Which could be true in case of this w but if we pump some substring of length i we should get $z = u\, v^i\, w\, x^i\, y \in L$ where $i \geq 0$

For example if we take z = abaaba

Let us group the input

$$z = (ab)\ (a)\ (a)\ (b)\ (a)$$

If we modify z by $u\, v^{i+1}\, w\, x^{i+1}\, y$ it becomes $z^l = (ab)\ (a)\ (a)\ (a)\ (b)\ (b)\ (a)$ and not

palindrome then $z^l \notin L$. Thus L is not a context free language.

Example 5.14

Show that the language L = {$a^i\, b^j\, c^k$ | i < j and i < k} is not a context free
             language.

Solution:

Let us prove this statement by method of contradiction. Assume L is a context free language. and z be any string such that $z \in L$.

Let, z be

$$z = u\, v\, w\, x\, y$$

Let, $|vx| \geq 1$ and $|vwx| \geq n$ and the vwx can contain at the most strings of a's and c's. The string $u\, v^i\, w\, x^i\, y$ contains additional occurrences of v's and x's.

As per the condition mentioned in a lemma, $|vx| \geq 1$ this means that either v or x contains at least one a then vx can not contain any c's and as per the language L the number of c's has to be more than number of a's. Therefore, the $z \notin L$.

Similarly if $z = u\, v^0\, w\, x^0\, y$ then also there are fewer occurrences of c's than a's which cannot be our L.

Hence the assumption which we had made that L is a context free language is wrong.

5.6 PROPERTIES OF CONTEXT-FREE LANGUAGES

Like regular languages context-free languages are closed under some operations i.e. on performing an operation on context-free languages, the resultant language is context-free language. The CFL's are closed under the following operations

1. The CFL's are closed under union.
2. The CFL's are closed under concatenation.
3. The CFL's are closed under kleen closure.

But the CFL's are not closed under intersection and complement. The CFL's are closed under union, concatenation and kleen closure.

If L_1 and L_2 be two context free languages then $L_1 \cup L_2$ is also context free language. Let $G_1 = (V_1, T, P_1, S_1)$ and $G_2 = (V_2, T, P_2, S_2)$ be two CFG's generating L_1 and L_2 respectively.

Here, P_1 is defined as

$$S_1 \to X_1 \, S_1 \, X_1 \mid Y_1 \, S_1 \, Y_1 \mid \varepsilon$$
$$X_1 \to a$$
$$Y_1 \to b$$

P_2 is defined as

$$S_2 \to a \, X_2 \, X_2 \mid b \, Y_2 \, Y_2$$
$$X_2 \to b$$
$$Y_2 \to a$$

Now let us perform the union operation on L_1 and L_2. Then we get $L = L_1 \cup L_2$ and $G \in L$, where G is given by

$$G = (V, T, P, S),$$
$$V = \{S_1, X_1, Y_1, S_2, X_2, Y_2\}$$
$$P = \{P_1 \cup P_2\}$$
$$P = \{S \to S_1 \mid S_2$$
$$S_1 \to X_1 S_1 X_1 \mid Y_1 S_1 Y_1 \mid \varepsilon$$
$$X_1 \to a$$
$$Y_1 \to b$$
$$S_2 \to a X_2 X_2 \mid b Y_2 Y_2$$
$$X_2 \to b$$
$$Y_2 \to a$$

S is a start symbol. Thus G is context-free grammar and L (G) is a context-free language. Similarly, we can show that the CFL's are closed under concatenation and kleen closure i.e. $L = L_1 L_2$ and $L = L_1^*$ is a CFL.

SOLVED PROBLEMS

Problem 1: *What is meant by ambiguous grammar?*
Solution:
A grammar is said to be ambiguous, if it produces more than one parse tree (or derivation tree) for some string generated by it.

(or)

A context free grammar G such that some word has two parse trees is said to be ambiguous.

(or)

For a given string, if there is more than one leftmost derivation or more than one rightmost derivation then the grammar is said to be ambiguous.

(or)

The terminal string 'w' belonging to L (G) is ambiguous, if there exist two or more derivation trees for 'w'.

An ambiguous grammar contains productions of the form **S→SαS** where 'α' is a string of terminals or non-terminals and the same non terminal appears twice on R.H.S of a production.

Problem 2: *Show that the following grammar is ambiguous.*

$S \rightarrow AB \mid C$
$A \rightarrow aAb \mid ab$
$B \rightarrow cBd \mid cd$
$C \rightarrow aCd \mid aDd$
$D \rightarrow bDc \mid bc$

Solution:
A grammar is said to be ambiguous, if it has more than one parse.

Left most derivation

1) $\quad\quad\quad S \Rightarrow AB$
$\quad\quad\quad\quad \Rightarrow aAbB$
$\quad\quad\quad\quad \Rightarrow aabbB$
$\quad\quad\quad\quad \Rightarrow aabbcBd$
$\quad\quad\quad\quad \Rightarrow aabbccdd$

$\quad \therefore$ **S ⇒ aabbccdd**

2) $\quad\quad\quad S \Rightarrow C$

$\Rightarrow$ aCd

$\Rightarrow$ aaDdd

$\Rightarrow$ aabDcdd

$\Rightarrow$ aabbccdd

$\therefore$ S $\Rightarrow$ **aabbccdd**

Parse trees:

1)

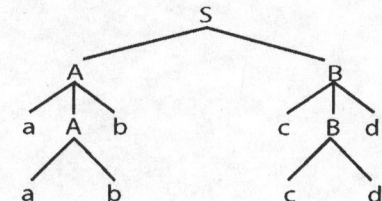

2)

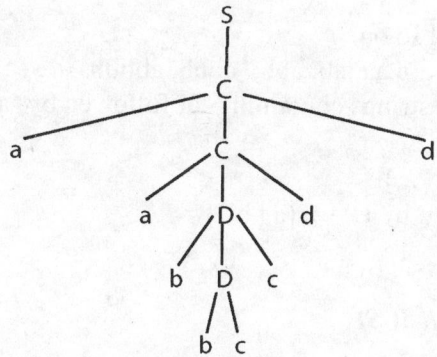

$\therefore$ For the given grammar, there are two distinct left parse trees.

$\therefore$ The given grammar is ambiguous.

Problem 3: *Identify the language generated by grammar*

G =({S, A, B}, {a, b}, P, S) where P = {(S→aA), (A→a), (A→B), (B→bB), (B→b)}

Solution:

Given grammar is

S → aA

A → a

A → B

$$B \rightarrow bB$$
$$B \rightarrow b$$

Let us derive the strings generated by G.

We start with start symbol 'S'

$$S \Rightarrow aA$$
$$\Rightarrow aa$$
$$\therefore S \Rightarrow aa$$

Let us derive another string from 'G'.

$$S \Rightarrow aA$$
$$\Rightarrow aB$$
$$\Rightarrow ab$$
$$\therefore S \Rightarrow ab$$

Let us derive another string from 'G'.

$$S \Rightarrow aA$$
$$\Rightarrow aB$$
$$\Rightarrow abB$$
$$\Rightarrow abbB$$
$$\Rightarrow abbbB$$
$$\Rightarrow abbbb \qquad \text{and so on.}$$

The set of strings generated by 'G' are aa, ab, abb, abbb, abbbb. . . . .

Clearly the grammar generates the set of strings containing 'a' followed by 'a' or more number of b's.

$$\therefore L = \{aa, ab, abb, abbb, abbbb. \ . \ . \ . \}$$

$\therefore$ The context-free language is given by $L = \{a(a \mid b^+)\}$.

Problem 4 : Simplify the grammar

$$G = (\{S, A, B, C, E\}, \{a, b, c\}, P, S)$$

Where P is $S \rightarrow AB$

$$A \rightarrow a$$
$$B \rightarrow b$$
$$B \rightarrow C$$
$$E \rightarrow C \mid \in$$

Solution:

1) Here the unit production is $B \rightarrow C$ and there is no production for 'C'. So we cannot substitute 'C' in R.H.S of $B \rightarrow C$

$\therefore$ Remove the unit production $B \rightarrow C$ from the grammar.

2) Consider the production $E \rightarrow C \mid \in$.

A variable $A \in V$ is said to be useful iff there is atleast one $w \in L(G)$ such that

$$S \overset{*}{\Rightarrow} xAy$$

$\overset{*}{\Rightarrow}$ w   with x, y in $(V \cup T)^*$ otherwise useless.
(or)

A variable is useful iff it occurs in at least one derivation, otherwise useless.
(or)

A variable is useless, either it cannot be reached from the start symbol or it cannot derive a terminal string.

Here the variable 'E' is useless, because 'E' does not involve in any derivation or it cannot be reached from the start symbol or there is no way to achieve $S \overset{*}{\Rightarrow} xEy$,

$S \overset{*}{\Rightarrow} w$ means 'w' can be derived from 'S' in zero or more steps.

We can conclude that 'E' is useless variable and the production $E \rightarrow C \mid \in$ is a useless production.

$\therefore$ We can remove the production $E \rightarrow C \mid \in$ from the grammar.

3) Consider the productions

$S \rightarrow AB$
$A \rightarrow a$
$B \rightarrow b$

Here S, A, B are useful variables.

$\therefore$ All these productions are useful productions. Here $\{S, A, B\}$ is the set of variables that can lead to a terminal string.

$$S \Rightarrow AB$$
$$\Rightarrow aB$$
$$\Rightarrow ab$$
$$\therefore S \Rightarrow ab$$

$\therefore$ The variables A and B are useful because they occur in at least one derivation, whereas the variable C and E are useless because they do not occur in atleast one derivation.

$\therefore$ The resulting simplified grammar is

G = (V, T, P, S) Where
$V = \{S, A, B\}$
$T = \{a, b\}$
P = {S→AB, A→a, B→b}
S = Start symbol

Problem 5:   *Find CFG with no useless symbols equivalent to*

$S \rightarrow AB \mid CA$
$A \rightarrow a$
$B \rightarrow BC \mid AB$
$C \rightarrow aB \mid b$

Solution:

Here the grammar does not contain ε - productions, unit productions so start with elimination of useless variables. A variable is useless if it cannot derive a terminal string and it cannot be reached from start symbol.

Elimination of useless symbols

Step 1: Eliminate the variables that cannot derive terminal string. i.e., useless variables.
In the given grammar, the variables S, A, C are deriving terminal strings.
Let us derive a terminal string using the variables S, A, C, we have to start from start symbol 'S'.

$$S \Rightarrow CA$$
$$\Rightarrow bA \qquad \text{(since C}\rightarrow\text{b)}$$
$$\Rightarrow ba \qquad \text{(since A}\rightarrow\text{a)}$$

$$\therefore S \Rightarrow ba$$

∴ The variables S, A, C are useful, because they involve in the derivation of terminal string 'ba'. (Since we know that a variable is useful iff it occurs in atleast one derivation)

∴ The productions S → CA, C → b, A → a are useful productions.
But the variable B is not deriving any terminal string.
For example

$$S \Rightarrow AB$$
$$\Rightarrow aBC \quad \text{(Since A}\rightarrow\text{a)}$$
$$\Rightarrow aABC \quad \text{(Since B}\rightarrow\text{AB)}$$
$$\Rightarrow aaBC \quad \text{(Since A}\rightarrow\text{a)}$$
$$\Rightarrow aaABC \quad \text{(Since B}\rightarrow\text{AB)}$$
$$\Rightarrow aaaBC \quad \text{(Since A}\rightarrow\text{a)}$$

(Here we cannot substitute a terminal for B)

The variable B is not deriving any terminating string. Therefore, B is a useless symbol.
∴ Removing B and its corresponding productions, the grammar becomes

$$S \rightarrow CA$$
$$A \rightarrow a$$
$$C \rightarrow b$$

Here we are removing the productions

$$S \rightarrow AB$$
$$B \rightarrow BC \mid AB$$
$$C \rightarrow aB$$

Because, we know that a production is useless if it involves any useless variable.

Step 2: Now, we find the variables that cannot be reached from the start variable. For this, draw a dependency graph for the variables.

Now the reduced grammar is

S→CA

A→a

C→b

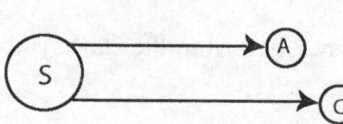

∴ In this graph, no variable is useless, since there is a path to 'A' and 'C' from start symbol 'S'.

∴ There is no further reduction.

The CFG after removing useless symbols is

S→CA

A→a

C→b

The reduced grammar is

G = ({S, A, C}, {a, b}, {S→CA, A→a, C→b}, S).

Problem 6: *Eliminate useless symbols from the following grammar.*

$S \rightarrow aAa$

$A \rightarrow bBB$

$B \rightarrow ab$

$C \rightarrow aB$

Solution:

A variable is useless, if it cannot derive terminal string or if it cannot be reached from the start symbol.

Step 1: Eliminate the variables that cannot derive the terminal string.

Here

(i)   B ⇒ ab

(ii)   C ⇒ aB

         ⇒ aab   (Since B→ab)

(iii)   A⇒ bBB

         ⇒ babB   (Since B→ab)

⇒ babab   (Since B→ab)

(iv)   S ⇒ aAa

⇒ abababa (Since A⇒babab)

∴ The variables S, A, B, C are deriving terminal strings.

∴ All the variables are useful.

∴ All the productions are useful at Step I.

Step 2: Eliminate the variables that cannot be reached from the start symbol.

Draw dependency graph for the productions,

S → aAa

A → bBB

B → ab

C → aB   (if A→αBγ then)

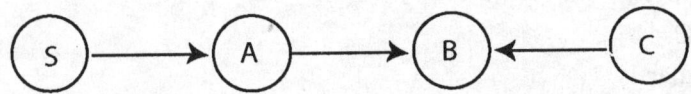

∴ There is no path to 'C' from start symbol 'S'.

∴ The variable 'C' cannot be reached from start symbol 'S'.

∴ 'C' is useless symbol. Eliminate C and its productions.

∴ The reduced grammar is

S→aAa

A→bBB

B→ab

Problem 7: *Convert the following grammar to Chomsky Normal Form (CNF)*

$$E → E+T \mid T$$
$$T → a \mid (E)$$

Solution:

To reduce the given grammar to CNF, first we have to check whether the given grammar contains any ε-productions or unit productions or useless symbols.

For that, use the following order.

1) Eliminate ε- productions

2) Eliminate unit productions

3) Eliminate useless productions

1) First check for ε or null productions.

In the given grammar if there are no null productions

∴ No need to apply (1)

2) Second, check for unit productions.

The given grammar contains one unit production.

The unit production is **E → T**

Substitute the production **T → a | (E)** in R.H.S of **E → T**

∴ **E → a | (E)**

∴ The resultant grammar after eliminating unit production is as follows

$$E \to E+T \mid a \mid (E)$$
$$T \to a \mid (E)$$

3) Third, check for useless variables or productions.

   i) The variables E and T are deriving terminal strings.

   ii) The dependency graph shows that the variables E and T can be reached from start symbol.

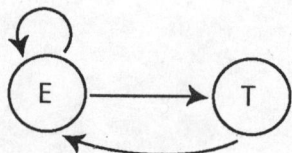

∴ There are no useless symbols in the above grammar.

∴ The reduced grammar is

$$E \to E+T \mid a \mid (E)$$
$$T \to a \mid (E)$$

Reduction to CNF: A grammar without ε-productions, unit productions, and useless productions can be reduced to Chomsky normal form. Reduce all productions in the form of A→BC or A→a.

Step 1: Eliminate all the terminals in R.H.S of the above productions by non-terminals by introducing new productions as follows

$$C_1 \to +$$
$$C_2 \to ($$
$$C_3 \to)$$

Now the grammar becomes

$$E \rightarrow EC_1T \mid a \mid C_2EC_3$$
$$T \rightarrow a \mid C_2EC_3$$
$$C_1 \rightarrow +$$
$$C_2 \rightarrow ($$
$$C_3 \rightarrow)$$

Step 2: Some productions are not in the form of A→BC
Now consider the productions

$$E \rightarrow E\ C_1T$$
$$E \rightarrow C_2E\ C_3$$
$$T \rightarrow C_2E\ C_3$$

Reduce the above productions in the form of A→BC
Introducing the new productions

$$A \rightarrow C_1T$$
$$B \rightarrow EC_3$$

∴ The final productions are

$$E \rightarrow EA$$
$$A \rightarrow C_1T$$
$$E \rightarrow C_2B$$
$$T \rightarrow C_2B$$
$$B \rightarrow EC_3$$
$$E \rightarrow a$$
$$T \rightarrow a$$
$$C_1 \rightarrow +$$
$$C_2 \rightarrow ($$
$$C_3 \rightarrow)$$

∴ All these productions are in the form of **A→BC or A→a**.

∴ The Chomsky normal form grammar is

$$G = (V, T, P, S)$$
where
$$V = \{E, A, B, T, C_1, C_2, C_3\}$$
$$T = \{a, +, (,)\}$$
$$P = \{ E \rightarrow EA \mid C_2B \mid a$$
$$T \rightarrow C_2B \mid a$$
$$A \rightarrow C_1T$$
$$B \rightarrow EC_3$$
$$C_1 \rightarrow +$$

$$C_2 \rightarrow ($$
$$C_3 \rightarrow) \}$$

S = E, start symbol.

Problem 8: Find a Griebach normal form grammar equivalent to the following CFG

$$S \rightarrow AA \mid 0$$
$$A \rightarrow SS \mid 1 \mid 0$$

Solution:

The given grammar does not contain any ε-productions, unit productions, useless variables.

Step 1: Rename the variables

Let 'S' and 'A' be renamed as 'A_1' and 'A_2' respectively.

The CFG becomes

$$A_1 \rightarrow A_2 A_2 \mid 0$$
$$A_2 \rightarrow A_1 A_1 \mid 1 \mid 0$$

Step 2: Apply the rule $A_i \rightarrow A_j \gamma$, j>i to the above productions.

i) Consider $A_1 \rightarrow A_2 A_2$ or $A_1 \rightarrow 0$

$A_1 \rightarrow A_2 A_2$ it is of the form $A_1 \rightarrow A_2 \gamma$, 2>1.

It satisfies the above rule, since the right hand side for A_1 starts with terminal or higher numbered variable.

ii) Consider $A_2 \rightarrow A_1 A_1$ or $A_2 \rightarrow 1$ or $A_2 \rightarrow 0$.

$A_2 \rightarrow A_1 A_1$, this production does not satisfy the above rule, since the right hand side of the production for A_2 starts with a lower-numbered variable.

So we substitute for the first occurrence of A_1 both $A_2 A_2$ and 0.

$\therefore A_2 \rightarrow A_1 A_1$ is replaced by $A_2 \rightarrow A_2 A_2 A_1 \mid 0 A_1 \mid 1 \mid 0$

$\therefore$ The new set is

$$A_1 \rightarrow A_2 A_2 \mid 0$$
$$A_2 \rightarrow A_2 A_2 A_1 \mid 0 A_1 \mid 1 \mid 0$$

iii) Now, consider $A_2 \rightarrow A_2 A_2 A_1 \mid 0 A_1 \mid 1 \mid 0$ and apply known lemma.

It is of the form $A_2 \rightarrow A_2 \alpha_1 \mid \beta_1 \mid \beta_2 \mid \beta_3$, which is a A_2 production.

Now introduce a new variable 'B_2' corresponding to A_2 and add 'B_2' to the set of non-terminals.   Replace all A_2 - productions using

1) $A_2 \rightarrow \beta_1 \mid \beta_1 B_2 \mid \beta_2 \mid \beta_2 B_2 \mid \beta_3 \mid \beta_3 B_2$

Since $A \rightarrow \beta_i B \mid \beta_i$, $1 \leq i \leq 3$ (since s =3)

2) $B_2 \rightarrow \alpha_1 \mid \alpha_1 B_2$

Since $B \rightarrow \alpha_i \mid \alpha_i B$, $1 \leq i \leq 1$ (since r = 1)

Substituting β_1, β_2, β_3 and α_1 in the above productions for A_2 and B_2, we have

$$A_2 \rightarrow 0A_1 \mid 0A_1B_2 \mid 1 \mid 1B_2 \mid 0 \mid 0B_2$$
$$B_2 \rightarrow A_2 A_1 \mid A_2 A_1 B_2$$

∴ The resulting productions are

$$A_1 \rightarrow A_2 A_2 \mid 0$$
$$A_2 \rightarrow 0A_1 \mid 0A_1B_2 \mid 1 \mid 1B_2 \mid 0 \mid 0B_2$$
$$B_2 \rightarrow A_2 A_1 \mid A_2 A_1 B_2$$

Now A_2 is in the form of GNF, i.e., $A \rightarrow a\alpha$ where 'a' is a terminal and 'α' is string of non-terminals.

Step 3: Substitute A_2 in A_1 and B_2

∴ $A_1 \rightarrow 0 A_1 A_2 \mid 0 A_1 B_2 A_2 \mid 1 A_2 \mid 1 B_2 A_2 \mid 0 A_2 \mid 0 B_2 A_2 \mid 0$

$A_2 \rightarrow 0 A_1 \mid 0 A_1 B_2 \mid 1 \mid 1B_2 \mid 0 \mid 0 B_2$

$B_2 \rightarrow 0 A_1A_1 \mid 0 A_1 B_2 A_1 \mid 1 A_1 \mid 1B_2 A_1 \mid 0 A_1 \mid 0 B_2 A_1 \mid 0 A_1A_1B_2 \mid$
$\qquad 0A_1B_2A_1B_2 \mid 1A_1B_2 \mid 1B_2 A_1 B_2 \mid 0 A_1 B_2 \mid 0 B_2 A_1 B_2$

which is in GNF.

Problem 9:   *Convert the following into Griebach normal form*
$$S \rightarrow AB$$
$$A \rightarrow BS \mid b$$
$$B \rightarrow SA \mid a$$

Solution:

The given grammar does not contain any ε-productions, unit productions, useless variables.

Reduction to GNF:

Step 1: Rename the variables

Let S, A and B be renamed as A_1, A_2 and A_3 respectively.
The CFG becomes

$$A_1 \rightarrow A_2 A_3 \qquad (2>1)$$
$$A_2 \rightarrow A_3 A_1 \mid b \qquad (3>2)$$
$$A_3 \rightarrow A_1 A_2 \mid a \qquad (1>3)$$

Step 2: The productions for A_1 and A_2 are in the form: $A_i \rightarrow A_j r, j>i$

The production for A_3 is not in the form
$$A_3 \rightarrow A_1 A_2 \mid a$$
$$A_3 \rightarrow A_2 A_3 A_2 \mid a \qquad \text{(by substituting } A_1 \text{ production)}$$
$$A_3 \rightarrow A_3 A_1 A_3 A_2 \mid b A_3 A_2 \mid a \quad \text{(by substituting } A_2 \text{ production)}$$
The above is a A_3 production
$\qquad$ Replace all A_3 productions by
$$A_3 \rightarrow \beta_1 \mid \beta_1 B_3 \mid \beta_2 \mid \beta_2 B_3$$

$$B_3 \rightarrow \alpha_1 \mid \alpha_1 B_3$$
$$\therefore A_3 \rightarrow bA_3 A_2 \mid b A_3 A_2 B_3 \mid a \mid a B_3$$
$$B_3 \rightarrow A_1 A_3 A_2 \mid A_1 A_3 A_2 B_3$$

Step 3: Now all the productions with 'A_3' on the left have right-hand sides that start with terminals.

These are used to replace 'A_3' in the production $A_2 \rightarrow A_3 A_1$ and then the productions with' A_2' on the left are used to replace 'A_2' in the production $A_1 \rightarrow A_2 A_3$.

The result is the following

$$A_3 \rightarrow bA_3 A_2 B_3 \qquad A_3 \rightarrow bA_3 A_2$$
$$A_3 \rightarrow aB_3 \qquad A_3 \rightarrow a$$
$$A_2 \rightarrow bA_3 A_2 B_3 A_1 \qquad A_2 \rightarrow bA_3 A_2 A_1$$
$$A_2 \rightarrow aB_3 A_1 \qquad A_2 \rightarrow a A_1$$
$$A_2 \rightarrow b$$
$$A_1 \rightarrow bA_3 A_2 B_3 A_1 A_3 \qquad A_1 \rightarrow bA_3 A_2 A_1 A_3$$
$$A_1 \rightarrow aB_3 A_1 A_3 \qquad A_1 \rightarrow aA_1 A_3$$
$$A_1 \rightarrow bA_3$$
$$B_3 \rightarrow A_1 A_3 A_2 \qquad B_3 \rightarrow A_1 A_3 A_2 B_3$$

Step 4: The two B_3 productions are converted to proper form; resulting in 10 more productions substitute the five productions of A_1 in both B_3's.

$$B_3 \rightarrow bA_3 A_2 B_3 A_1 A_3 \ A_3 \ A_2 \qquad B_3 \rightarrow bA_3 A_2 B_3 A_1 A_3 \ A_3 \ A_2 B_3$$
$$B_3 \rightarrow aB_3 A_1 A_3 A_3 \ A_2 \qquad B_3 \rightarrow aB_3 A_1 A_3 A_3 \ A_2 B_3$$
$$B_3 \rightarrow bA_3 A_3 A_2 \qquad B_3 \rightarrow bA_3 A_3 A_2 B_3$$
$$B_3 \rightarrow bA_3 A_2 A_1 A_3 \ A_3 \ A_2 \qquad B_3 \rightarrow bA_3 A_2 A_1 A_3 \ A_3 \ A_2 B_3$$
$$B_3 \rightarrow aA_1 A_3 A_3 \ A_2 \qquad B_3 \rightarrow aA_1 A_3 A_3 \ A_2 B_3$$

$\therefore$ The final set of productions is

$$A_3 \rightarrow bA_3 A_2 B_3 \qquad A_3 \rightarrow bA_3 A_2$$
$$A_3 \rightarrow aB_3 \qquad A_3 \rightarrow a$$
$$A_2 \rightarrow b A_3 A_2 B_3 A_1 \qquad A_2 \rightarrow b A_3 A_2 A_1$$
$$A_2 \rightarrow aB_3 A_1 \qquad A_2 \rightarrow aA_1$$
$$A_2 \rightarrow b$$
$$A_1 \rightarrow bA_3 A_2 B_3 A_1 A_3 \qquad A_1 \rightarrow bA_3 A_2 A_1 A_3$$
$$A_1 \rightarrow aB_3 A_1 A_3 \qquad A_1 \rightarrow aA_1 A_3$$
$$A_1 \rightarrow bA_3$$
$$B_3 \rightarrow bA_3 A_2 B_3 A_1 A_3 \ A_3 \ A_2 B_3 \qquad B_3 \rightarrow bA_3 A_2 B_3 A_1 A_3 \ A_3 \ A_2$$
$$B_3 \rightarrow aB_3 A_1 A_3 A_3 A_2 \ B_3 \qquad B_3 \rightarrow aB_3 A_1 A_3 A_3 \ A_2$$
$$B_3 \rightarrow bA_3 A_3 A_2 \ B_3 \qquad B_3 \rightarrow bA_3 A_3 A_2$$
$$B_3 \rightarrow bA_3 A_2 A_1 \ A_3 \ A_3 \ A_2 \ B_3 \qquad B_3 \rightarrow bA_3 A_2 A_1 \ A_3 \ A_3 \ A_2$$

$B_3 \rightarrow aA_1A_3A_3A_2B_3$ $B_3 \rightarrow aA_1A_3A_3A_2$

Problem 10: Convert the following grammar to Griebach normal form (GNF)

$\quad\quad S \rightarrow Ba \mid ab$
$\quad\quad A \rightarrow aAB \mid a$
$\quad\quad B \rightarrow ABb \mid b$

Solution:

The given grammar does not contain null or unit or useless productions.

Step 1: Out of the given productions

$\quad\quad A \rightarrow aAB$
$\quad\quad A \rightarrow a$
$\quad\quad B \rightarrow b$ are in the required form.

Step 2: Replace the terminals which appear in the second and subsequent places of the R.H.S of

$\quad\quad S \rightarrow Ba$
$\quad\quad S \rightarrow ab$
$\quad\quad B \rightarrow ABb$

Introduce the new variables 'C' and 'D' and add productions $C \rightarrow a$, $D \rightarrow b$

$\quad\quad \therefore\ S \rightarrow BC$
$\quad\quad\quad S \rightarrow aD$
$\quad\quad\quad B \rightarrow ABD$

Renaming S, B, A, C, D as A_1, A_2, A_3, A_4, and A_5 respectively

Step 3: The modified productions are

$\quad\quad A_1 \rightarrow A_2 A_4 \quad\quad A_1 \rightarrow aA_5$
$\quad\quad A_3 \rightarrow aA_3 A_2 \quad\quad A_3 \rightarrow a$
$\quad\quad A_2 \rightarrow A_3 A_2 A_5 \quad\quad A_2 \rightarrow b$
$\quad\quad A_4 \rightarrow a$
$\quad\quad A_5 \rightarrow b$

Step 4: Substitute A_3 in A_2

$\quad\quad \therefore A_2 \rightarrow aA_3 A_2A_2 A_5 \mid aA_2 A_5 \mid b$

Substitute A_2 in A_1

$\quad\quad A_1 \rightarrow aA_3 A_2A_2 A_5A_4 \mid aA_2 A_5 \mid bA_4 \mid a A_5$

$\therefore$ The resulting grammar in Griebach Normal Form has the productions of the form

$\quad\quad A_1 \rightarrow aA_3 A_2A_2 A_5A_4 \mid aA_2 A_5 A_4 \mid bA_4 \mid aA_5$
$\quad\quad A_2 \rightarrow aA_3 A_2A_2 A_5 \mid aA_2 A_5 \mid b$
$\quad\quad A_3 \rightarrow aA_3 A_2 \mid a$
$\quad\quad A_4 \rightarrow a$

$$A_5 \rightarrow b$$

Problem 11: ***Show that*** $L = \{a^n \ b^n \ c^n \mid n \geq 0\}$ ***is not a context free language.***

Solution:

Let us assume that
$$L = a^n \ b^n \ c^n \text{ is a context free language.}$$
Let, z be any string such that $z \in L$.
Let,
$$z = uvwxy$$
Let, $|vx| \geq 1$ and $|vwx| \geq n$
Now let us consider
$$z = u \ v^i \ w \ x^i \ y$$
i.e. we have additional occurrences of v and x. Consider various cases.

Case 1: Consider i=0 then,
$$z = u \ v^i w \ x^i \ y$$
$$z = aa \ bb \ cc \qquad here \qquad aa - u, b - v, b - w, c - x, c - y$$
If i=0 then v^0 and x^0. That means v and x absent then,
$$z = uwy$$
$$= a^2 \ bc$$
$$= aabc \notin L$$
Hence our assumption of L being CFG is wrong.

Case 2: Consider i=2 then,
$$z = u \ v^i \ w \ x^i \ y$$
$$= uvvwxxy$$
Consider, $\qquad z = aa \ bb \ cc \qquad here \qquad aa - u, b - v, b - w, c - x, c - y$
Now with i =2 we get,
$$z = aa \ bbb \ ccc \quad here \ aa - u, bb - v, \ b - w, cc - x, c - y$$
$$= a^2 \ b^3 \ c^3 \notin L$$
Thus our assumption of L being CFG is wrong. This proves that the given language L is not a context free.

EXERCISE PROBLEMS

1) Reduce the grammar given by

    S→aAa

    A→Sb | bcc | DaA

    C→abb |DD

    E→aC

    D→aDA

into an equivalent grammar by removing useless symbols and useless productions from it.

2) Give the CNF of the following grammar

    S→AB | CA

    B→BC | AB

    A→a

    C→aB | b

3) Convert to Chomsky normal form for the following grammar

    G = (V, T, P, S) where

    V = {S, A, B}

    T = {a, b}

    P = {S→aB

        S→bA

        A→aS

        A→bAA

        A → a

        B→bS

        B→aBB

        B→b}

4) Convert the following grammar to Chomsky Normal Form (CNF)

    S→aAbB

    A→aA | a

    B→bB | b

5) Reduce the following grammar and convert to CNF

    S→AB1 | 0

    B→00A | B

    B→1A1

6) Generate GNF for the following

    S→bA | aB

    A→bAA | aS.| a

    B→aBB | bS | b

7) Convert the following grammar to Greiback Normal Form (GNF)

$S \rightarrow \sim S \mid [SS] \mid p \mid q$

8) Convert the following grammar toGreiback Normal Form (GNF)

$S \rightarrow AB$

$A \rightarrow a$

$B \rightarrow C \mid b$

$C \rightarrow a$

9) Convert the following grammar to Griebach Normal Form (GNF).

$E \rightarrow E+T \mid T$

$T \rightarrow T*F \mid F$

$F \rightarrow (E) / a$

10) Convert the following grammar to Griebach normal form (GNF).

a) $S \rightarrow AA \mid a$

$A \rightarrow SS \mid b$

b) $S \rightarrow aA \mid B \mid C \mid a$

$A \rightarrow aB \mid \varepsilon$

$B \rightarrow aA$

$C \rightarrow cCd$

$D \rightarrow abd$

11) Show that $L = \{a^n b^{2n} c^n / n \geq 1\}$ is not context free language.

12) Show that $L \{ a^{2n} b^{2n} c^n / n \geq 1\}$ is not context free language.

Chapter 6

Push Down Automata

6.1 INTRODUCTION

Regular languages are accepted by finite automata. A finite automaton is a machine which accepts limited set of languages. Since it is a machine without memory, it cannot remember the symbol which it has seen so far. With the help of pumping lemma, the language $L = \{a^n b^n \mid n \geq 1\}$ have been proved to be not regular. This is a context-free language $(S \rightarrow a S b \mid ab)$

A finite automata does not accept the string of the form $a^n b^n$ since it cannot remember the number of a's it has seen to check with equal number of b's it is going to scan at. This can be overcome by pushdown automata (PDA), where auxiliary memory is added to the machine in form of a stack. For the language $a^n b^n$, initially all **a's** are pushed on to the stack and for one **b** on the input tape, one **a** is popped up from the stack. When the tape is read and the stack is empty, then the string is accepted by the machine.

In this chapter, we discuss the model and acceptance of PDA. The relationship between PDA and CFL is discussed. An overview of deterministic CFL's and deterministic PDA has been given.

6.2 DEFINITION AND MODEL OF PDA

Finite automata cannot recognize every context-free language, since some context-free languages are not regular.

We add some extra features to finite automata to accept any context-free language. So, we can say that pushdown automata is an improvement over finite automata. Finite automata with a stack memory can be viewed as pushdown automata. The operation of stack is based on **LIFO** principle (i.e. last symbol pushed on the stack will be popped first).

A PDA Consists of:

i)   A finite non-empty set of states denoted by **Q**

ii) A finite non-empty set of input symbols denoted by Σ

iii) A finite non-empty set of pushdown symbols denoted by Γ

iv) A special state called initial state denoted by $\mathbf{q_0}$

v)   A special pushdown symbol called initial symbol on the pushdown store denoted by $\mathbf{Z_0}$

vi) The set of final states, a subset of Q denoted by **F**

vii) The transition function δ from $Q \times (\Sigma \cup \varepsilon) \times \Gamma$ to the set of finite subsets of $Q\Gamma^*$

The model of PDA is as follows

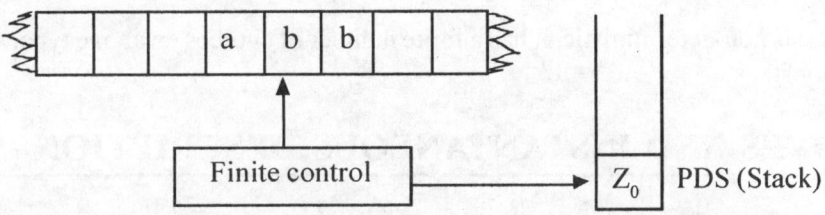

The PDA in some state and on reading an input symbol from the tape and the topmost symbol in the push down store(PDS) moves to a new state and writes a string of symbols in PDS.

Auxiliary Push Down Automata (APDA)

Auxiliary pushdown automata is shown below:

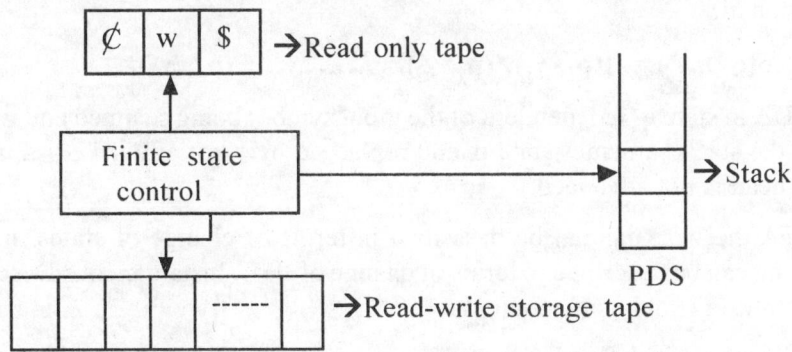

It consists of:

1) A read-only input tape, surrounded by the end markers ¢ and $
2) A finite state control
3) A read-write storage tape of length S(n), where **'n'** is the length of the input string **'w'**
4) A stack

A move of APDA is determined by the state of the finite control, along with the symbols scanned by the input, storage tape and stack heads.

In one move, the APDA may do any or all of the following:

1) Changes state
2) Move its input head one position left or right but not off the input
3) Print a symbol on the cell scanned by the storage head and move the head one position left or right
4) Push a symbol onto the stack or pop the top symbol of the stack

If the device is non-deterministic, it has a finite number of choices on above type. Acceptance is by empty stack.

6.3 MOVES AND INSTANTANEOUS DESCRIPTION

The interpretation of .....

$$\delta(q, a, Z) = \{(p_1, \gamma_1), (p_2, \gamma_2), \ldots\ldots\ldots\ldots (p_m, \gamma_m)\}$$

.....where **q** and p_i, $1 \le i \le m$ are states, **a** is in Σ, **Z** is a stack symbol and γ_i is in Γ^*, $1 \le i \le m$, is that the PDA in state **q**, with input symbol **a** and **Z** the top symbol on the stack can, for any i, enter state p_i, replace symbol by string γ_i, and advance the input head one symbol.

The interpretation of....

$$\delta(q, \varepsilon, Z) = \{(p_1, \gamma_1), (p_2, \gamma_2), \ldots\ldots\ldots\ldots(p_m, \gamma_m)\}$$

.... is that the PDA in state **q**, independent of the input symbol being scanned and with **Z** the top symbol on the stack can enter state p_i and replace **Z** by γ_i for any i, $1 \le i \le m$. In this case, the input head is not advanced.

For a FA the working can be described in terms of change of states. In case of PDA, the working can be described in terms of change of ID's (Instantaneous Description). An ID records the state and stack contents.

Formally, an ID is defined as a triple **(q, w, γ)**, where q is a state in Q, w is string input symbol from Σ^* and γ a string of stack symbol in Γ^*.

If $M = (Q, \Sigma, \Gamma, \delta, q_0, Z_0, F)$ is a PDA, then $(q, aw, Z\alpha) \vdash_M (p, w, \beta\alpha)$ if (q, a, Z) contains (p, β), where 'a' may be ε or an input symbol.

A move relation (denoted as $\vdash$) between ID's is defined as

$$(q, a_1 a_2 \ldots\ldots a_n, z_1 z_2 \ldots\ldots\ldots z_n) \vdash (q^l, a_2 \ldots\ldots a_n, \beta z_2 \ldots\ldots\ldots z_n)$$

if $\delta(q, a_1, z_1)$ contains (q^l, β)

Example 6.1

Design a PDA that accepts L =$\{$ $a^n b^n \mid n \geq 1$ $\}$

Solution:

Let $M = (Q, \Sigma, \Gamma, \delta, q_0, Z_0, F)$ be a PDA , where δ is given as

$\delta(q_0, a, Z_0) = (q_0, aZ_0)$

$\delta(q_0, a, a) = (q_0, aa)$

$\delta(q_0, b, a) = (q_1, \varepsilon)$

$\delta(q_1, b, a) = (q_1, \varepsilon)$

$\delta(q_1, \varepsilon, Z_0) = (q_f, Z_0)$

Consider a string **aaabbb**, the ID's are given as

$(q_0, aaabbb, Z_0) \vdash (q_0, aabbb, aZ_0)$

$\vdash (q_0, abbb, aaZ_0)$

$\vdash (q_0, bbb, aaaZ_0)$

$\vdash (q_1, bb, aaZ_0)$

$\vdash (q_1, b, aZ_0)$

$\vdash (q_1, \varepsilon, Z_0)$

$\vdash (q_f, Z_0)$.

Hence the string is accepted.

Example 6.2

Design a PDA for the following language $L(G) = \{w \mid w \in (0, 1)^*$, w consists of equal number of 0's and 1's$\}$

Solution:

Given that $L(G) = \{w \mid w \in (0, 1)^*$, w consists of equal number of 0's and 1's$\}$

We can construct a PDA for accepting the above language by empty stack.

Let $M = (\{q\}, \{0, 1\}, \{R, B, G\}, q, R, \{\emptyset\})$, assume that the stack symbol are R, B, G i.e. $\Gamma = \{R, B, G\}$. **R** is the initial stack top symbol

Input alphabet is $\Sigma = \{0, 1\}$

q is the start state

To accept equal number of 0's and 1's, we have to construct moves as follows:

a) From starting state, reading an input symbol **0**. At that time top of the stack is **R** and then add **B** to the stack, without changing the state, if the input symbol **1** is read then add **G** to the stack.

b) From the same state reading symbol **0** if the top of the stack is **B**, then again add **B** to the stack. If the top of the stack is **G** then remove topmost element from the stack.

c) Similarly readings symbol **1**, if the top of the stack is **G**, and then add **G** to the stack. If the top of the stack is **B** then remove topmost element from the stack.

The moves of PDA are as follows

Let M = ({q}, {0, 1}, {R, B, G}, δ, q, R, Ø) here final state in null, because this is acceptance of a string by PDA by empty stack.

R – Initial stack symbol

B – 0

G – 1

δ is defined as follows

$$\delta(q, 0, R) = (q, BR) \qquad \text{initial}$$
$$\delta(q, 1, R) = (q, GR)$$
$$\delta(q, 0, B) = (q, BB) \qquad \text{no matching}$$
$$\delta(q, 1, G) = (q, GG)$$
$$\delta(q, 0, G) = (q, \varepsilon) \qquad \text{matching}$$
$$\delta(q, 1, B) = (q, \varepsilon)$$
$$\delta(q, \varepsilon, R) = (q, \varepsilon)$$

Matching means for an input symbol **0** if topmost stack symbol is **G** and for an input symbol **1** if topmost stack symbol is **B**

i.e.   B — 0

G — 1

No matching means, if input symbol is **0** and stack top symbol is **B** or if input symbol is **1** and stack top symbol is **G**

B ⤬ 0
G ⤬ 1

Consider the string with equal number of 0's and 1's i.e. 0101

$$
\begin{aligned}
(q, 0101, R) \quad &\vdash (q, 101, BR) \\
&\vdash (q, 01, R) \\
&\vdash (q, 1, BR) \\
&\vdash (q, \varepsilon, R) \\
&\vdash (q, \varepsilon, \varepsilon)
\end{aligned}
$$

$$\therefore (q, 0101, R) \quad \vdash (q, \varepsilon, \varepsilon)$$

∴ The string '0101' is accepted by PDA by empty stack.

The moves of PDA are diagrammatically as follows

We can write the moves of PDA for the string 1010, 1001, 0011, 1100, 010010,etc

    1)Consider the string '0101'

Move 1:

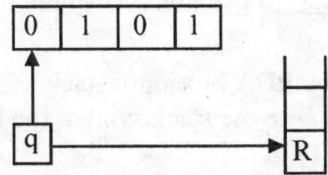

Move 2: Since δ(q, 0, R) = (q, BR) – initial

    ∴ Add a blue plate 'B' to the stack.

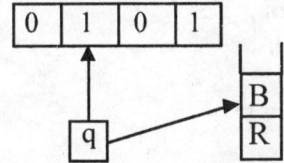

Move 3: Since δ(q, 1, B) = (q, ε) –matching

    ∴ Remove 'B' from the stack, here matching is found i.e. for input symbol '1' there is a 'B' in the stack.

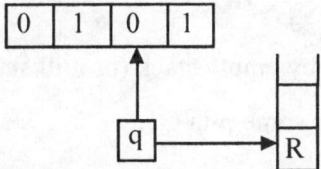

Move 4: Since δ(q, 0, R) = (q, BR)

    ∴ Add 'B' to the stack.

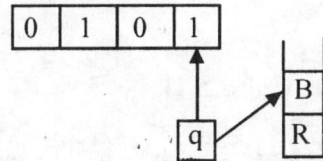

Move 5: Since δ(q, 1, B) = (q, ε) – matching

    ∴ Remove 'B' from the stack.

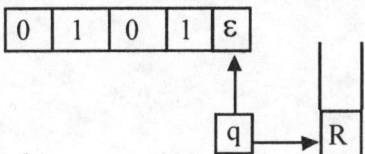

Move 6: Since $(q, \varepsilon, R) = (q, \varepsilon)x$

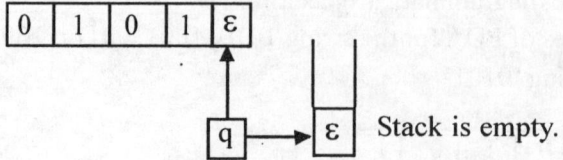

Stack is empty.

$\therefore$ The string '0101' is accepted by PDA by empty stack.

Note: If the input symbol does not match with the stack symbol the PDA halts.

6.4 ACCEPTANCE OF CFL

Acceptance of a PDA for a given CFL's can be in any one of the following ways:

- Acceptance by Final State or
- Acceptance by Empty Stack

6.4.1 ACCEPTANCE BY FINAL STATE

For PDA $M = (Q, \Sigma, \Gamma, \delta, q_0, Z_0, F)$ we define L(M), the language accepted by final state to be

$$\{w \mid (q_0, w, Z_0) \mid \stackrel{*}{-} (p, \varepsilon, \gamma) \text{ for some p in F and } \gamma \text{ in } \Gamma^*\}$$

6.4.2 ACCEPTANCE BY EMPTY STACK

We define N(M), the language accepted by empty stack (or null stack) to be

$$\{w \mid (q_0, w, Z_0) \mid \stackrel{*}{-} (p, \varepsilon, \varepsilon) \text{ for some p in Q}\}.$$

When acceptance is by empty stack, the set of final states is irrelevant, and, in this case, we usually let the set of final states be the empty set.

Example 6.3
Obtain a PDA to accept the language
$$L = \{a^n b^{2n} \mid n \geq 1\} \text{ or } L = \{0^n 1^{2n} \mid n \geq 1\}$$
Solution:
We construct a PDA, to accept the language $L = \{a^n b^{2n} \mid n \geq 1\}$ by **empty stack.**
        Taking $a = 0$, $b = 1$ in $a^n b^{2n}$, we get $0^n 1^{2n}$
Let $M = (\{q_0, q_1, q_2\}, \{0, 1\}, \{R, B\}, \delta, q_0, R, \emptyset)$, δ is defined as below
$$\delta(q_0, 0, R) = (q_1, BR)$$
$$\delta(q_1, 0, B) = (q_1, BB)$$
$$\delta(q_1, 1, B) = (q_2, B)$$

$$\delta(q_2, 1, B) = (q_1, \epsilon)$$
$$\delta(q_1, \epsilon, R) = (q_1, \epsilon)$$

Consider the string '001111' and check whether it is accepted by PDA or not.

$$(q_0, 001111, R) \vdash (q_1, 01111, BR) \text{ (since } \delta(q_1, 0, B) = (q_1, B))$$
$$\vdash (q_1, 111, BBR)$$
$$\vdash (q_2, 111, BBR)$$
$$\vdash (q_1, 11, BR)$$
$$\vdash (q_2, 1, BR)$$
$$\vdash (q_1, \epsilon, R)$$
$$\vdash (q_1, \epsilon, \epsilon)$$

$$(q_0, 001111, R) \quad \vdash^* (q_1, \epsilon, \epsilon)$$

∴ The string is accepted by empty stack

Consider the string "00111"

$$(q_0, 00111, R) \vdash (q_1, 0111, BR)$$
$$\vdash (q_1, 111, BR)$$
$$\vdash (q_2, 11, BR)$$
$$\vdash (q_1, 1, R)$$

Here stack is empty, but the input tape is non-empty

∴ The string "00111" is not accepted by PDA, because the string does not belong to L.

Example 6.4

Construct a pushdown automata to accept the language
L = {wcw^R | w ∈ (0+1)*} by empty stack.

Solution:

Let $M = (\{q_1, q_2\}, \{0, 1, c\}, \{R, B, G\}, \delta, q, R, \emptyset)$

Given language $L = \{w\,cw^R \mid w \in (0+1)^*\}$

The string before 'c' is exactly reversed and is followed by 'c'.

In this we assume two states $\{q_1, q_2\}$

- State 'q_1' is continued before reading 'c'. Once we read 'c' we enter into state 'q_2'

- In state 'q_1', reading symbol '0' with any top stack element, 'B' is added to the stack.

- In state 'q_1', reading symbol '1', with any top stack element, G is added to the stack.

- Similarly in q_2 state, if matching is found, the topmost element from the stack is removed.

δ is defined as follows

$$\delta(q_1, 0, R) = (q_1, BR)$$
$$\delta(q_1, 0, B) = (q_1, BB)$$
$$\delta(q_1, 0, G) = (q_1, BG)$$

$$\delta(q_1, 1, R) = (q_1, GR)$$
$$\delta(q_1, 1, B) = (q_1, GB)$$
$$\delta(q_1, 1, G) = (q_1, GG)$$
$$\delta(q_1, c, R) = (q_2, R)$$
$$\delta(q_1, c, B) = (q_2, B)$$
$$\delta(q_1, c, G) = (q_2, G)$$
$$\delta(q_2, 0, B) = (q_2, \varepsilon)$$
$$\delta(q_2, 1, G) = (q_2, \varepsilon)$$
$$\delta(q_2, \varepsilon, R) = (q_2, \varepsilon)$$

Consider the string "01c10"

$(q_1, 01c10, R)$	$\vdash (q_1, 1c10, BR)$
	$\vdash (q_1, c10, GBR)$
	$\vdash (q_2, 10, GBR)$
	$\vdash (q_2, 0, BR)$
	$\vdash (q_2, \varepsilon, R)$
	$\vdash (q_2, \varepsilon, \varepsilon)$
$\therefore (q_1, 01c10, R)$	$\vdash \underline{*} (q_2, \varepsilon, \varepsilon)$

$\therefore$ The string "01c10" is accepted by PDA

Consider the string "101c101"

$(q_1, 101c101, R)$	$\vdash (q_1, 01c101, GR)$
	$\vdash (q_1, 1c101, BGR)$
	$\vdash (q_1, c101, GBGR)$
	$\vdash (q_2, 101, BGR)$
	$\vdash (q_2, 01, BGR)$
	$\vdash (q_2, 1\varepsilon, GR)$
	$\vdash (q_2, \varepsilon, R)$
	$\vdash (q_2, \varepsilon, \varepsilon)$
$\therefore (q_1, 101c101, R)$	$\vdash^* (q_2, \varepsilon, \varepsilon)$

$\therefore$ The string "101c101" is accepted

Consider the string "01c101"

$(q_1, 01c101, R)$	$\vdash (q_1, 1c101, BR)$
	$\vdash (q_1, c101, GBR)$
	$\vdash (q_2, 101, GBR)$
	$\vdash (q_2, 01, BR)$
	$\vdash (q_2, 1, R)$
$(q_1, 01c101, R)$	$\vdash (q_2, 1, R)$

Here stack is empty, but input string is non-empty.

$\therefore$ "01c101" is not accepted by PDA, because $01c101 \notin L$

6.4.3 EQUIVALENCE OF FINAL STATE AND EMPTY STACK

1) If $L = N(P_1)$ for some PDA P_1 then there is a PDA P_2 such that $L = L(P_2)$. That means the language accepted by empty stack PDA will also be accepted by final state PDA ie., $N(P_1) = L(P_2)$.

2) If there is a language $L = L(P_1)$ for some PDA P_1 then there is a PDA P_2 such that $L = N(P_2)$. That means language accepted by final state PDA is also acceptable by empty stack PDA ie., $L(P_1)=N(P_2)$.

6.5 PUSH DOWN AUTOMATA AND CFG

Regular expressions and regular languages are accepted by FA. Similarly CFL's are accepted by PDA. The language accepted by PDA can be represented in form of CFG. And, a CFG can be designed to be accepted by a PDA. PDA accepts CFL's; PDA can be represented in form of deterministic PDA's and non-deterministic PDA's. The CFG is accepted either by DPDA or NPDA.

6.5.1 NON-DETERMINISTIC PDA (NPDA)

Similar to NFA, there exist NPDA for context-free grammars. The CFG accepted by DPDA is also accepted by NPDA. But some CFG's which are accepted by NPDA need not be accepted by DPDA.

Example 6.5

Construct a NPDA for the language $L = \{ww^R \mid w \in \{a, b\}^*\}$

Solution:

The language consists of all the strings of the form $L = \{\varepsilon, aa, bb, aaaa, abba, bbbb, baab,...\}$

The δ functions are given as

$$\delta(q_0, a, Z_0) = (q_0, aZ_0)$$
$$\delta(q_0, b, Z_0) = (q_0, bZ_0)$$
$$\delta(q_0, \varepsilon, Z_0) = (q_1, Z_0)$$
$$\delta(q_0, a, a) = \{(q_0, aa), (q_1, \varepsilon)\}$$
$$\delta(q_0, a, b) = (q_0, ab)$$
$$\delta(q_0, b, a) = (q_0, ba)$$
$$\delta(q_0, b, b) = \{(q_0, bb), (q_1, \varepsilon)\}$$
$$\delta(q_1, a, a) = (q_1, \varepsilon)$$
$$\delta(q_1, b, b) = (q_1, \varepsilon)$$

$$\delta(q_1, \varepsilon, Z_0) = (q_f, Z_0)$$

Where q_f is the final state.

Let w= aba

$\therefore ww^R = abaaba$

$(q_0, abaaba, Z_0)$	$\vdash$	$(q_0, baaba, aZ_0)$
	$\vdash$	$(q_0, aaba, baZ_0)$
	$\vdash$	$(q_0, aba, abaZ_0)$
	$\vdash$	$\{(q_0, ba, aabaZ_0), (q_1, ba, baZ_0)\}$
	$\vdash$	$\{(q_0, a, baabaZ_0), (q_1, a, aZ_0)\}$
	$\vdash$	$\{(q_0, \varepsilon, abaabaZ_0), (q_1, \varepsilon, Z_0)\}$
	$\vdash$	(q_f, ε, Z_0)

Hence accepted

6.5.2 CONSTRUCTION OF PDA FROM CFG

Theorem 1: If L is a context-free language, then we can construct a PDA M accepted L by empty store or empty stack.

Proof :   We construct M by making use of productions in 'G'.

Construction of M:

Let $G = (V, T, P, S)$ is a context-free grammar

We construct a PDA 'M' as $M = (\{q\}, \Sigma, V \cup \Sigma, \delta, q, S, \emptyset)$

where δ is defined by the following rules:

R_1: $\delta(q, \varepsilon, A) = \{(q, \alpha) \mid A \rightarrow \alpha \text{ is in P}\}$

or

If $A \rightarrow \alpha$ then $\delta(q, \varepsilon, A) = \{(q, \alpha)\}$

If the PDA reads a variable A on the top of PDS, it makes 'ε' move by placing the R.H.S any A-production (after erasing A)

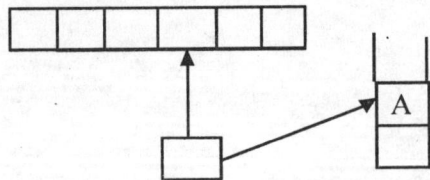

R_2: $\delta(q, a, a) = \{(q, \varepsilon)\}$ for every a in Σ

$\forall a \in \Sigma, \delta(q, a, a) = \{(q, \varepsilon)\}$

The pushdown symbols in 'M' are variables and terminals.

If the PDA reads a terminal 'a' and if it matches the current input symbol, then the PDA erases 'a'.

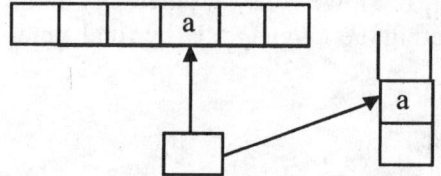

In other cases the PDA halts.

R_1 can be written as if A $\rightarrow$ α then (q, ϵ, A) = {(q, α)}

Theorem 2: If L is a context-free language, then we can construct a PDA 'M' accepting 'L' by final state or accepting state.

Proof: We can construct 'M' by making use of productions in 'G'.

Construction of M:

If L is a ϵ-free context-free language, then there exists a context-free grammar in Greibach Normal Form for it. We then construct a non-deterministic PDA which simulates left most derivations in this grammar.

Let G = (V, T, P, S) be a context-free grammar in Greibach normal from.

In Greibach normal form, each production should be in the form of either A $\rightarrow$ aα or A $\rightarrow$ a

A $\rightarrow$ a can be written as A $\rightarrow$ aϵ (if α = ϵ)

We construct PDA 'M' as

$$M = (Q, \Sigma, \Gamma, \delta, q_0, Z_0, F)$$

where 'δ' is defined by the following rules:

R_1: If A $\rightarrow$ aα then δ(q, a, A) = {(q, α)}

R_2: δ(q, a, a) = {(q, ϵ)}, $\forall$ a $\in$ Σ

If the PDA reads an input symbol 'a' in input alphabet and a variable 'A' on the top of the stack then it goes to the next input symbol by placing the string 'α' on the top of the stack (after erasing A)

Example 6.6

Design automata for the following CFG

$$\begin{aligned}
S &\rightarrow 0A \\
A &\rightarrow 0AB \mid 1 \\
B &\rightarrow 1
\end{aligned}$$

Solution:

The given grammar is context-free grammar and also in Greibach normal form, because all productions are of the form A $\rightarrow$ aα or A $\rightarrow$ a, α is a string of variables.

For the given CFG, G = (V, T, P, S), we construct the PDA in the following way. Here we construct the PDA 'M' accepting the language 'L' by final state.

Let $M = (Q, \Sigma, \Gamma, \delta, q_0, Z_0, F)$

Where
$$Q = \{q_0, q_1, q_2\}$$
$$\Sigma = \{0, 1\}$$
$$\Gamma = \{S, A, B, Z_0\}$$
q_0 = initial state
Z_0 = initial stack symbol
$F = \{q_2\}$ = final state.

δ is calculated using the following rules:

R$_1$: If $A \to a\alpha$ then $\delta(q, a, A) = \{(q, \alpha)\}$

R$_2$: $\forall\ a \in \Sigma,\ \delta(q, a, a) = \{(q, \varepsilon)\}$

- The initial move of PDA is to place 'S' on the stack and move to 'q_1',
  hence $\delta(q_0, \varepsilon, Z_0) = \{(q_1, SZ_0)\}$
- The only move to the accepting state 'q_2' is from 'q_1', when the stack is empty except for 'Z_0', hence $\delta(q_1, \varepsilon, Z_0) = \{(q_2, Z_0)\}$

By applying these rules to the above given productions, we get the transition function as follows

(start)	$\Rightarrow$	$\delta(q_0, \varepsilon, Z_0)$	$\Rightarrow \{(q_1, SZ_0)\}$	—————— R$_1$
$S \to 0A$	$\Rightarrow$	$\delta(q_1, 0, S)$	$\Rightarrow \{(q_1, A)\}$	—————— R$_2$
$A \to 0AB$	$\Rightarrow$	$\delta(q_1, 0, A)$	$\Rightarrow \{(q_1, AB)\}$	—————— R$_3$
$A \to 1$	$\Rightarrow$	$\delta(q_1, 1, A)$	$\Rightarrow \{(q_1, \varepsilon)\}$	—————— R$_4$
$B \to 1$	$\Rightarrow$	$\delta(q_1, 1, B)$	$\Rightarrow \{(q_1, \varepsilon)\}$	—————— R$_5$
(final)	$\Rightarrow$	$\delta(q_1, \varepsilon, Z_0)$	$\Rightarrow \{(q_2, Z_0)\}$	—————— R$_6$

One additional production for terminal '1' is

$$\delta(q_1, 1, 1) = \{(q_1, \varepsilon)\}$$

Derivation of a string using the grammar

$$S \Rightarrow 0A$$
$$\Rightarrow 00AB \qquad \text{(since } A \to 0AB)$$
$$\Rightarrow 001B \qquad \text{(since } A \to 1)$$
$$\Rightarrow 0011 \qquad \text{(since } B \to 1)$$
$$\therefore S \Rightarrow 0011$$

Acceptance of the string using PDA by final state

The sequences of moves are as follows. Start from the initial state 'q_0'.

$(q_0, \varepsilon\ 0011\varepsilon, Z_0)$ |- $(q_1, 0011\varepsilon, SZ_0)$ .... R_1

|- $(q_1, 011\varepsilon, AZ_0)$ .... R_2

|- $(q_1, 11\varepsilon, ABZ_0)$ .... R_3

|- $(q_1, 1\varepsilon, BZ_0)$ .... R_4

|- $(q_1, \varepsilon, Z_0$................................ R_5

|- (q_2, Z_0)................................ R_6

$\therefore (q_0, 0011, Z_0)$|-* (q_2, Z_0)

Since 'q_2' is final state, the string "0011" is accepted by PDA.

Here, input tape is:

ε	0	0	1	1	ε

Finite control is:

q_0, q_1, q_2

Contents of stack are:

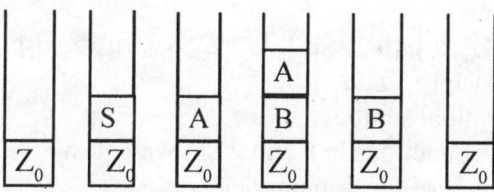

6.5.3 CONSTRUCTION OF CFG FOR A GIVEN PDA

If M = $(Q, \Sigma, , \delta, q_0, Z_0, F)$ is a PDA, then there exists a CFG 'G' such that
L(G) = N(M)
L(G) is a CFL

Construction of G

We define G = (V, T, P, S) where
V = $\{S\} \cup \{[q, Z, q^1] \mid q, q^1 \in Q, Z \in \Gamma\}$
    S – start symbol for G
    q, q^1 – states
    Z – push down symbol
The productions in P are induced by moves of PDA as follows:
R_1: S – productions are given by
$$\forall\ q \in Q, S \rightarrow [q_0, Z_0, q]$$
R_2: Each move erasing a push down symbol given by $\delta(q, a, Z) = (q^1, \varepsilon)$ induces a production
$$[q, Z, q^1] \rightarrow a\ \text{ i.e. } \delta(q, a, Z) = (q^1, \varepsilon) \implies [q, Z, q^1] \rightarrow a$$

R_3: Each move not erasing a push down symbol given by $\delta(q, a, Z) = (q_1, Z_1 Z_2 \ldots\ldots Z_m)$ induces many productions of the form

$$[q, Z, q^1] \rightarrow a[q_1, Z_1, q_2] [q_2, Z_2, q_3]\ldots [q_m Z_m q^1],$$

where each of the states $q^1, q_2, q_3, \ldots q_m$ can be any state in Q.

Example 6.7

Construct a context-free grammar G which accepts N(M), where M = ({q_0, q_1}, {a, b}, {Z_0, Z}, δ, q_0, Z_0, Ø)

δ **is given by**

$$\delta(q_0, b, Z_0) = (q_0, ZZ_0)$$
$$\delta(q_0, \varepsilon, Z_0) = (q_0, \varepsilon)$$
$$\delta(q_0, b, Z) = (q_0, ZZ)$$
$$\delta(q_0, a, Z) = (q_1, Z)$$
$$\delta(q_1, b, Z) = (q_1, \varepsilon)$$
$$\delta(q_1, a, Z_0) = (q_0, Z_0)$$

Solution:

Let G = (V, T, P, S) where

V = {S, [q_0, Z_0, q_0], [q_0, Z, q_0], [q_0, Z_0, q_1], [q_0, Z, q_1], [q_1, Z_0, q_0], [q_1, Z, q_0], [q_1, Z_0, q_1], [q_1, Z, q_1]}

T – set of terminal symbols   {q_0, q_1}

P – find the productions in P and is shown below

S – start symbol of the grammar G

The productions are   Q = {q_0, q_1}

$$\left. \begin{array}{l} P_1 : S \rightarrow [q_0, Z_0, q_0] \\ P_2 : S \rightarrow [q_0, Z_0, q_1] \end{array} \right\} \text{ S - productions}$$

$\delta(q_0, b, Z_0) = \{(q_0, ZZ_0)\}$ yields

$$P_3 : [q_0, Z_0, q_0] \rightarrow b [q_0, Z, q_0] [q_0, Z_0, q_0]$$
$$P_4 : [q_0, Z_0, q_0] \rightarrow b [q_0, Z, q_1] [q_1, Z_0, q_0]$$
$$P_5 : [q_0, Z_0, q_1] \rightarrow b [q_0, Z, q_0] [q_0, Z_0, q_1] \qquad (R_3)$$
$$P_6 : [q_0, Z_0, q_1] \rightarrow b [q_0, Z, q_1] [q_1, Z_0, q_1]$$

Since $\delta(q, a, Z) = (q_1, Z_1 Z_2 \ldots\ldots Z_n)$

$$\Rightarrow [q, Z, q^1] \rightarrow a [q_1, Z_1, q_2] [q_2, Z_2, q_3]\ldots[q_m, Z_m, q^1]$$

Here q^1 is q_0 or q_1

- $\delta(q_0, \varepsilon, Z_0) = \{(q_0, \varepsilon)\}$ gives

$\rightarrow$ $P_7 : [q_0, Z_0, q_0] \rightarrow \varepsilon$ (R_2)

- $\delta(q_0, b, Z) = \{(q_0, ZZ)\}$ gives

$\rightarrow$ $P_8 : [q_0, Z, q_0] \rightarrow b [q_0, Z, q_0] [q_0, Z, q_0]$

$\rightarrow$ $P_9 : [q_0, Z, q_0] \rightarrow b [q_0, Z, q_1] [q_1, Z, q_0]$ (R_3)

→ P_{10} : $[q_0, Z, q_1]$ → b $[q_0, Z, q_0]$ $[q_0, Z, q_1]$
→ P_{11}: $[q_0, Z_0, q_1]$ → b $[q_0, Z, q_1]$ $[q_1, Z_0, q_1]$
- $\delta(q_0, a, Z) = \{(q_1, Z)\}$ yields
→ P_{12}: $[q_0, Z, q_0]$ → a $[q_1, Z, q_0]$ (R_3)
→ P_{13}: $[q_0, Z, q_1]$ → a $[q_1, Z, q_1]$
- $\delta(q_1, b, Z) = \{(q_1, \varepsilon)\}$ gives
→ P_{14}: $[q_1, Z, q_1]$ → b   (R_2)
- $\delta(q_1, a, Z_0) = \{(q_0, Z_0)\}$
→ P_{15}: $[q_1, Z_0, q_0]$ → a $[q_0, Z_0, q_0]$ (R_3) $q^1 = q_0$ or q_1
→ P_{16}: $[q_1, Z_0, q_1]$ → a $[q_0, Z_0, q_1]$

$P_1 - P_{16}$ give the productions in P. We can reduce the number of variables and productions.

6.6 DETERMINISTIC PUSH DOWN AUTOMATA

A Push Down Automata M = $(Q, \Sigma, \Gamma, \delta, q_0, Z_0, F)$ is deterministic if

1) For each **'q'** in Q and Z in Γ, whenever $\delta(q, \varepsilon, Z)$ is non-empty, then $\delta(q, a, Z)$ is empty for all **'a'** in .

2) For no **q** in Q, Z in Γ and **'a'** in $\Sigma \cup \{\varepsilon\}$ does $\delta(q, a, Z)$ contain more than one element.

Here deterministic in the sense that at most one move is possible from any instantaneous description.

Condition (1) prevents the possibility of a choice between a move independent of the input symbol (ε -move) and a move involving an input symbol. Condition (2) prevents a choice of move for any (q, a, Z) or (q, ε, Z)

For example, $\{wcw^R \mid w \in (0+1)^*\}$ is deterministic PDA, note that unlike the finite automata, a PDA is assumed to be non-deterministic unless we state otherwise.

For finite automata, the deterministic and non-deterministic are equivalent with respect to the languages accepted. The same is not true for PDA. In fact ww^R is accepted by a non-deterministic PDA but not by any deterministic PDA.

6.7 DETERMINISTIC CONTEXT-FREE LANGUAGE

This language is used to describe syntax for various other languages. It's properties lies between the available regular sets and also the context-free language. Also the deterministic context-free language can be easily generated using L-R grammar.

Few closure properties of DFCL:

1. DCFL's are closed under complementation
2. DCFL's are closed under union, concatenation and Kleene closure.

SOLVED QUESTIONS

Problem 1: -Differentiate PDA by empty stack and final state by giving their definitions. Are they same. If so explain their equivalence with the help of an example.

Solution:

A pushdown automata has final states similar to non-deterministic or deterministic finite automata and with additional feature of stack memory. Stack feature enhances the capability of pushdown automata when compared to finite automata.

A pushdown automata accepts the input strings in two ways:
1. Accepting by entering to final state
2. Accepting by emptying its stack contents

Acceptance of input string by entering into final state

For PDA $M = (Q, \Sigma, \Gamma, \delta, q_0, Z_0, F)$, we define L(M), the language accepted by final state, to be

$$\{w \mid (q_0, w, Z_0)|\text{-}^* (P, \varepsilon, \gamma) \text{ for some P in F and } \gamma \text{ in } \Gamma^*\}$$

Acceptance of input string by emptying the stack contents

For PDA $M = (Q, \Sigma, \Gamma, \delta, q_0, Z_0, F)$, we define N(M), the language accepted by empty stack or null stack, to be

$$\{w \in \Sigma^* \mid (q_0, w, Z_0) |\text{-}^* (P, \varepsilon, \varepsilon) \text{ for some P in Q}\}$$

When acceptance is by empty stack, the set of final states are irrelevant. In this case we take the set of final states as empty set.

➢ The above two methods of accepting an input string by the PDA are equivalent i.e., suppose that a language L has a PDA, P_1 that accepts the given language L by final state iff L has a PDA, P_2 that accepts the same language L by empty stack.

➢ The language accepted by pushdown automata P through final state and the language accepted by the same PDA P through empty stack are different.

➢ A language that is accepted by a PDA by empty stack can be accepted by other PDA by final state.

For example:

1. **Consider the CFG**

$$S \rightarrow 0A$$
$$A \rightarrow 0AB \mid 1$$
$$B \rightarrow 1$$

The language generated by this grammar is L = {0011, 000111, 00001111, ...}
We can show that this language is accepted by two different PDA's i.e., one PDA by empty stack and the other PDA by final state.

2. **Consider the CFG**

$S \rightarrow aAA$

$A \rightarrow aS \mid bS \mid a$

The pushdown automata (PDA), M_1 accepting the language generated by the given grammar by final state is as follows

$$\delta(q_0, \varepsilon, Z_0) = (q_1, SZ_0)$$
$$\delta(q_1, a, S) = (q_1, AA)$$
$$\delta(q_1, a, A) = (q_1, S)$$
$$\delta(q_1, b, A) = (q_1, S)$$
$$\delta(q_1, a, A) = (q_1, \varepsilon)$$
$$\delta(q_1, \varepsilon, Z_0) = (q_2, Z_0)$$

The string which is accepted by this PDA M_1 is **w = 'aabaaa'**

The pushdown automata (PDA), M_2 accepting the language generated by this given grammar by empty stack is as follows

$$\delta(q, \varepsilon, S) = (q, aAA)$$
$$\delta(q, \varepsilon, A) = (q, aS)$$
$$\delta(q, \varepsilon, A) = (q, bS)$$
$$\delta(q, \varepsilon, A) = (q, a)$$
$$\delta(q, a, a) = (q, \varepsilon)$$
$$\delta(q, b, b) = (q, \varepsilon)$$

The string which is accepted by this PDA M_2 is **w = 'aabaaa'**

∴ The same string **w = 'aabaaa'** is accepted by both the PDA's M_1 and M_2 i.e., M_1 is by final state and M_2 is by empty stack. Similarly we can show that the set of all strings i.e,. the language generated by the grammar is also accepted by M_1 and M_2.

Problem 2: Obtain PDA to accept all strings generated by the language $\{a^n b^m a^n \mid m, n \geq 1\}$

Solution:

We construct a pushdown automata (PDA) accepting $\{a^n b^m a^n \mid m, n \geq 1\}$ by empty stack.

∴ The PDA which accepts the language $\{a^n b^m a^n \mid m, n \geq 1\}$ is

$$M = (\{q_0, q_1\}, \{a, b\}, \{a, Z_0\}, \delta, q_0, Z_0, F)$$

We define the transition function δ as follows

$$\delta(q_0, a, Z_0) = \{(q_0, aZ_0)\} \qquad \rightarrow R_1$$
$$\delta(q_0, a, a) = \{(q_0, aa)\} \qquad \rightarrow R_2$$
$$\delta(q_0, b, a) = \{(q_1, a)\} \qquad \rightarrow R_3$$
$$\delta(q_1, b, a) = \{(q_1, a)\} \qquad \rightarrow R_4$$
$$\delta(q_1, a, a) = \{(q_1, \varepsilon)\} \qquad \rightarrow R_5$$
$$\delta(q_0, \varepsilon, Z_0) = \{(q_1, \varepsilon)\} \qquad \rightarrow R_6$$

We start by storing a's on the stack until we encounter the symbol b. As soon as the input symbol 'b' is encountered, the current state changes but the stack contents remain same. As all the b's are processed in the input string, the remaining a's are erased one by one.

Finally, we erase the symbol Z_0 using rule 6.

$\therefore (q_0, a^n b^m a^n, Z_0) \vdash (q_1, \varepsilon, Z_0) \vdash \delta(q_1, \varepsilon, \varepsilon)$.

Let us consider whether the string 'aabaa' is accepted by the above PDA by empty stack or not.

$$
\begin{array}{lll}
(q_0, aabaa, Z_0) & \vdash (q_0, abaa, aZ_0) & (\text{Using } R_1) \\
& \vdash (q_0, baa, aaZ_0) & (\text{Using } R_2) \\
& \vdash (q_1, aa, aaZ_0) & (\text{Using } R_3) \\
& \vdash (q_1, a, aZ_0) & (\text{Using } R_5) \\
& \vdash (q_1, \varepsilon, Z_0) & (\text{Using } R_5) \\
& \vdash (q_0, \varepsilon, \varepsilon) & (\text{Using } R_6) \quad \textbf{(Empty stack)}
\end{array}
$$

$\therefore$ The string 'aabaa' is accepted by the PDA, by empty stack.

Problem 3: *Construct a PDA to accept the language $L = \{w \mid w \in (a, b)^*$, $n_a(w) = 2n_b(w)\}$*

Solution:

The PDA that will accept the language is given by

$$M = (\{q_0, q_1, q_2\}, \{a, b\}, \{a, b, Z_0\}, \delta, q_0, Z_0, \{q_2\})$$

where 'δ' is given by

$$\delta(q_0, a, Z_0) = (q_0, Z_0)$$
$$\delta(q_0, b, Z_0) = \{(q_0, bZ_0)\}$$
$$\delta(q_0, a, b) = (q_0, \varepsilon)$$
$$\delta(q_0, b, b) = (q_0, bb)$$
$$\delta(q_1, a, Z_0) = (q_2, Z_0)$$
$$\delta(q_1, b, Z_0) = (q_0, Z_0)$$
$$\delta(q_2, a, Z_0) = (q_2, aZ_0)$$

$$\delta(q_2, a, a) \quad = \quad (q_2, aa)$$
$$\delta(q_2, b, a) \quad = \quad (q_2, \varepsilon)$$
$$\delta(q_2, b, Z_0) \quad = \quad (q_1, Z_0)$$

The transition diagram for the PDA is as follows

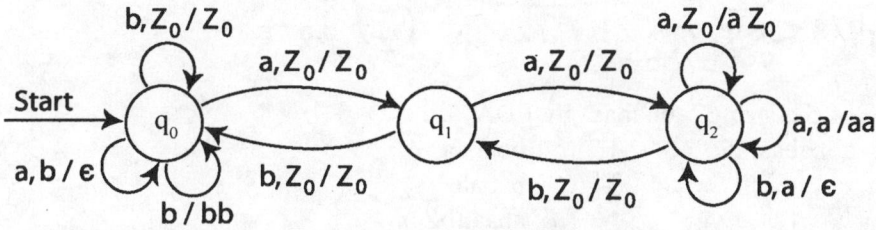

Problem 4: ***Obtain the PDA accept the language $L = \{w \mid w \in (a, b)^* \text{ and } n_a(w) > n_b(w)\}$***
i.e., number of a's in 'w' is greater than number of b's in 'w'

Solution:

Given that $L = \{w \mid w \in (a, b)^* \text{ and } n_a(w) > n_b(w)\}$.

The transition function δ is defined as follows

$$\delta(q_0, a, Z_0) \quad = \quad (q_1, Z_0)$$
$$\delta(q_0, b, Z_0) \quad = \quad (q_0, bZ_0)$$
$$\delta(q_0, a, b) \quad = \quad (q_0, \varepsilon)$$
$$\delta(q_0, b, b) \quad = \quad (q_0, bb)$$
$$\delta(q_1, a, Z_0) \quad = \quad (q_0, aZ_0)$$
$$\delta(q_1, b, Z_0) \quad = \quad (q_0, Z_0)$$
$$\delta(q_1, a, a) \quad = \quad (q_1, aa)$$
$$\delta(q_1, b, a) \quad = \quad (q_1, \varepsilon)$$

The transition table for the PDA:

State	Input	Stack symbol	Move
q_0	a	Z_0	(q_1, Z_0)
q_0	b	Z_0	(q_0, bZ_0)
q_0	a	b	(q_0, ε)
q_0	b	b	(q_0, bb)
q_1	a	Z_0	(q_1, aZ_0)
q_1	b	Z_0	(q_0, Z_0)
q_1	a	a	(q_1, aa)
q_1	b	a	(q_1, ε)

The transition diagram for PDA is as follows

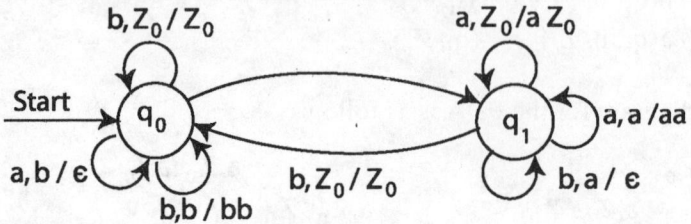

Acceptance of the string 'abbabaa' by PDA

$(q_0,\ abbabaa,\ Z_0)$ |- $(q_1,\ bbabaa,\ Z_0)$

                         |- $(q_0,\ babaa,\ Z_0)$

                         |- $(q_0,\ abaa,\ bZ_0)$

                         |- $(q_0,\ baa,\ Z_0)$

                         |- $(q_0,\ aa,\ bZ_0)$

                         |- $(q_0,\ a,\ Z_0)$

                         |- $(q_0,\ \varepsilon,\ Z_0)$

$\therefore (q_0,\ abbabaa,\ Z_0)$ |-* $(q_0,\ \varepsilon,\ Z_0)$

$\therefore$ The string 'abbabaa' is accepted by PDA.

$\therefore$ The language given is

        L = {aabbbaa, abaab, aaaba, ababa...............}

$\therefore$ The PDA accepts all the strings, which are in the given language L.

Problem 5: Design a non-deterministic pushdown for the following language

        L = {ww^R | w in (0+1)^*}

Solution:

     Here we consider two states in the design. The machine is in q_0 state until the middle of the string is reached. If the middle of the string is reached, then M enters into state q_1 and tries to match remaining input symbols with contents of the stack.

     If the input is of the form **ww^R** then M will empty its stack and accepts the input string. Here input symbols are {0, 1} and stack symbols are {R, B, G}, R is the initial top of the stack.

     Let M = ({q_0, q_1}, {0, 1}, {R, B, G}, δ, q_0, R, $\emptyset$) and δ is defined as

        $\delta(q_0,\ 0,\ R)$ $= (q_0,\ BR)$

        $\delta(q_1,\ 1,\ R)$ $= (q_0,\ GR)$

        $\mathbf{\delta(q_0,\ 0,\ B)}$ $\mathbf{= \{(q_0,\ BB),\ (q_1, \varepsilon)\}}$

        $\delta(q_0,\ 0,\ G)$ $= (q_0,\ BG)$

$$\delta(q_0, 1, B) \qquad = (q_0, GB)$$
$$\delta(\mathbf{q_0, 1, G}) \qquad = \{(\mathbf{q_0, GG}), (\mathbf{q_1}, \varepsilon)\}$$
$$\delta(q_1, 0, B) \qquad = (q_1, \varepsilon)$$
$$\delta(q_1, 1, G) \qquad = (q_1, \varepsilon)$$
$$\delta(\mathbf{q_0, \varepsilon, R}) \qquad = (q_1, \varepsilon)$$
$$\delta(\mathbf{q_1, \varepsilon, R}) \qquad = (q_1, \varepsilon)$$

Acceptance of the strings by PDA

The language $L = \{011110, 001100, 00111100, 1001, 0110\ldots\ldots\ldots\ldots\ldots\}$
We can show that this language is accepted by PDA by taking strings from this language.
Consider the string **'011110'**

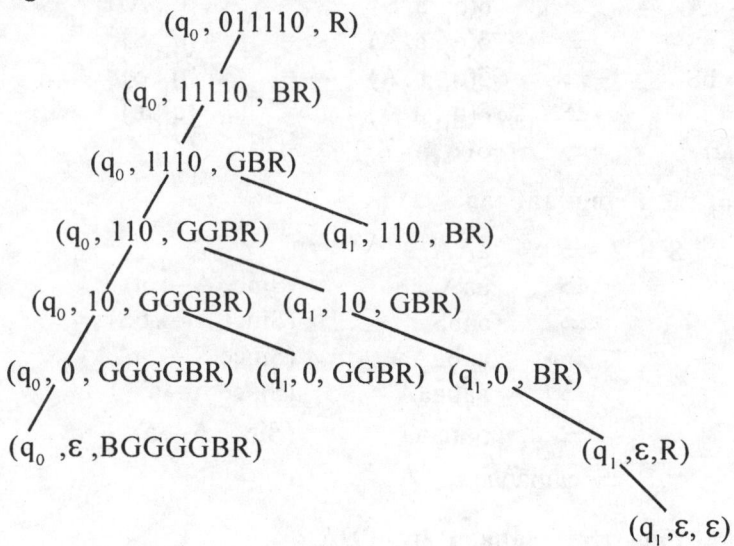

$\therefore$ There exists one path showing stack empty.
$\therefore$ The given string '011110' is accepted by PDA.

Problem 6: Construct a PDA equivalent to the following grammar and verify the result for 'aabaaa'

$$S \to aAA$$
$$A \to aS \mid bS \mid a$$

Solution:

The given grammar is CFG and also in Greibach normal form.

For the given CFG, $G = (V, T, P, S)$, we construct the PDA as follows. Here we construct the   PDA 'M' accepting the language 'L' by final state.

Let $M = (Q, \Sigma, \Gamma, \delta, q_0, Z_0, F)$, where

$Q = \{q_0, q_1, q_2\}$

$\Sigma = \{a, b\}$

$\Gamma = \{S, A, Z_0\}$

q_0 = initial state

Z_0 = initial stack symbol

$F = \{q_2\}$-Final state

δ is calculated as follows

R_1: If $A \rightarrow a\alpha$ then $\delta(q, a, A) = \{(q, \alpha)\}$.

By applying this rule to the given productions, we get

(Start)	$\Rightarrow$	$\delta(q_0, \varepsilon, Z_0)$	=	$\{q_1, SZ_0\}$	$\rightarrow R_1$
$S \rightarrow aAA$	$\Rightarrow$	$\delta(q_1, a, S)$	=	$\{q_1, AA\}$	$\rightarrow R_2$
$A \rightarrow aS$	$\Rightarrow$	$\delta(q_1, a, A)$	=	$\{q_1, S\}$	$\rightarrow R_3$
$A \rightarrow bS$	$\Rightarrow$	$\delta(q_1, b, A)$	=	$\{q_1, S\}$	$\rightarrow R_4$
$A \rightarrow a$	$\Rightarrow$	$\delta(q_1, a, A)$	=	$\{q_1, \varepsilon\}$	$\rightarrow R_5$
(Final)	$\Rightarrow$	$\delta(q_1, \varepsilon, Z_0)$	=	$\{q_2, Z_0\}$	$\rightarrow R_6$

Derivation of the string 'aabaaa'

S	$\Rightarrow$	aAA	
	$\Rightarrow$	aaA	(Since $A \rightarrow a$)
	$\Rightarrow$	$aabS$	(Since $A \rightarrow bS$)
	$\Rightarrow$	$aabaAA$	(Since $S \rightarrow aAA$)
	$\Rightarrow$	$aabaaA$	(Since $A \rightarrow a$)
	$\Rightarrow$	$aabaaa$	(Since $A \rightarrow a$)

$\therefore S \Rightarrow aabaaa$

Acceptance of the string 'aabaaa' by 'PDA'

The sequences of moves are as follows. Start from the initial state q_0

$(q_0, \varepsilon aabaaa\varepsilon, Z_0)$	$\vdash$	$(q_1, aabaaa\varepsilon, SZ_0)$	$\rightarrow R_1$
	$\vdash$	$(q_1, abaaa\varepsilon, AAZ_0)$	$\rightarrow R_2$
	$\vdash$	$(q_1, baaa\varepsilon, AZ_0)$	$\rightarrow R_5$
	$\vdash$	$(q_1, aaa\varepsilon, SZ_0)$	$\rightarrow R_4$
	$\vdash$	$(q_1, aa\varepsilon, AAZ_0)$	$\rightarrow R_2$
	$\vdash$	$(q_1, a\varepsilon, AZ_0)$	$\rightarrow R_5$
	$\vdash$	(q_1, ε, Z_0)	$\rightarrow R_5$
$\therefore (q_0, aabaaa, Z_0)$	$\vdash^*$	(q_2, ε, Z_0)	$\rightarrow R_6$

Since 'q_2' is final state, the string 'aabaaa' is accepted by PDA.
For this PDA
The input tape is:

ε	a	a	b	a	a	a	ε

The finite control is:

q_0	q_1	q_2

The contents of the stack are as follows

Problem 7: Show that the set of all string over {a, b} consisting of equal number of a's and b's accepted by a 'PDA'.

Solution:

The CFG for equal number of a's and b's is as follows

$S \rightarrow aAS|bBS|\varepsilon$	$S \rightarrow aSbS$
$A \rightarrow aAA \mid b$ **(OR)**	$S \rightarrow bSaS$
$B \rightarrow bBB \mid a$	$S \rightarrow \varepsilon$

The PDA for the above CFG is as follows

$M = (\{Q\}, \Sigma, \Gamma, \delta, q, S, \emptyset)$ where
$\Sigma = \{a, b, \varepsilon\}$
$\Gamma = \{S, A, B, a, b\}$ and
δ is defined as follows.

Here we construct the PDA, to accept the language generated by empty stack

$\therefore$
$\delta(q, \varepsilon, S) = (q_1, aAS)$
$\delta(q, \varepsilon, S) = (q, bBS)$
$\delta(q, \varepsilon, S) = (q, \varepsilon)$
$\delta(q, \varepsilon, A) = (q, aAA)$
$\delta(q, \varepsilon, A) = (q, b)$
$\delta(q, \varepsilon, B) = (q, bBB)$

$$\delta(q, \varepsilon, B) = (q, a)$$
$$\delta(q, a, a) = (q, \varepsilon)$$
$$\delta(q, b, b) = (q, \varepsilon)$$

Now, using the above transition functions we check whether the given string is accepted by PDA or not.

Derivation of the string using CFG

$$S \Rightarrow aAS$$
$$\Rightarrow aaAAS$$
$$\Rightarrow aabaAS$$
$$\Rightarrow aabbab$$

Acceptance of the string 'aabbab' by PDA by empty stack

$$
\begin{aligned}
(q,\varepsilon aabbab,S) \quad &|\text{-} (q, aabbab , aAS) \\
&|\text{-} (q, \varepsilon abbab , AS) \\
&|\text{-} (q, abbab , aAAS) \\
&|\text{-} (q, \varepsilon bbab, AAS) \\
&|\text{-} (q, bbab , bAS) \\
&|\text{-} (q, \varepsilon bab, AS) \\
&|\text{-} (q, bab , bS) \\
&|\text{-} (q, \varepsilon ab , S) \\
&|\text{-} (q, ab , aAS) \\
&|\text{-} (q, \varepsilon b , AS) \\
&|\text{-} (q, b\varepsilon , bS) \\
&|\text{-} (q, \varepsilon, S) \\
&|\text{-} (q, \varepsilon) \text{ or } (q, \varepsilon,\varepsilon)
\end{aligned}
$$

$\therefore (q, aabbab, S) |\text{-}^* (q, \varepsilon,\varepsilon)$

$\therefore$ The string 'aabbab' is accepted by 'PDA' by empty stack.

Consider the string 'aba'.

$$
\begin{aligned}
(q, \varepsilon aba, S) \quad &|\text{-} (q, aba, aAS) \\
&|\text{-} (q, \varepsilon ba, AS) \\
&|\text{-} (q, ba , bS) \\
&|\text{-} (q, \varepsilon a, S) \\
&|\text{-} (q, a, aAA) \\
&|\text{-} (q, \varepsilon, AA)
\end{aligned}
$$

$\therefore$ The string contains 'AA' it is not empty.

$\therefore$ The string 'aba' is not accepted by PDA, by empty stack, because $aba \notin L$

Problem 8: *Construct a CFG which accepts N(A) and simplify the same where*
$A = (\{q_0, q_1\}, \{a, b\}, \{z_0, z\}, \delta, z_0, \emptyset)$ *where δ is given by*

$$\delta(q_0, b, z_0) = \{(q_0, zz_0)\}$$
$$\delta(q_0, \in, z_0) = \{(q_0, \in)\}$$
$$\delta(q_0, b, z) = \{(q_0, zz)\}$$
$$\delta(q_0, a, z) = \{(q_1, z)\}$$
$$\delta(q_1, b, z) = \{(q_1, \in)\}$$
$$\delta(q_1, a, z_0) = \{(q_0, z_0)\}$$

Solution:

Let $G = (V, T, P, S)$ where $V = \{S, [q_0, z, q_0], [q_0, z, q_0], [q_0, z_0, q_1], [q_0, z, q_1], [q_1, z_0, q_0], [q_1, z, q_0], [q_1, z_0, q_1], [q_1, z, q_1]\}$

where P is the set of productions and is given below

$\left. \begin{array}{l} P_1: S \rightarrow [q_0, z_0, q_0] \\ P_2: S \rightarrow [q_0, z_0, q_1] \end{array} \right\}$ R1

$\delta(q_0, b, z_0) = \{(q_0, zz_0)\}$ gives

$P_3: [q_0, z_0, q_0] \rightarrow b [q_0, z, q_0] [q_0, z_0, q_0]$
$P_4: [q_0, z_0, q_0] \rightarrow b [q_0, z, q_1] [q_1, z_0, q_0]$
$P_5: [q_0, z_0, q_1] \rightarrow b [q_0, z, q_0] [q_0, z_0, q_1]$
$P_6: [q_0, z_0, q_1] \rightarrow b[q_0, z, q_1] [q_1, z_0, q_1]$

since $\delta(q, a, z) = (q_1, z_1 z_2 \ldots \ldots z_m)$, we have

$[q, z, q^1] \rightarrow a[q_1, z_1, q_2] [q_2, z_2, q_3]\ldots\ldots\ldots\ldots[q_m, z_m, q^1]$

Here 'q^1' is 'q$_0$' or 'q$_1$'

$\delta(q_0, \in, z_0) = \{(q_0, \in)\}$ gives

$P_7: [q_0, z_0, q_0] \rightarrow \in$ ————— R2

$\delta(q_0, b, z) = \{(q_0, zz)\}$ gives

$P_8: [q_0, z, q_0] \rightarrow [q_0, z, q_0] [q_0, z, q_0]$
$P_9: [q_0, z, q_0] \rightarrow [q_0, z, q_1] [q_1, z, q_0]$
$P_{10}: [q_0, z, q_1] \rightarrow [q_0, z, q_0] [q_0, z, q_1]$ (R3) q^1 =q$_0$ or q$_1$
$P_{11}: [q_0, z, q_1] \rightarrow [q_0, z, q_1] [q_1, z, q_1]$

$\delta(q_0, a, z) = \{(q_1, z)\}$ gives

$P_{12}: [q_0, z, q_0] \rightarrow a [q_1, z, q_0]$ (R3) q^1= q$_0$ or q$_1$
$P_{13}: [q_0, z, q_1] \rightarrow a [q_1, z, q_1]$

$\delta(q_1, b, z) = \{(q_1, \in)\}$ gives

$P_{14}: [q_1, z, q_1] \rightarrow b$ (R2)

$\delta(q_1, a, z_0) = \{(q_0, z_0)\}$ gives

$P_{15}: [q_1, z_0, q_0] \rightarrow a [q_0, z_0, q_0]$
$P_{16}: [q_1, z_0, q_1] \rightarrow a [q_0, z_0, q_1]$ (R3), q^1= q$_0$, q$_1$

$P_1 - P_{16}$ give the productions in p

We can reduce the number of variables and productions using procedure for minimization of context free grammars.

Problem 9: *Construct a CFG which accepts N(A) where A is ({q_0, q_1, q_2}, {a, b, c}, {a, b, z_0}, δ, q_0, z_0, q_2)where δ is given by*

$$δ(q_0, a, z_0) = (q_0, az_0)$$
$$δ(q_0, b, z_0) = (q_0, bz_0)$$
$$δ(q_0, a, a) = (q_0, aa)$$
$$δ(q_0, b, a) = (q_0, ba)$$
$$δ(q_0, a, b) = (q_0, ab)$$
$$δ(q_0, b, b) = (q_0, bb)$$
$$δ(q_0, c, z_0) = (q_1, z_0)$$
$$δ(q_0, c, a) = (q_1, a)$$
$$δ(q_0, c, b) = (q_1, b)$$
$$δ(q_1, a, a) = (q_1, \in)$$
$$δ(q_1, b, b) = (q_1, \in)$$
$$δ(q_1, \in, z_0) = (q_2, z_0)$$

Solution:

Let CFG be G = (V, T, P, S)

where T = {a, b, c}

V = {S, [q_0, a, q_0], [q_0, a, q_1], [q_0, a, q_2], [q_0, b, q_0], [q_0, b, q_1], [q_0, b, q_2], [q_0, z_0, q_0], [q_0, z_0, q_1], [q_0, z_0, q_2], [q_1, a, q_0], [q_1, a, q_1], [q_1, a, q_2], [q_1, b, q_0], [q_1, b, q_1], [q_1, b, q_2], [q_1, z_0, q_0], [q_1, z_0, q_1], [q_1, z_0, q_2], [q_2, a, q_0], [q_2, a, q_1], [q_2, a, q_2], [q_2, b, q_0], [q_2, b, q_1], [q_2, b, q_2], [q_2, z_0, q_0], [q_2, z_0, q_1], [q_2, z_0, q_2]}

Using the rules (1), (2) & (3). We find out the productions of the CFG.

The S – productions are

P_1: S → [q_0, z_0, q_0]
P_2: S → [q_0, z_0, q_1]
P_3: S → [q_0, z_0, q_2]

Since the S – productions are given by S → [q_0, z_0, q], ∀q ∈ Q.

If δ [q, a, z] = [q_1, $z_1 z_2$........z_m] then [q, z, q^l] → a [q_1, z_1, q_2] [q_2, z_2, q_3] .....[q_m, z_m, q^l]

where each of the state q^l, q_2...........q_m can be any state in Q.

$δ(q_0, a, z_0) = (q_0, az_0)$ gives

P_4: [q_0, z_0, q_0] → a [q_0, a, q_0] [q_0, z_0, q_0]
P_5: [q_0, z_0, q_0] → a [q_0, a, q_1] [q_1, z_0, q_0]
P_6: [q_0, z_0, q_0] → a [q_0, a, q_2] [q_2, z_0, q_0]
P_7: [q_0, z_0, q_1] → a [q_0, a, q_0] [q_0, z_0, q_1]
P_8: [q_0, z_0, q_1] → a [q_0, a, q_1] [q_1, z_0, q_1]
P_9: [q_0, z_0, q_1] → a [q_0, a, q_2] [q_2, z_0, q_1]

P_{10}: $[q_0, z_0, q_2] \rightarrow a [q_0, a, q_0] [q_0, z_0, q_2]$
P_{11}: $[q_0, z_0, q_2] \rightarrow a [q_0, a, q_1] [q_1, z_0, q_2]$
P_{12}: $[q_0, z_0, q_2] \rightarrow a [q_0, a, q_2] [q_2, z_0, q_2]$

Similarly, we can calculate for

$\delta[q_0, b, z_0] = [q_0, bz_0]$ - 9 productions

$\delta[q_0, a, a] = [q_0, aa]$ - 9 productions

$\delta[q_0, b, a] = [q_0, ba]$ - 9 productions

$\delta[q_0, a, b] = [q_0, ab]$ - 9 productions

$\delta[q_0, b, b] = [q_0, bb]$ - 9 productions

Up to this, we get '57' productions. For the remaining transitions, we will write the productions

$\delta(q_0, c, z_0) = (q_1, z_0)$ gives

P_{58}: $[q_0, z_0, q_0] \rightarrow c [q_1, z_0, q_0]$
P_{59}: $[q_0, z_0, q_1] \rightarrow c [q_1, z_0, q_1]$

P_{60}: $[q_0, z_0, q_2] \rightarrow c [q_1, z_0, q_2]$

Similarly we can find for

$\delta [q_0, c, a] = [q_1, a]$ - 3 productions

$\delta [q_0, c, b] = [q_1, b]$ - 3 productions

$\delta [q_1, a, a] = [q_1, \in]$ gives

P_{67}: $[q_1, a, q_1] \rightarrow a$

$\delta [q_1, b, b] = [q_1, \in]$ gives

P_{68}: $[q_1, b, q_1] \rightarrow b$

$\delta [q_1, \in, z_0] = [q_2, z_0]$ gives

P_{69}: $[q_1, z_0, q_2] \rightarrow \in$

The productions 1 to 69 are in p.

$\therefore$ The grammar is context- free.

Problem 10: *Construct a CFG to the PDA, $M = (\{q_0, q_1\}, \{0, 1\}, \{x, z_0\}, \delta, q_0, z_0, \emptyset)$*
        where δ is given by

$\delta(q_0, 0, z_0) = \{(q_0, xz_0)\}$

$\delta(q_0, 0, x) = \{(q_0, xx)\}$

$\delta(q_0, 1, x) = \{(q_1, \in)\}$

$\delta(q_1, 1, x) = \{(q_1, \in)\}$

$\delta(q_1, \in, x) = \{(q_1, \in)\}$

$\delta(q_1, \in, z_0) = \{(q_1, \in)\}$

Solution:

To construct a CFG G = (V, T, P, S) generating N(M) let v ={s, $[q_0, x, q_0]$, $[q_0, x, q_1]$, $[q_1, x, q_0]$, $[q_1, x, q_1]$, $[q_0, z_0, q_0]$, $[q_0, z_0, q_1]$, $[q_1, z_0, q_0]$, $[q_1, z_0, q_1]$} and T ={0, 1}.

To construct the set of productions easily, we must realize that some variables may not appear in any derivation starting from the symbols S.

Thus we can save some effort, if we start with the productions for 'S', then add productions only for these variables that appear on the right of some production already in the set.

The productions for S are

P_1: S → $[q_0, z_0, q_0]$
P_2: S → $[q_0, z_0, q_1]$

Next we add productions for the variables $[q_0, z_0, q_0]$ these are

P_2: $[q_0, z_0, q_0]$ → $0[q_0, x, q_0] [q_0, z_0, q_0]$
P_3: $[q_0, z_0, q_0]$ → $0[q_0, x, q_1] [q_1, z_0, q_0]$

These productions are required by $\delta(q_0, o, z_0) = (q_0, x, z_0)$

Next, the productions for $[q_0, z_0, q_1]$ are

P_4: $[q_0, z_0, q_1]$ → $0[q_0, x, q_0] [q_0, z_0, q_1]$
P_5: $[q_0, z_0, q_1]$ → $0[q_0, x, q_1] [q_1, z_0, q_1]$

These are also required by $\delta(q_0, 0, z_0) = \{(q_0, xz_0)\}$

The productions for the remaining variables and the relevant moves of the PDA are

P_6: $[q_0, x, q_0]$ → $0[q_0, x, q_0] [q_0, x, q_0]$
P_7: $[q_0, x, q_0]$ → $0[q_0, x, q_1] [q_1, x, q_0]$
P_8: $[q_0, x, q_1]$ → $0[q_0, x, q_0] [q_0, x, q_1]$
P_9: $[q_0, x, q_1]$ → $0[q_0, x, q_1] [q_1, x, q_1]$

Since $\delta(q_0, 0, x) = \{(q_0, xx)\}$

P_{10}: $[q_0, x, q_1]$ →1   since $\delta(q_0, 1, x) = \{(q_1, \in)\}$
P_{11}: $[q_1, z_0, q_1]$ → ∈   since $\delta(q_0, \in, z_0) = \{(q_1, \in)\}$
P_{12}: $[q_1, x, q_1]$ →∈   since $\delta(q_1, \in, x) = \{(q_1, \in)\}$
P_{13}: $[q_1, x, q_1]$ →1   since $(q_1, 1, x) = \{(q_1, \in)\}$

It should be noted that there are no productions for the variables $[q_1, x, q_0]$ and $[q_1, z_0, q_0]$.

As the productions for $[q_0, x, q_0]$ and $[q_0, z_0, q_0]$ have $[q_1, x, q_0]$ or $[q_1, z_0, q_0]$ on the right, no terminal string can be derived from $[q_0, x, q_0]$ or $[q_0, z_0, q_0]$ either. Deleting all productions involving one of these four variables on either the right or left, we end up with the following productions

$S \rightarrow [q_0, z_0, q_1]$ —— P_1

$[q_0, z_0, q_1] \rightarrow 0[q_0, x, q_1] [q_1, z_0, q_1]$ —— P_2

$[q_1, x, q_1] \rightarrow 0[q_0, x, q_1] [q_1, x, q_1]$ —— P_3

$[q_0, x, q_1] \rightarrow 1$ —— P_4

$[q_1, z_0, q_1] \rightarrow \in$ —— P_5

$[q_1, x, q_1] \rightarrow \in$ —— P_6

$[q_1, x, q_1] \rightarrow 1$ —— P_7

This is the minimized context free grammar.

REVIEW QUESTIONS

1. Construct PDA that recognizes the languages $L = \{ x = x^R \mid x \in (a, b)^* \}$.

2. Obtain PDA to accept all strings generated by the language

   $\{ a^n b^m a^n \mid m, n >= 1 \}$.

3. Let G be a CFG that generates the set of palindromes given by

   $S \rightarrow aSa \mid bSb \mid a \mid b$ Find PDA that accepts L(G).

4. Let G be the grammar given by

   $S \rightarrow aABB \mid aAA$

   $A \rightarrow aBB \mid a$

   $B \rightarrow bBB \mid A$

   Find the PDA that accepts the language generated by 'G'.

5. Design a PDA to accept the following CFG

   $S \rightarrow AA \mid a$

   $A \rightarrow SA \mid b$

6. Design a PDA to accept

   $S \rightarrow 0A$

   $A \rightarrow 0AB \mid 1$

   $B \rightarrow 1BA \mid 0$

7. Construct PDA for the grammar

   $S \rightarrow aA$

   $A \rightarrow aABC \mid bB \mid a$

$B \rightarrow b$

$C \rightarrow c$

8. Construct a PDA to accept the following CFG and verify by a suitable example.

   $S \rightarrow aABC$

   $A \rightarrow aB \mid C$

   $B \rightarrow bA \mid b$

   $C \rightarrow a$

9. Find CFG for the PDA whose transition mapping is as follows

   $\delta(S, a, X) = (s, AX)$

   $\delta(S, b, A) = (s, AA)$

   $\delta(S, a, A) = (s, \in)$

10. Construct the equivalent grammar for the following PDA

    $M = (\{q_0, q_1\}, \{0, 1\}, \{R, Z_0\}, \delta, q_0, Z_0, \phi)$ and δ is given by

    $\delta(q_0, 0, Z_0) = (q_0, RZ_0)$

    $\delta(q_0, 0, R) = (q_0, RR)$

    $\delta(q_0, 1, R) = (q_1, R)$

    $\delta(q_1, 1, R) = (q_1, R)$

    $\delta(q_1, 0, R) = (q_1, \in)$

    $\delta(q_1, 0, Z_0) = (q_1, \in)$

Chapter 7

Turing Machines

7.1 INTRODUCTION

The different machines which have been studied so far are FA and PDA. Finite automata is used for recognizing specific class of languages only. Regular expressions can be modelled by finite automata. The limitations of FA is it can remember the current input symbol only, but cannot remember the past sequence of input symbols as there is no memory. This limitation has been overcome in pushdown automata (PDA) machine. PDA makes use of external stack to store the sequence of input symbols the machine has seen so far. It accepts a large class of languages which a FA does not accepts, and these are called as context free languages. But PDA also has certain limitations. These limitations can be overcome by Turing Machines (TM). The TM is a simple machine.

Alan Turing is father of such a model which has computing capability of general purpose computer. Hence this model is popularly known as Turing machine. This machine has following features-

1. It has external memory which remembers arbitrarily long sequence of input.
2. It has unlimited memory capability.
3. The model has a facility by which the input at left or right on the tape can be read easily.
4. The machine can produce certain output based on its input. Sometimes it may be required that the same input has to be used to generate the output. So in this machine the distinction between input and output has been removed. Thus a common set of alphabets can be used for the Turing machine.

7.2 BASIC MODEL OF A TURING MACHINE

A formal model for an effective procedure should be finitely describable. Second, the procedure should consist of discrete steps, each of which can be carried out mechanically.

Such a model was introduced by Alan Turing in 1963.

The Basic model is as follows

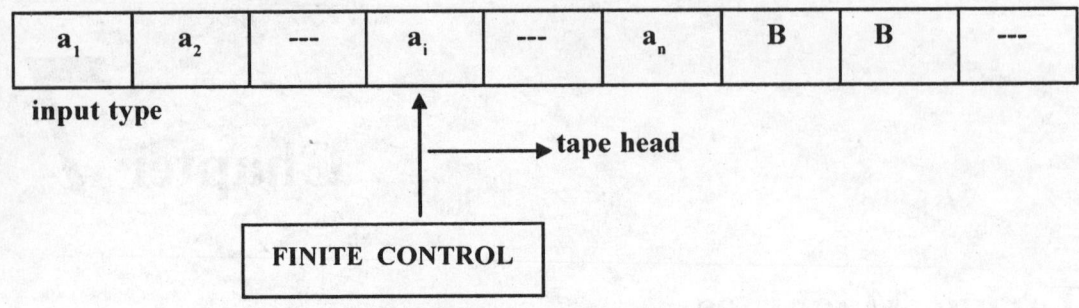

Basic Turing Machine

It has a finite control, an input tape that is divided into cells, and a tape head that scans one cell of the tape at a time. The tape has a left most cell but is infinite to the right. Each cell of the tape may hold exactly one of a finite number of tape symbols.

Initially, the **n** leftmost cells for some finite **n ≥ 0**, hold the input, which is a string of symbols chosen from a subset of tape symbols called the input symbols.

The remaining infinity of cells can hold a blank, which is a special tape symbol that is not an input symbol.

In one move the Turing machine, depending upon the symbol scanned by the tape head and the state of the finite control.

i.   Changes state.

ii. Prints a symbol on the tape cell scanned, replacing what was written there and

iii. Moves its head left or right one cell.

Note: The difference between a Turing machine and a two was finite automation lies in the former's ability to change symbols on its tape.

Definition

A Turing machine (TM) is represented by a 7-tuple, M = (Q, Σ, Γ, δ, q_0, B, F) whose components have the following meanings.

    Q: The finite set of states of the finite control.

    Σ: The finite set of input symbols.

    Γ: The finite set or complete set of tape symbols, Σ is always a subset of Γ.

    δ: The delta function or transition function.

         The arguments of δ(**q, x**) are a state **q** and a tape symbol **x**.

         δ: Q x Γ to Q x Γ x {L, R}.

The value of δ(**q, x**), if it is defined, is a triple (P, Y, D) where

        i.   **P** is the next state, in **Q**.

       ii.   **Y** is the symbol, in Γ, written in the cell being scanned, replacing whatever symbol was there

      iii. **D** is the direction, either **L** or **R**, standing for "left" or "right" , respectively and telling us the direction in which the head moves.

q_0:   The start state, a member of **Q**, in which the finite control is found initially.

B: The blank symbol. This symbol is in Γ but not in Σ, i.e., it is not an input symbol. The blank appears initially in all but the finite number of initial cells that hold input symbols.

F: The set of all final or accepting states, **F** is a subset of **Q**.

7.3 DESIGN OF TM

We can describe a Turing Machine using Instantaneous Descriptions (ID) using move relations.

Instantaneous Description:

We denote an instantaneous description (ID) of the Turing machine **M** by $\alpha_1 \, q \, \alpha_2$ where **q** is the current state of **M**, which is in **Q**, $\alpha_1 \alpha_2$ is the string in Γ^* that is the contents of the tape up to the right most non-blank symbol or the symbol to the left of the head, whichever is the rightmost. (Observe that the blank may occur in $\alpha_1 \alpha_2$)

We assume that **Q** and Γ are disjoint to avoid confusion.

Finally the tape head is assumed to be scanning the leftmost symbol of α_2 or if $\alpha_2 = \varepsilon$, the head is scanning a blank.

A move of the Turing machine is a function of the state of the finite control and the tape symbol scanned.

In one move, the Turing machine will:

1.   Change state. The next state optionally may be same as the current state.

2.   Write a tape symbol in the cell scanned. This tape symbol replaces whatever symbol was in that cell. Optionally, the symbol may be the same as the symbol currently there.

3.   Move the tape head left or right. In our formalism we require a move and do not allow the head to remain stationary. This restriction does not constrain what a Turing Machine can compute, since any sequence of moves with a stationary head could be condensed, along with the next tape-head move, into a single state change, a new tape symbol and a move left or right.

The TM is studied for

1. The class of languages it defines, called recursively enumerable sets.

2. The class of integer functions it computes called partial recursive functions.

    Example: n!, $f(x, y) = x + y$, $f(x) = x^2 =$ square of an integer.

(or) An ID of a Turing Machine **M** is a string $\alpha B \gamma$, where **B** is the present state of **M**, the entire input string is split as $\alpha \gamma$, the first R/W head and γ has all the subsequent symbols of the input string, and the string α is the substring of the input string formed by all symbols to the left of **a**.

Example: A snapshot of Turing Machine is shown in the following figure. Obtain the instantaneous description.

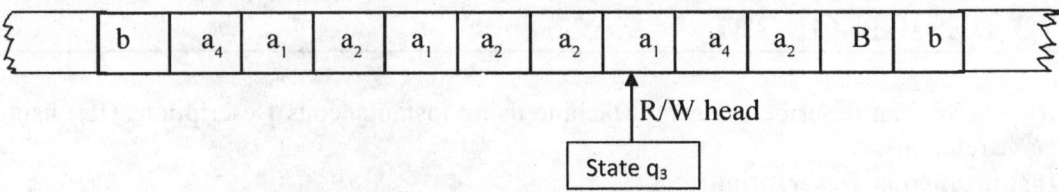

A snapshot of Turing Machine

Solution: The present symbol under R/W head is a_1. The present state is q_3.
So a_1 is written to the right of q_3.
The non-blank symbols to the left of a_1 form the string "$a_4 a_1 a_2 a_1 a_2 a_2$" which is written to the left of q_3
The sequence of non-blank symbols to the right of a_1 is $a_4 a_2$.
Thus the ID is as follows

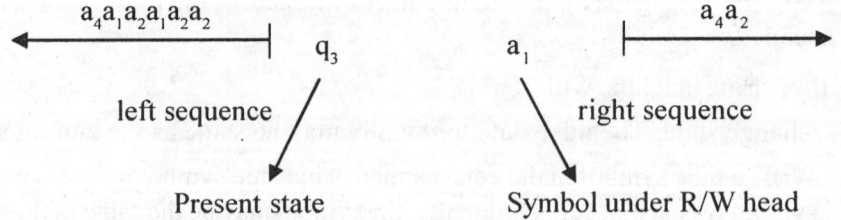

Representation of ID

Note:

a) For constructing the ID, we simply insert the current state in the input string to the left of the symbol under R/W head.

b) We observe that a blank symbol may occur as part of the left or right sudstring.

Moves in Turing Machine

As in case of pushdown automata, $\delta(q, x)$ induces a change in ID of the Turing Machine. We call this change in ID move.

We define a move of **M** as follows

Let $x_1 x_2 \ldots\ldots x_{i-1}\, q\, x_i \ldots\ldots x_n$ be an ID

Suppose $\delta(q, x_i) = (p, y, L)$

The input string to be processed is $x_1 x_2 \ldots\ldots x_n$ and the present symbol under R/W head is x_i.

So the ID before processing x_i is $x_1 x_2 \ldots\ldots x_{i-1}\, q\, x_i \ldots\ldots x_n$ After processing of x_i, the resulting ID is

$$x_1 x_2 \ldots\ldots x_{i-2}\, p\, x_{i-1}\, y\, x_{i+1} \ldots\ldots x_n.$$

This change of ID is represented by

$$x_1 x_2 \ldots\ldots x_{i-1}\, q\, x_i \ldots\ldots x_n \;\vert\!-\; x_1 x_2 \ldots\ldots x_{i-2}\, p\, x_{i-1}\, y\, x_{i+1} \ldots\ldots x_n$$

If $\delta(q, x_i) = (p, y, R)$ then the change of ID is represented by

$$x_1 x_2 \ldots\ldots x_{i-1}\, q\, x_i \ldots\ldots x_n \;\vert\!-\; x_1 x_2 \ldots\ldots x_{i-1}\, yp\, x_{i+1} \ldots\ldots x_n$$

We can denote an ID by I_j, for some **j**. $I_j \;\vert\!-\; I_k$ define a relation among ID's. So the symbol $\vert\!-^*$ denotes the reflexive transitive closure of the relation $\vert\!-$.

In particular $I_j \;\vert\!-\; I_j$.

Also if $I_1 \;\vert\!-^*\; I_n$, then we can split this as $I_1 \;\vert\!-\; I_2 \;\vert\!-\; I_3 \;\vert\!- \ldots\ldots \vert\!-\; I_n$.

Note: The description of moves by IDs is very mush useful to represent the processing of input strings

If two IDs are related by $\vert\!-_M$, then we say that the second results from the first by one move.

If one ID results from another by some finite number of moves, including zero moves, they are related by the symbol $\vert\!-_M^*$

We drop the subscript **M** $\vert\!-_M$ or $\vert\!-_M^*$ when no confusion results.

The language accepted by **M** denoted by **L(M)**, is the set of those words in Σ^* that cause **M** to enter a final state when placed, justified at the left on the tape of **M**, with **M** in state q_0, and the tape head of **M** at he left most cell.

Formally the language accepted by $M = (Q, \Sigma, \Gamma, \delta, q_0, B, F)$ is

$$\{ w/w \text{ in } \Sigma^* \text{ and } q_0 w \;\vert\!-^*\; \alpha_1 p \alpha_2 \text{ for some p in F and } \alpha_1, \alpha_2 \text{ in } \Gamma^* \}$$

Given a TM recognizing a language **L**, we assume without loss of generality that the **TM** halts i.e., has no next move, whenever the input is accepted.

However, for words not accepted is possible that the **TM** will never halt.

Representations of TM:

A TM can be represented in form of

     1.   Transition table

     2.   Transition diagram

     3.   Change of IDs

Example: Design a Turing Machine to accept the strings having equal number of a's and 1's.

Solution: The strings are of the form aababb, aabbababba, ...
         The transition diagram for the TM is as follows

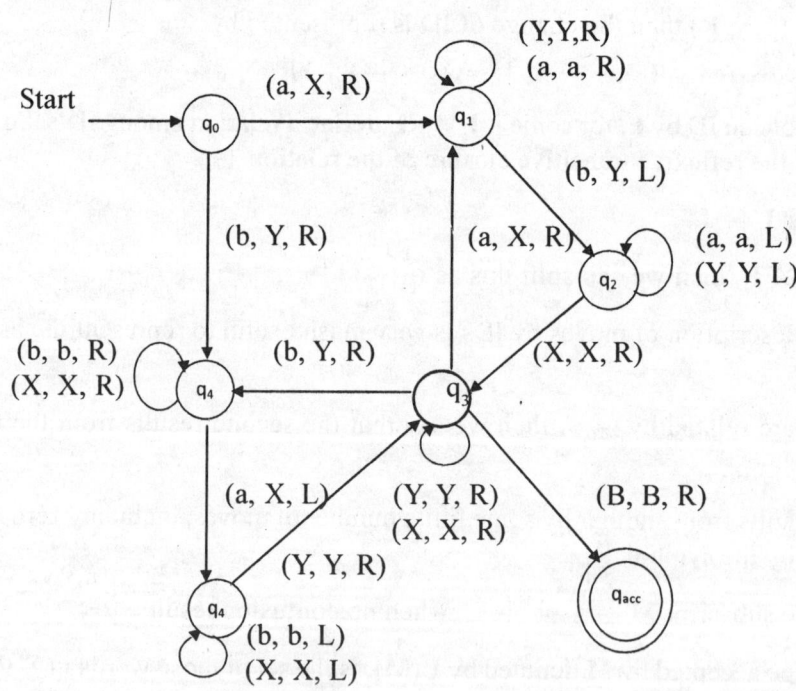

Example Strings   1. aabbabB
                 2. bbaabaB

Using these two strings we can prepare the table.
First prepare for 1 then for 2.

Transition table for the TM is as follows x

State	a	b	X	Y	B
q_0	(q_1, X, R)	(q_4, Y, R)	—	—	—
q_1	(q_1, a, R)	(q_2, Y, L)	—	(q_1, Y, R)	—
q_2	(q_2, a, L)	—	(q_3, X, R)	(q_2, Y, L)	—
q_3	(q_1, X, R)	(q_4, Y, R)	(q_3, X, R)	(q_3, Y, R)	(q_{acc}, B, R)
q_4	(q_5, X, L)	(q_4, b, R)	(q_4, X, R)	—	—
q_5	—	(q_5, b, L)	(q_5, X, L)	(q_3, Y, R)	—
q_{acc}	—	—	—	—	—

Example Design a Turing Machine that recognizes any palindrome of digits $\{0, 1\}$. Give its state transition diagram and table also.

Solution: Design of TM for any palindrome: (odd or even)

Let $M = (Q, \Sigma, \Gamma, \delta, q_0, B, F)$ where
$Q = \{q_0, q_1, q_2, q_3, q_4, q_5, q_6\}$
$\Sigma = \{0, 1\}$
$\Gamma = \{0, 1, B\}$
$F = \{q_6\}$

Procedure for constructing of TM:

First TM scans first symbol of input tape 0, 1 erases it and changes states q_1, q_2.

Next, machine scans remaining part without changing input symbols until encounters **B**. Then read/write head moves left.

If right most symbol tallies with the left most symbol, the right most symbol is erased. Otherwise machine halts. The read/write head moves to the left until **B** is encountered. Same is repeated with changes of states as shown in the table.

Transition table

State	0	1	B
q_0	(q_1, B, R)	(q_4, B, R)	(q_6, B, R)
q_1	$(q_1, 0, R)$	$(q_1, 1, R)$	(q_2, B, L)
q_2	(q_3, B, L)		(q_6, B, R)
q_3	$(q_3, 0, L)$	$(q_3, 1, L)$	(q_0, B, R)
q_4	$(q_4, 0, R)$	$(q_4, 1, R)$	(q_5, B, L)
q_5		(q_3, B, L)	(q_6, B, R)
q_6			

Example 3: Design a Turing Machine to recognize the language L = {$0^n 1^n$ / n ≥ 1}

Solution: The moves of the Turing machine are as follows

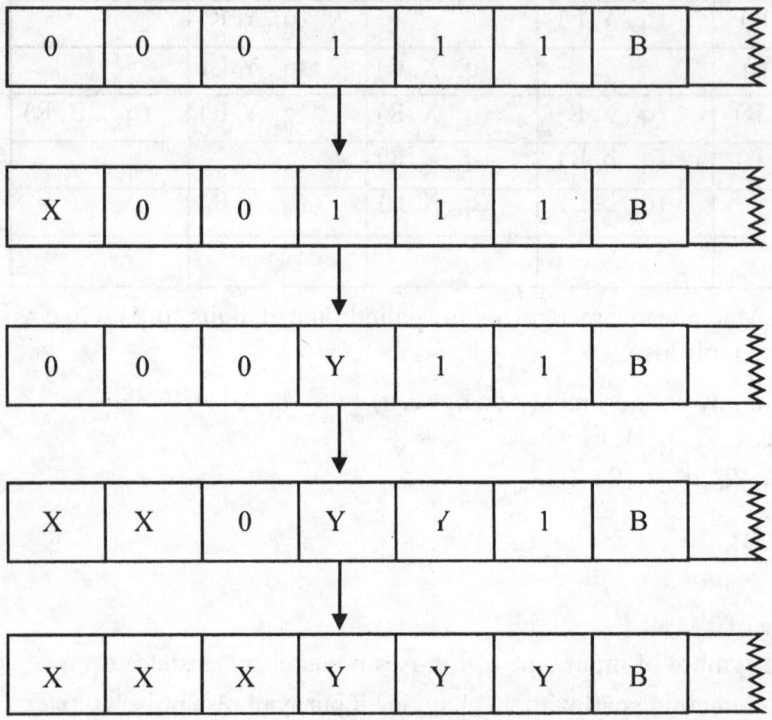

Repeatedly, **M** replaces the leftmost **0** by **X** moves right to the leftmost **1**, replacing it by **Y** moves left to find the rightmost **X**, then moves one cell right to the leftmost and repeats the cycle.

If however when searching for a **1 M** finds a blank instead, then **M** halts without accepting.

If after changing a **1** to a **Y**, **M** finds no more **0**'s, then M checks that no more **1**'s remains accepting if there are none.

Step-I: State q_0 is entered initially and also immediately prior to each replacement of leftmost **0** by an **X**.

Step-II: State q_1 is used to search right, skipping over **0**'s and **Y**'s until it finds the leftmost **1**. If **M** finds a **1** it changes it to **Y**, entering the state q_2.

Step-III: State q_2 searches left for an **X** and enters state q_0 upon finding it, moving right to the leftmost **0**, as it changes state.

As **M** searches right instate q_1, if a **B** or **X** is encountered before a **1** then the input is rejected; either there are too many **0**'s or the input is not in 0^*1^*.

State q_0 has another role:

If after state q_2 finds the rightmost **X** there is a **Y** immediately to its right, then the **0**'s are exhausted. From q_0, scanning **Y**, state q_3 is entered to scanner **Y**'s and check that no is remain.

If the **Y**'s are followed by a **B**, state q_4 is entered and acceptance occurs, otherwise the string is rejected.

We can prove that for some rejected inputs such as 001101, 001, 011 etc.

The δ-function of $\{0^n1^n \,/\, n \geq 1\}$ is defined as follows

1. $\delta(q_0, 0) = (q_1, X, R)$
2. $\delta(q_1, 0) = (q_1, 0, R)$
3. $\delta(q_1, 1) = (q_2, Y, L)$
4. $\delta(q_2, 0) = (q_2, 0, L)$
5. $\delta(q_2, X) = (q_0, X, R)$
6. $\delta(q_1, Y) = (q_1, Y, R)$
7. $\delta(q_2, Y) = (q_2, Y, L)$
8. $\delta(q_0, Y) = (q_3, Y, R)$
9. $\delta(q_3, Y) = (q_3, X, R)$
10. $\delta(q_3, B) = (q_4, B, R)$

Instantaneous description:

Input string - 0011B

$q_0 0011 \vdash Xq_1 001 \vdash X0q_1 11 \vdash Xq_2 0Y1 \vdash q_2 X0Y1 \vdash Xq_0 0Y1 \vdash XXq_1 Y1 \vdash XXYq_1 1 \vdash XXq_2 YY \vdash Xq_2 XYY \vdash XXq_0 YY \vdash XXq_2 YY \vdash Xq_2 XYY \vdash XXq_0 YY \vdash XXYq_3 Y \vdash XXYYq_3 \vdash XXYYBq_4$

Procedure for $\{0^n1^n \,/\, n \geq 1\}$

The TM, $M = (Q, \Sigma, \Gamma, \delta, q_0, B, F)$ where

$Q = \{q_0, q_1, q_2, q_3, q_4\}$

$\Sigma = \{0, 1\}$

$\Gamma = \{0, 1, X, Y, B\}$

$F = \{q_4\}$.

Informally each state represents a statement or group of statements in a program. Transition table for $\{0^n 1^n / n \geq 1\}$

State	0	1	X	Y	B
q_0	(q_1, X, R)			(q_3, Y, R)	
q_1	$(q_1, 0, R)$	(q_2, Y, L)		(q_1, Y, R)	
q_2	$(q_2, 0, L)$		(q_0, X, R)	(q_2, Y, L)	
q_3				(q_3, Y, R)	(q_4, B, R)
q_4	accepted				

Transition diagram for $\{0^n 1^n / n \geq 1\}$

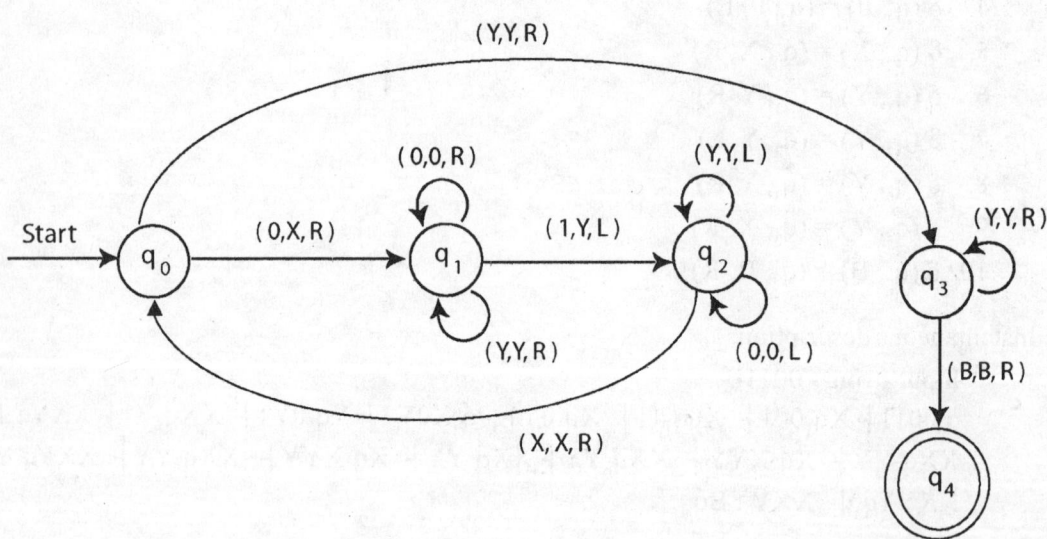

Importance of Turing Machines

1. The Turing Machine is an abstract computing machine with the power of both real computers and other mathematical definitions of what can be computed.

2. TM consists of a finite state control and an infinite tape divided into cells.

3. TM makes moves based on its current state and the tape symbol at the cell scanned by the tape head.

4. The blank is one of the tape symbol, but not an input symbol.

5. TM accepts its input if it ever enters an accepting state.

6. The languages accepted by a Turing machine are called recursively enumerable languages.

7. Instantaneous description of a Turing Machine describes current configuration of a TM by finite length string.

8. Storage in the finite control helps to design a TM for a particular language.

9. A TM can simulate the storage and control of a real computer by using one tape to store all the locations and their contents.

7.4 COMPUTABLE LANGUAGES AND FUNCTIONS

A language that is accepted by a TM is said to be recursively enumerable. The term "enumerable" derives from the fact that it is precisely these languages where strings can be enumerated (listed) by a TM.

The TM also accepts the computable functions, such as addition, multiplications, division, power function, square function, logarithmic function and many more.

7.5 TECHNIQUES FOR TURING MACHINE CONSTRUCTION

Construction of Turing machine is a process of writing out the complete set of states and next move function. This is totally a conceptual phenomenon. The Turing machine can be designed with the help of some conceptual tools. Let us discuss some of these tools.

7.5.1 STORAGE IN FINITE CONTROL

As shown in the figure the model of Turing machine has a finite control. This finite control can be used to hold some amount of information. The finite automata stores the information in pair of elements such as the current state and the current symbol pointed by the tape head. This is just a conceptual arrangement.

Example: The function δ can be written as follows.

$$\delta([q_0, 0], 1) \rightarrow ([q_1, 1], B, R)$$

This means that if finite control shows the initial state is q_0 and stores the current symbol **0** if it reads the symbol **1** then the machine goes to next state q_1 replace that **1** by **B** and moves to right. This helps in building the transition graph of the language.

Example: Construct a Turing machine **M** for $\Sigma = \{a, b\}$ which will convert lower case letters to upper case.

Solution:

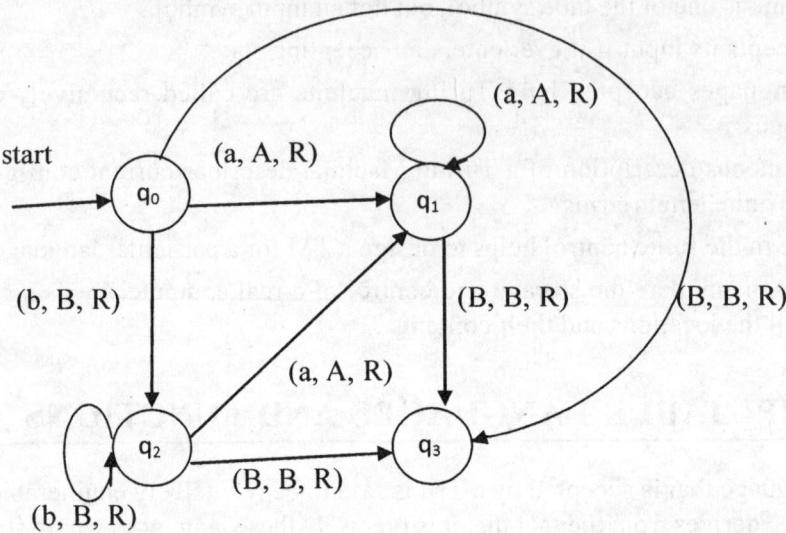

The Turing machine is designed such that we will get the equivalent upper case string to the input. The q_0 is an initial state and q_3 is final state. The δ function is written along each edge. For the sake of understanding consider the transition between q_0 and q_1, if the finite control reads a on the tape **A** will be printed there and the move will be in right direction. After converting the given string to upper case we reach to the final state q_3. The q_3 is a halt state.

7.5.2 MULTIPLE TRACKS

If the input tape is divided into multiple tracks then the input tape will be as follows

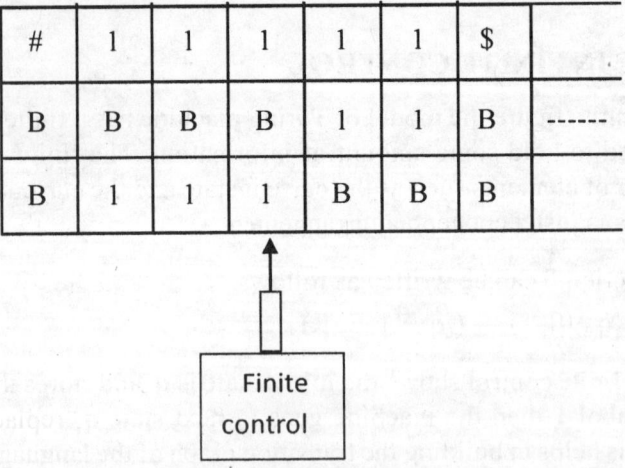

Fig 7.4: Multiple tracks

For example:

As shown in figure the input tape has multiple tracks on the first track the input which is placed is surrounded by # and $. The unary number equivalent to **5** is placed on the input tape, on the first track. On the second track unary **2** is placed. If we construct a TM which subtracts **2** from **5** we get the answer on the third track and that is **3**, in unary form. Thus this TM is for subtracting two unary numbers with the help of multiple tracks.

7.5.3 CHECKING OF SYMBOLS

Checking of symbols is an effective way of recognizing the language by TM. The symbols are to be placed on the input tape. The symbol which is read is marked by any special character. The tape head then can be moved to the right or left. Let us take some example and we will see how to build Turing machine by checking of symbols.

Example :   Construct a Turing machine M =(Q, Σ, Γ, δ, q_0, B, F) which recognizes the language L = {w c w/w∈(a+b)$^+$}

Solution :   In the language the input set is Σ = {a, b, c}. The string which, when placed on the input tape it will have two distinct parts separated by letter c, such as

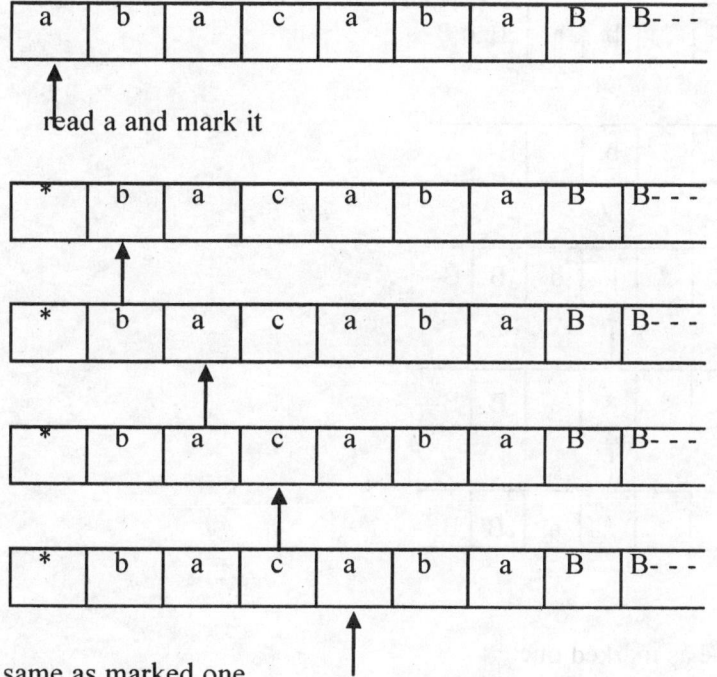

read a and mark it

it is same as marked one

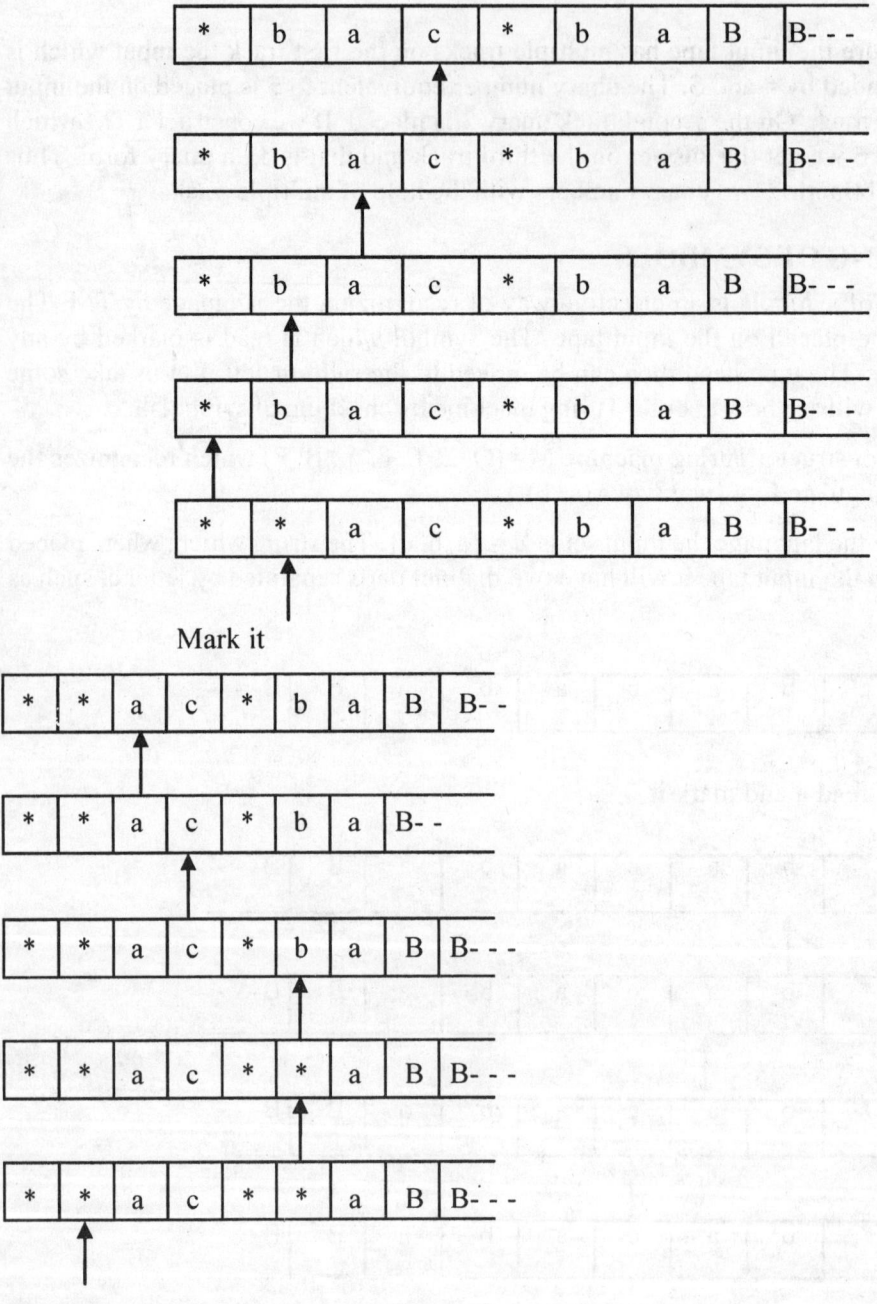

Mark it

Mark if it is same as marked one

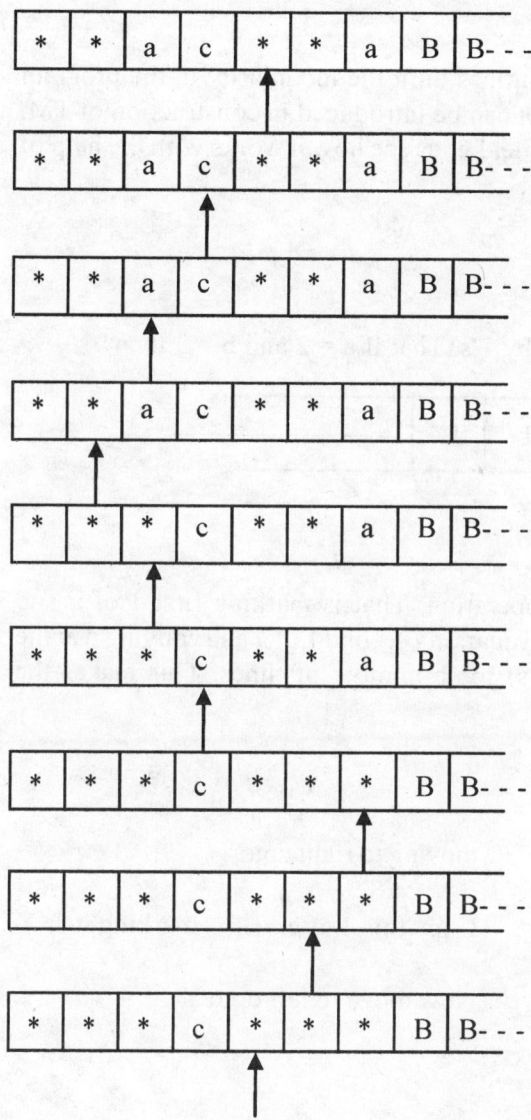

In the checking off symbols, each symbol is marked by special character. The simple logic in construction of this TM will be we mark the first letter and then move to the right till we get **c**, the very first letter after **c** will be compared with the marked letter. If it is the same as which we have marked this symbol otherwise go to reject state.

Now machine goes to accept state. Thus, in this TM we are scanning each symbol and trying to recognize the string.

7.5.4 SUBROUTINE

In the high level languages use of subroutines built the modularity in the program development process. The same type of concept can be introduced in construction of TM. We can write the subroutines as a Turing machine. Let us see how it works with the help of some example.

Example: Construct a TM for the subroutine

$$f(a, b) = a*b$$

Where a and b are unary numbers.

Solution: The unary numbers are represented by 1's. That if a = 2 and b = 3 then

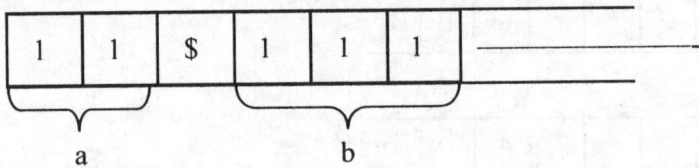

For this subroutine we will perform a copy operation. That is marking first 1 of **a** and copying it after **B** at **b** number of times, similarly marking second **1** of **a** and copying it at the rightmost end (i.e. after previously copied 1's) for **b** number of times. This makes the multiplication function complete

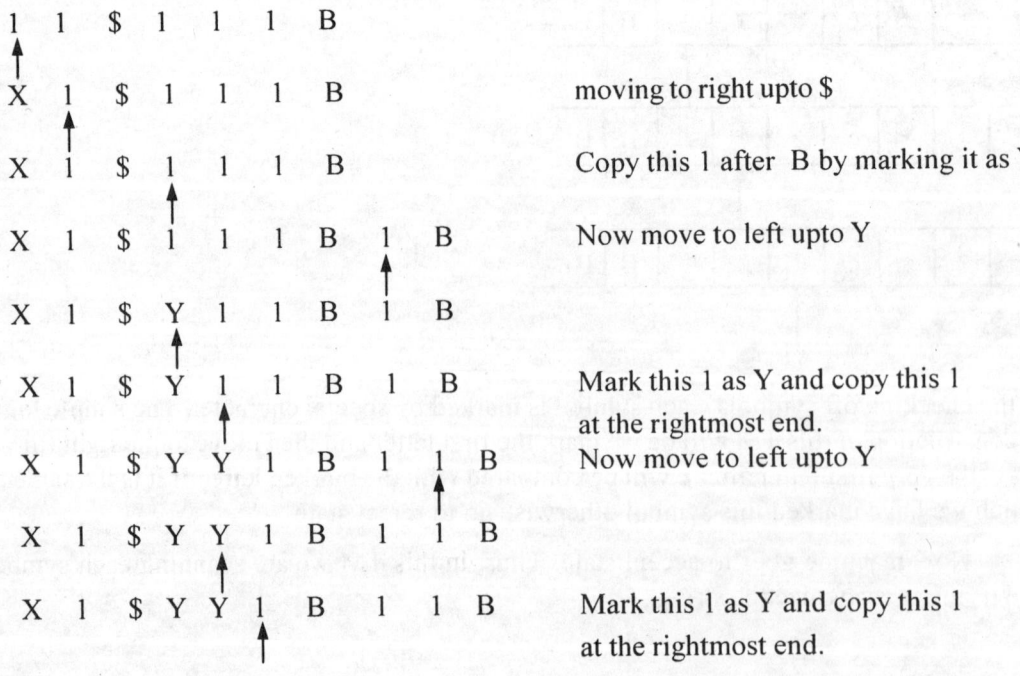

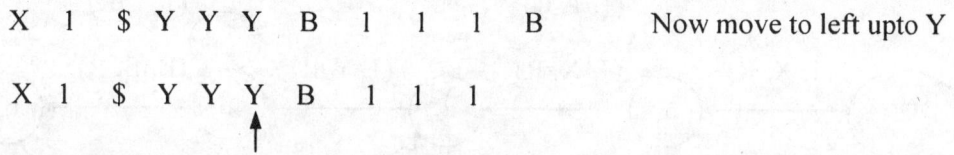

X 1 $ Y Y Y B 1 1 1 B   Now move to left upto Y

X 1 $ Y Y Y B 1 1 1

The head is pointing to **Y** next **Y** is **B** that means all the **1**'s are over we will convert these **Y** to **1**'s and repeat the same procedure for the **1** which is left to $.

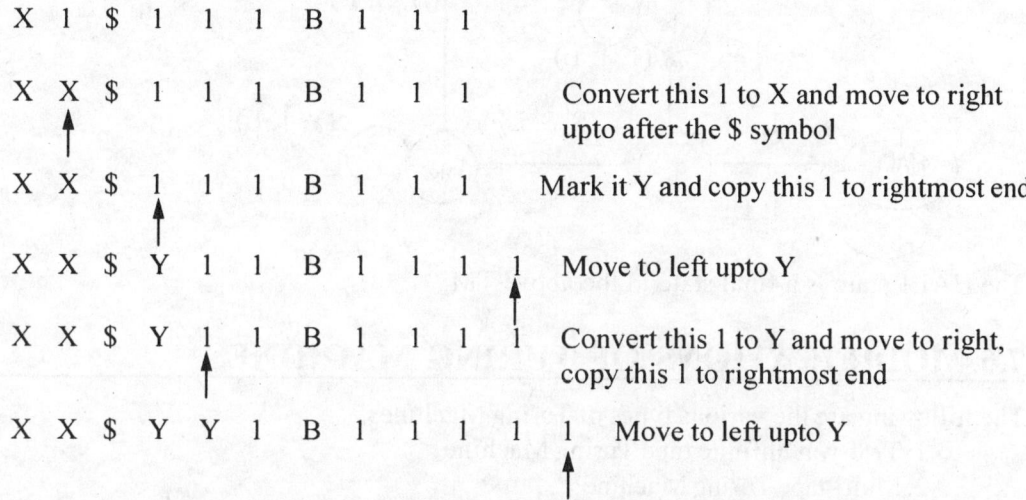

X 1 $ 1 1 1 B 1 1 1

X X $ 1 1 1 B 1 1 1   Convert this 1 to X and move to right upto after the $ symbol

X X $ 1 1 1 B 1 1 1   Mark it Y and copy this 1 to rightmost end

X X $ Y 1 1 B 1 1 1   1   Move to left upto Y

X X $ Y 1 1 B 1 1 1   1   Convert this 1 to Y and move to right, copy this 1 to rightmost end

X X $ Y Y 1 B 1 1 1 1   1   Move to left upto Y

The above steps will be repeated and all the **1**'s between **$** and **B** are copied to the rightmost end. This will be like this

X X $ Y Y Y B 1 1 1 1 1 1

Now we will move to left tape will have by converting all Y's and X's to 1's. Finally input tape will have

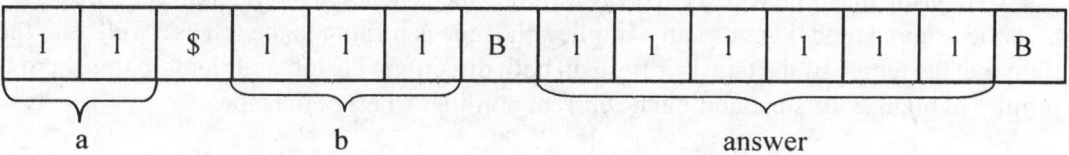

We put these actions together and draw the transition graph and model out the Turing machine.

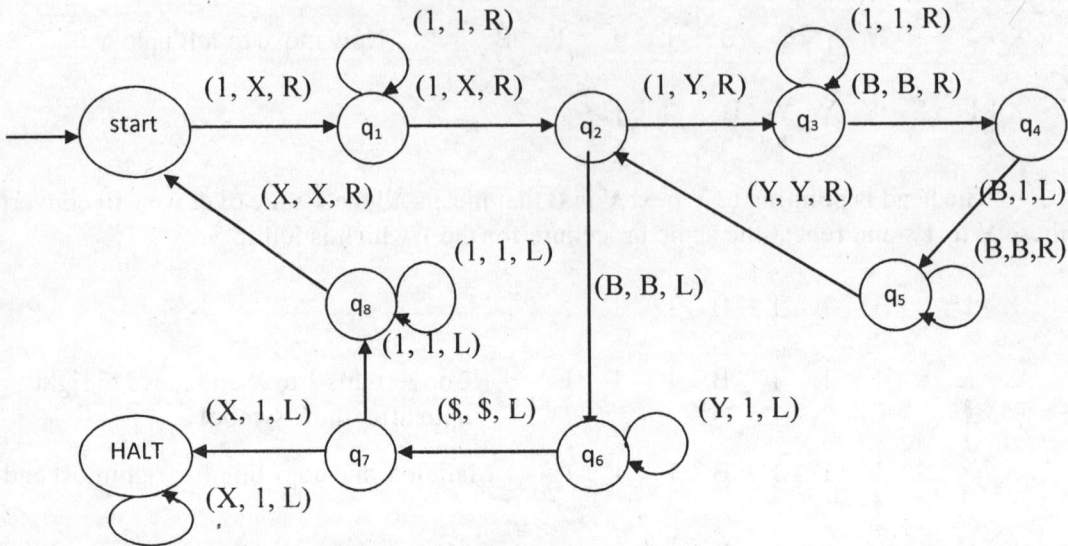

The HALT state is a final state in the above TM.

7.6 MODIFICATIONS OF TURING MACHINES

The following are the various types of Turing Machines.
1. Two-way infinite tape Turing Machine
2. Multi-tape Turing Machine
3. Non-deterministic Turing Machine
4. Multi-head Turing Machine
5. Multi-dimensional Turing Machine
6. Off-line Turing Machine

7.6.1. TWO-WAY INFINITE TAPE TURING MACHINE

A Turing machine with a two-way infinite tape is denoted by $M = (Q, \Sigma, \Gamma, \delta, q, B, F)$ as in the original model. As its name implies, the tape is infinite to the left as a well as to the right, i.e. the length of the tape is infinite in both directions i.e. left and right. In this there is infinity of blank cells surrounding the current non-blank part of the tape.

7.6.2. MULTI-TAPE TURING MACHINE

It is a modification of the two-way infinite tape Turing Machine. It consists of **n** tapes and **n** tape heads. Similar to the two-way infinite tape, each tape is infinite in both directions.

Each tape is divided into cells and each cell hold any symbol of the finite tape alphabet.

We definite an **n**-tape machine by

$M = (Q, \Sigma, \Gamma, \delta, q_0, B, F)$ where $Q, \Sigma, \Gamma, q_0, F$ are same as standard Turing Machine.

But the mapping function δ is defined as

$$Q \times \Gamma \rightarrow Q \times \Gamma \times \{ L, R \}$$

Depending on the state of the finite control and the symbol read by each tape heads the machine can change the state, and independently move each of its tape heads

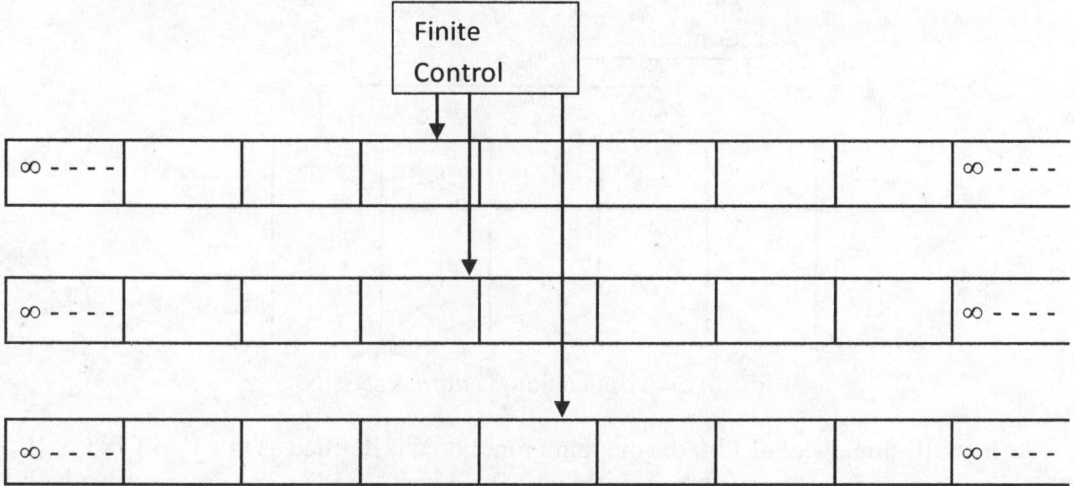

Multiple-tape Turing Machine

7.6.3. NON-DETERMINISTIC TURING MACHINE

A non-deterministic Turing Machine is a device with a finite control and a single, one way infinite tape. For a given state and tape symbol scanned by the tape head, the machine has infinite number of choices for the next move.

Each choice consists of a new state, a tape symbol to print, and a direction of head motion. The non-deterministic TM accepts its input if any sequence of choices of moves leads to an accepting state.

As with finite automation, the addition of non-deterministic to the Turing Machine does not allow the device to accept new languages.

7.6.4. MULTI-HEAD TURING MACHINE

As its name implies, it consists of multiple heads. Let the number of heads be **n**. The heads are numbered **1** to **n**. The movement of Turing Machine depends on the state of the finite control and on the symbol read by each head. In one move, the heads may each move independently left, right or remain stationary.

7.6.5. MULTI-DIMENSIONAL TURING MACHINE

It consists of a finite control, a tape. The tape consists of n-dimensional array of cells infinite in all 2n directions.

A two-dimensional Turing Machine is shown below

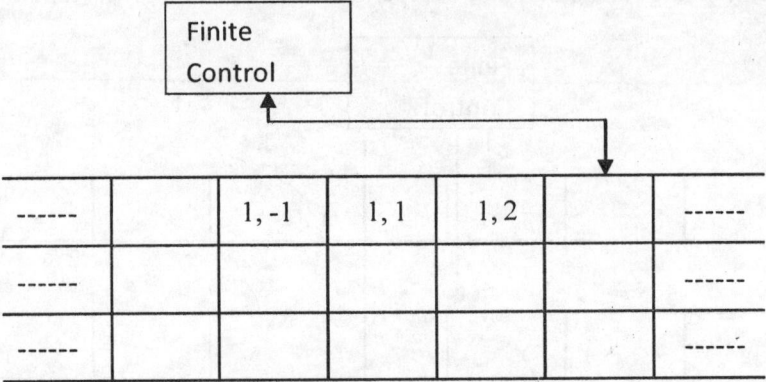

Fig 7.6: Two dimensional address scheme

In multi-dimensional TM, the mapping function δ is defined as $Q \times \Gamma \rightarrow Q \times \Gamma \times \{L, R, U, D\}$ where **U** and **D** specify the movement of read/write head up and down respectively.

7.6.6 OFF-LINE TURING MACHINE

To make the Turing Machine concept simple, we put the input file back into the picture; we get what is known as an off-line Turing Machine.

In this each move is governed by the internal state, what is currently read from the input file, and what is seen by the read/write head. A schematic representation of an off-line Turing Machine is shown below.

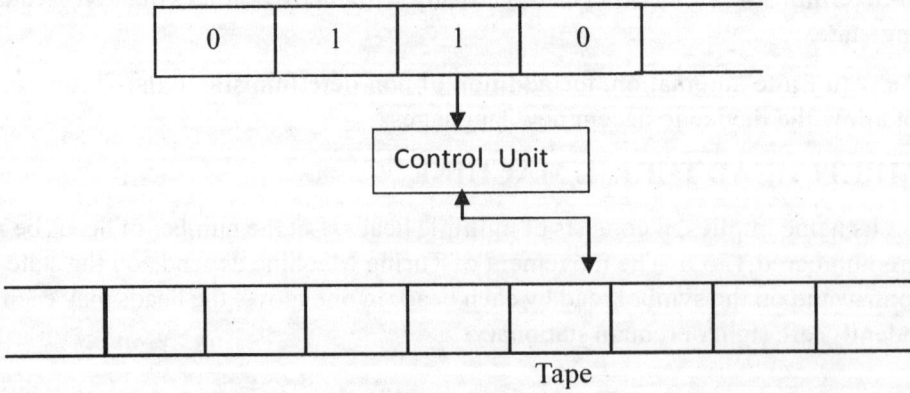

Tape

7.7 CHURCH'S HYPOTHESIS

All the proposals for a model of computation have the same power, i.e. they compute the same functions or recognize the same languages. The unprovable assumption that any general way to compute will allow us to compute only the partial recursive functions (or equivalently, what TM's or modern day computers can compute) is known as church's hypothesis or the church Turing thesis.

According to church's hypothesis, all the functions which can be defined by human beings can be computed by Turing Machines. The TM is believed to be ultimate computing machine.

The church's original statement was slightly different because he gave his thesis before machines were actually developed. He said that any machine that can do certain list of operations will be able to perform all algorithms.

Turing machines can perform what church's asked, so they are possibly the machines which church's described.

Church tied together recursive functions and computable functions together.

Every partial recursive function is computable on TM. Computer models such as RAM also give rise to partial recursive functions. So they can be simulated on TM which confirms the validity of church's hypothesis.

Importance of Church's Hypothesis:

1. First we will prove certain problems which cannot be solved using Turing Machine.

2. If church's thesis is true this implies that problems cannot be solved by any computer or any programming languages we might ever develop.

3. Thus in studying the capabilities and limitations of Turing Machines, we are indeed · studying the fundamental capabilities and limitations of any computational device we might ever construct.

It provides a general principle for algorithmic computation and, while not provable, gives strong evidence that no more powerful models can be found.

Simulation of random access by TM:

The following figure suggests how a "Turing Machine" would simulate a "RAM COMPUTER"

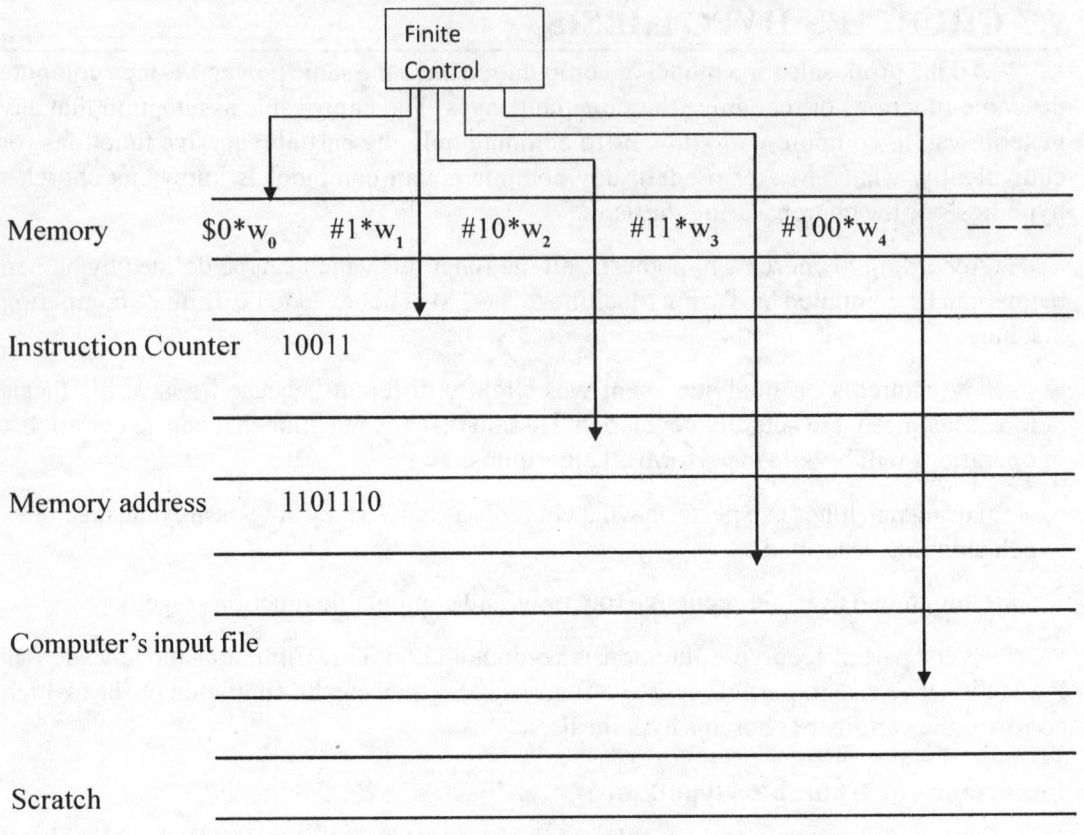

A Turing machine that simulates a typical computer

The above Turing machine uses several tapes, but it could be converted to a one-tape TM using the construction procedure.

The first tape represents the entire memory of the computer. We have used a code in which addresses of memory words, in numerical order, alternate with the contents of those memory words. But addresses and contents are written in binary.

The marker symbols * and # are used to make it easy to find the ends of addresses and contents, and to tell whether a binary string is an addresses or content. Another marker $ indicates the beginning of the sequence of addresses and contents.

The second tape is the "instruction counter". This tape holds one integer in binary, which represents one of the memory locations on tape 1. The value stored in this location will be interpreted as the next computer instruction to be executed.

The third tape holds a "Memory Addresses" or the contents of that address after the address has been located on tape 1. To execute an instruction, the TM must find the contents of one or more memory addresses that hold data involved in the computation. First the desired address is copied onto tape 3 and compared with the addresses on tape 1, until a match is found. The contents of this address is copied on to the third tape and moved to wherever it is needed, typically to one of the low-numbered addresses that represent the registers of the computer.

7.8 COUNTER MACHINE

A counter machine can be represented by following model

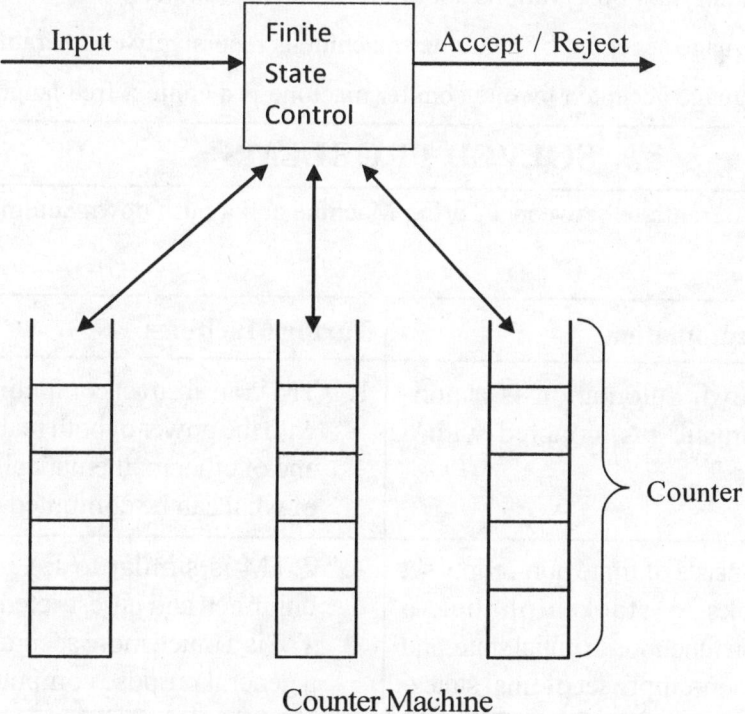

Counter Machine

There are two ways of representing counter machine –

1. The counter machine is similar to a multi-stack Turing Machine, but the only difference between them is that in place of each stack there is a counter. The counters contain non negative integers. Each move of counter machine depends on its state, input symbol. In one move counter machine can :

   i. Changes state

   ii. Add or subtract **1** from any of its counters. The negative counters are not allowed at all.

2. The counter machine is similar to restricted multi-stack machine. These restrictions are –

    i.   There are only two stack symbols: Z_0 and X.

    ii. The Z_0 is the bottom of stack marker. It is initially on each stack.

    iii. Replace Z_0 only by string of the form $X^i Z_0$ where $i \geq 0$.

    iv. Replace X only by X^i for $i \geq 0$. i.e., the Z_0 appears on the bottom of each stack, and all other stack symbols are X.

There are two important observations about the counter machine.

1. Every language accepted by a counter machine is recursively enumerable.

2. Every language accepted by one counter machine is a context free language.

SOLVED PROBLEMS

Problem 1: Differentiate between a Turing Machine and a push down automation?

Solution:

Push down automation	Turing Machines
1. Push down automation is a non-deterministic FA, coupled with a stack.	1. TM is an abstract computing machine with the power of both real computers and of other mathematical definitions of what can be computed.
2. PDA consists of finite non-empty set of stacks, a stack alphabet, a transition function, an initial state, and a finite non-empty set of final states.	2. TM is similar to FA but with an unlimited and unrestricted memory, a TM is a much more accurate model of a general purpose computer.
3. Abstractly a PDA consists of a finite tape, a reading head (which reads from the stack), a stack memory operating in LIFO fashion.	3. A TM consists of a tape (which is finite at the left-end and infinite at right-end), a finite control, and a read/write head.
4. PDA chooses its next state based on its current state, the next input symbol, and the symbol on the top of the stack.	4. TM makes moves based on its current state and the tape symbol scanned by the tape head.

5.	There are two ways in which the PDA is allowed to single acceptance. One is by entering an accepting state, the other by emptying its stack.	5.	TM accepts its input if it enters an accepting state.
6.	The languages accepting by PDA's are called context-free languages.	6.	The languages accepted by TM's are called recursively enumerable languages.
7.	PDA can only read from the tape	7.	TM can both write on the tape and read from it.
8.	The read/write head can only move to the right	8.	The read/write head can move both to the left and the right.

Problem 2: Compare and contrast finite automation, push down automation and Turing Machine?

Solution:

S. no	Finite automaton	Push-down automaton	Turing Machine
1.	Finite automatons are good examples for computers with an extremely limited amount of memory.	Push-down automaton is a non-deterministic FA, coupled with a stack.	TM is an abstract computing machine with the power of both real computers and of other mathematical definitions of what can be computed.
2.	FA consists of finite non-empty set of states, an input alphabet, a transition function, an initial state, and a finite non-empty set of final states.	PDA consists of finite non-empty set of states, a stack alphabet, a transition function, an initial state, and a finite non-empty set of final states.	TM is similar to FA but with an unlimited and unrestricted memory, a TM is a much more accurate model of a general purpose computer.

3.	Abstractly a FA consists of a finite tape, a finite control and read/write head.	Abstractly a PDA consists of a finite tape, a reading head (which reads from the stack), and a stack memory operating on LIFO fashion.	A TM consists of a tape (which is finite at the left end and infinite at the right end), a finite control and a read/write head.
4.	FA chooses its next state based on its current state and the next input symbols.	PDA chooses its next state based on its current state, the next input symbol, and the symbol at the top of the stack.	TM makes moves based on its current state and the tape symbol scanned by the tape head.
5.	FA accepts its input if it enters an accepting state.	There are two ways in which the PDA is allowed to signal acceptance. One is by entering an accepting state, the other by emptying its stack.	TM accepts its input if it enters an accepting state.
6.	The languages accepting by FA's are called regular languages.	The languages accepted by PDA's are called context-free languages.	The languages accepted by TM's are called recursively enumerable languages.
7.	FA can only read from tape.	PDA can only read from the tape.	Turing machine can both read from the tape and write on the tape.
8.	The read/write head can only move to the right.	The read/write head can only move to the right.	The read/write head can move both to the left and to the right.

Problem 3:Construct a Turing machine to recognize the language.
$$\{ ww^R / w \in (a+b)^* \}$$

The transition diagram is as follows

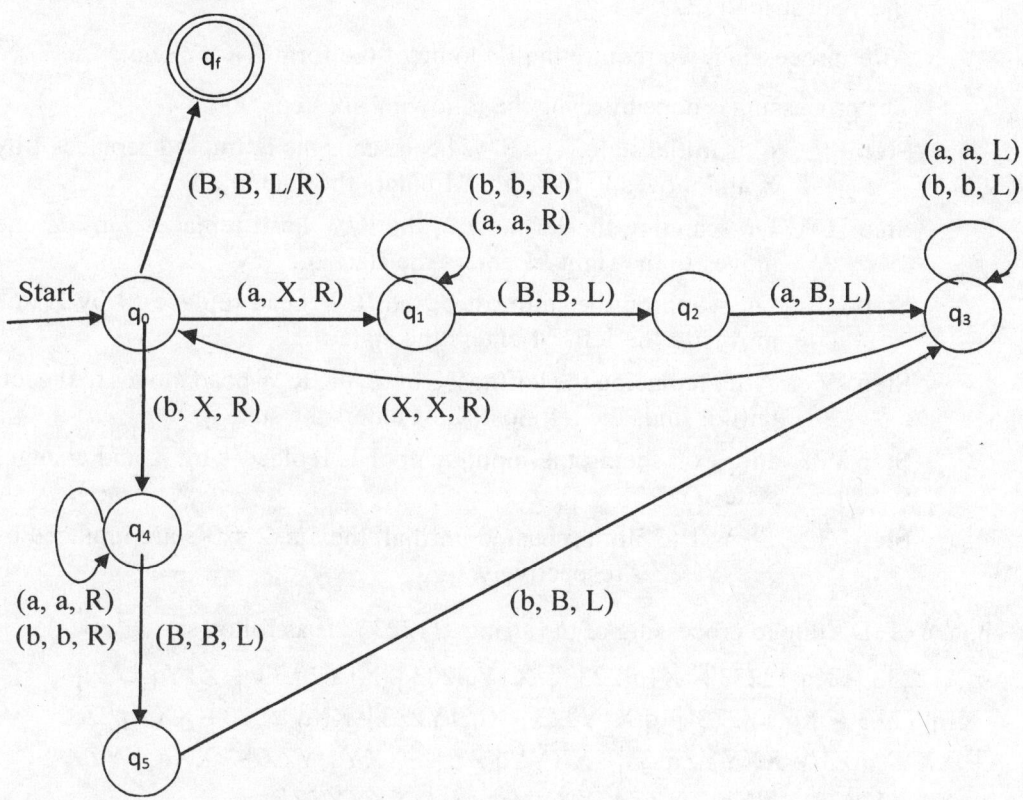

The transition table is as follows

State	A	B	X	B
q_0	(q_1, X, R)	(q_4, X, R)	-	$(q_f, B, L/R)$
q_1	(q_1, a, R)	(q_1, b, R)	-	(q_2, B, L)
q_2	(q_3, B, L)	-	-	-
q_3	(q_3, a, L)	(q_3, b, L)	(q_0, X, R)	-
q_4	(q_4, a, R)	(q_4, b, R)	-	(q_5, B, L)
q_5	-	(q_3, B, L)	-	-
q_f	-	-	-	-

Problem 4: Design a Turing Machine to recognize the language.
$$L = \{ 1^n 2^n 3^n / n \geq 1 \}$$

Solution: Before designing a Turing Machine M, let us evolve a procedure for processing the input string 112233.

After processing, we require the ID to be of the form $XXYYZZq_5$.

The processing is done by using the following six steps.

Step I : q_0 is initial state. The R/W head scans the leftmost **1** replaces **1** by **X** and moves to the right **M** enters the state q_1.

Step II :   On scanning the leftmost **2**, the R/W head replaces **2** by **Y** and moves to the right **M** enters the state q_2.

Step III :   On scanning the leftmost **3**, the R/W head replaces **3** by **Z** and moves to the left. **M** enters the state q_3.

Step IV :   After replacing the leftmost **3** by **Z**, the R/W head moves to the left until it finds the leftmost **X**, **M** enters the state q_4.

Step V :   In q_4, on seeing the input symbol **1**, replace **1** by **X** and changes state q_1.

Step VI :   Steps 1 to 5 are repeated until all the 1's, 2's, 3's are replaced by X's, Y's, Z's respectively.

The change of ID's due to processing of the string "112233" is as follows.

$q_0 112233 \vdash Xq_1 12233 \vdash X1q_1 2233 \vdash X1Yq_2 233 \vdash X1Y2q_2 33 \vdash X1Yq_3 2Z3 \vdash$
$X1q_3 Y2Z3 \vdash Xq_3 1Y2Z3 \vdash q_3 X1Y2Z3 \vdash Xq_4 1Y2Z3 \vdash XXq_1 Y2Z3 \vdash XXYq_1 2Z3$
$\vdash XXYYq_2 Z3 \vdash XXYYZq_2 3 \vdash XXYYq_3 ZZ \vdash XXYq_3 YZZ \vdash XXq_3 YYZZ$
$\vdash Xq_3 XYYZZ \vdash XXq_4 YYZZ \vdash XXYYq_4 ZZ \vdash XXYYZZB$

Steps to remember

Present State	Input Symbol	Replacement	Next State	Move
q_0	1	X	q_1	R
q_1	2	Y	q_2	R
q_2	3	Z	q_3	L
q_3	X	X	q_4	R
q_4	1	X	q_1	R
q_4	B	B	q_5	R

The transition diagram is as follows

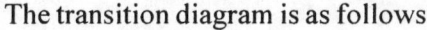

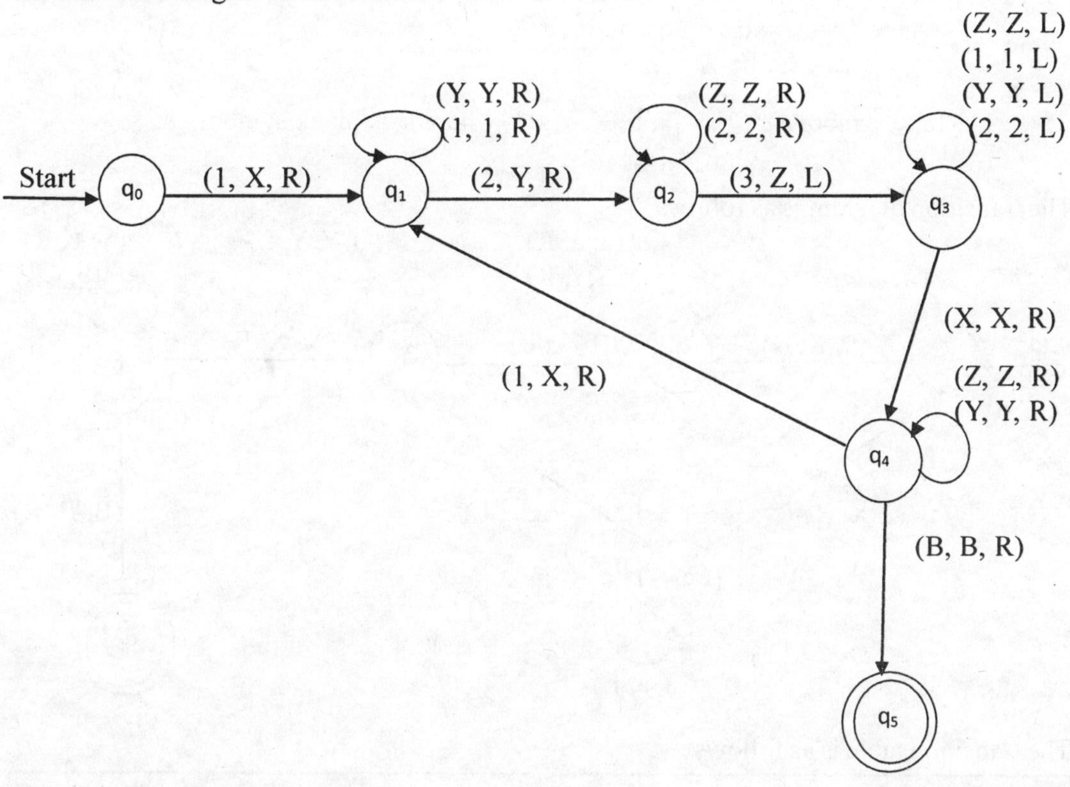

Start → q_0 — (1, X, R) → q_1 — (2, Y, R) → q_2 — (3, Z, L) → q_3

q_1 self-loop: (Y, Y, R), (1, 1, R)

q_2 self-loop: (Z, Z, R), (2, 2, R)

q_3 self-loop: (Z, Z, L), (1, 1, L), (Y, Y, L), (2, 2, L)

q_3 — (X, X, R) → q_4

q_4 self-loop: (Z, Z, R), (Y, Y, R)

q_4 — (1, X, R) → q_1

q_4 — (B, B, R) → q_5

The transition table is as follows

State	1	2	3	X	Y	Z	B
q_0	(q_1, x, R)	–	–	–	–	–	–
q_1	$(q_1, 1, R)$	(q_2, Y, R)	–	–	(q_1, Y, R)	–	–
q_2	–	$(q_2, 2, R)$	(q_3, Z, L)	–	–	(q_2, Z, R)	–
q_3	$(q_3, 1, L)$	$(q_3, 2, L)$	–	(q_4, X, R)	(q_3, Y, L)	(q_3, Z, L)	–
q_4	(q_1, X, R)	–	–	–	(q_4, Y, R)	(q_4, Z, R)	(q_5, B, R)
q_5	-	-	-	-	-	-	-

Problem 5: Design a Turing machine to recognize the language and give its state transition table and diagram. $L = \{ a^n b^m : n \geq 1 \text{ and } n \leq m \}$

Solution : Let the TM **M** be M = (Q, Σ, Γ, δ, q_0, B, F)

where Q = {q_0, q_1, q_2, q_3, q_4, q_5}

Σ = {a, b}

Tape symbols are Γ = {a, b, X, Y, B}, where B is blank symbol.

F = {q_5}.

The transition diagram is as follows

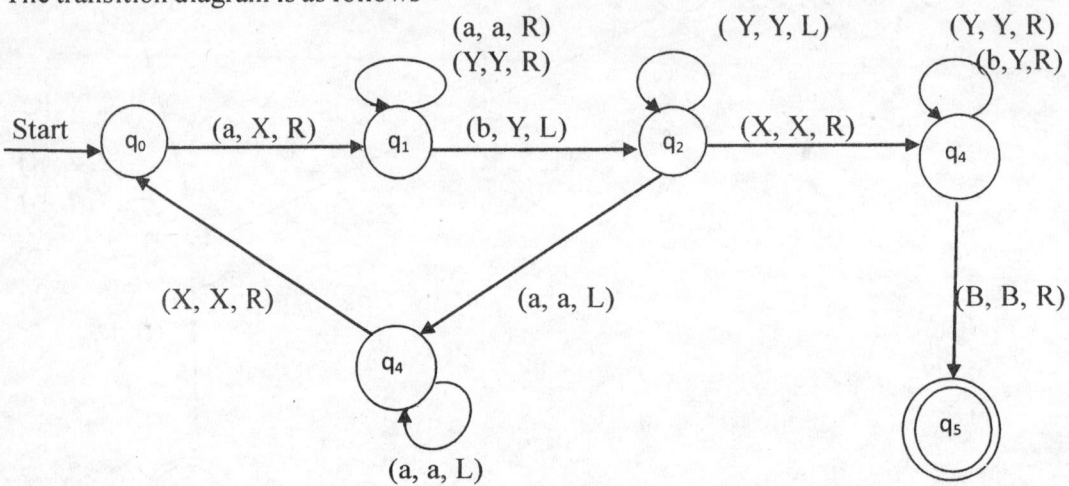

The transition table is as follows

Present state	a	b	X	Y	B
q_0	(q_1, X, R)	-	-	-	-
q_1	(q_1, a, R)	(q_2, Y, L)	-	(q_1,Y, R)	-
q_2	(q_3, a, L)	-	(q_4, X, R)	(q_2,Y, L)	-
q_3	(q_3, a, L)	-	(q_0, X, R)	-	-
q_4	-	(q_4, Y, R)	-	(q_4, Y, R)	(q_5, B, R)
q_5	-	-	-	-	-

Problem 6: Construct a Turing machine for the following

That shifts the input string, often the alphabet (a, b) by one position right by inserting # as the first character.

Solution: aaaaB   bbbbB   aaabB   bbbaB
              # aaaa   # bbbb   # aaab   # bbba

Transition diagram is as follows:

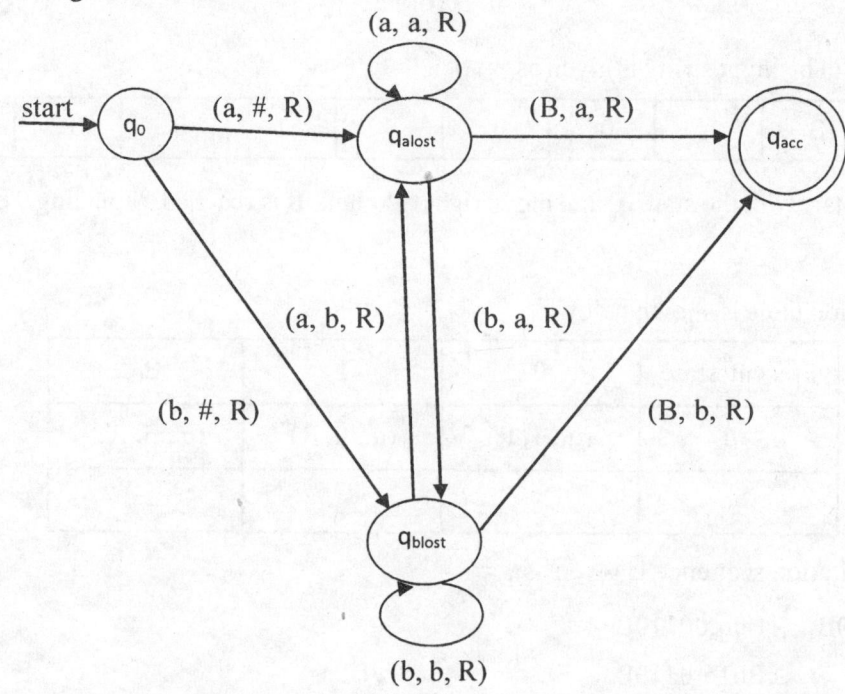

Transition table is as follows

	a	b	#	B
q_0	$(q_{alost}, \#, R)$	$(q_{blost}, \#, R)$	-	-
q_{alost}	(q_{alost}, a, R)	(q_{blost}, a, R)	-	(q_{acc}, a, R)
q_{blost}	(q_{alost}, b, R)	(q_{blost}, b, R)	-	(q_{acc}, b, R)
q_{acc}	-	-	-	-

Problem 7: Design Turing Machines for the following

     i. To compliment a given binary number

     ii. To compute $f(x, y) = x + y$ for x and y positive integers represented in unary

Solution :

i.   To compliment a given binary number, we simply replace 1 by 0 and 0 by 1. For
      example the complement of 100110 is

                        011001

      The input string is given as

....	B	1	0	0	1	1	0	B	....

We start with the state q_0 and move right till blank **B** is reached, replacing **0** by **1** and
1 by **0**.

The transition table is shown below

Present state	0	1	B
$\to q_0$	$(q_0, 1, R)$	$(q_0, 0, R)$	(q_f, B, R)
q_f	-	-	-

The computation sequence is given as

$B\,q_0 100110B$ $\vdash B0q_0 00110B$

$\vdash B01q_0 0110B$

$\vdash B011q_0 110B$

$\vdash B0110q_0 10B$

$\vdash B01100q_0 0B$

$\vdash B011001q_0 B$

$\vdash B011001q_1 B$

$\therefore B\,q_0 100110B \vdash B011001q_1 B$

ii.   Suppose the two unary numbers are 11 and 11.
       Now, to find the addition of these two numbers, the input tape is given as

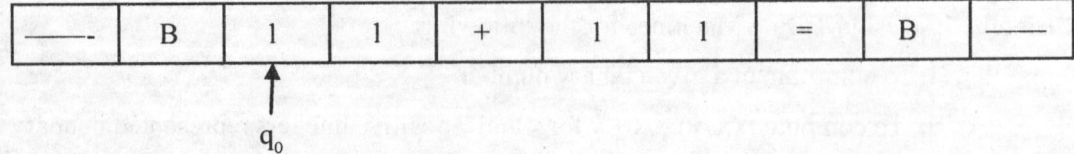

Now, to design the TM, consider the initial state as q_0. In state q_0, replace 1 by x and move towards right till blank, there we will print 1 then come back for second one.

When we get + sign, we jump to new state q_2 and move right.

When we get = sign, we jump to new state q_3 and move right.

In state q_3, we replace blank and print here 1 and move left in a new state say q_4.

The transition table is as follows

State	1	+	=	x	y	B
$\rightarrow q_0$	(q_1, x, R)	$(q_5, +, R)$	–	–	–	–
q_1	$(q_1, 1, R)$	$(q_2, +, R)$	–	–	–	–
q_2	(q_2, x, R)	–	$(q_3, =, R)$	–	–	–
q_3	$(q_3, 1, R)$	–	–	–	–	$(q_4, 1, L)$
q_4	$(q_4, 1, L)$	$(q_4, +, L)$	$(q_4, =, L)$	(q_0, x, R)	–	–
q_5	(q_6, y, R)	–	$(q_9, =, R)$	–	–	–
q6	$(q_6, 1, R)$	–	$(q_7, =, R)$	–	–	–
q_7	$(q_7, 1, R)$	–	–	–	–	$(q_8, 1, L)$
q_8	$(q_8, 1, L)$	–	$(q_8, =, L)$	–	(q_5, y, R)	–
q_9	$(q_9, 1, R)$	–	–	–	–	–

The computation sequence is given as

$Bq_0 11+11=B$ $\vdash Bxq_0 1 +11 = B$

$\vdash Bxq_1 1 +11 = B$

$\vdash Bx1q_1 +11 = B$

$\vdash Bx1+q_2 11 = B$

$\vdash Bx1 +11 = q_3 B$

$\vdash Bx1 +11 = q_4 1B$

$\vdash Bx1 +11q_4 = 1B$

$\vdash Bx1 +1q_4 1 =1 B$

$\vdash Bx1+q_4 11 =1B$

$\vdash Bx1q_4 +11 =1B$

$$|- Bxq_41 +11 = 1B$$

$$|- Bq_4x1 +11 = 1B$$

$$|- Bxq_01 +11 =1 B$$

$$|- Bxxq_1 +11 =1B$$

$$|- Bxx+q_211 = 1B$$

$$|- Bxx+1q_21 =1B$$

$$|- Bxx+11q_2 =1B$$

$$|- Bxx +11 =1q_3B$$

$$|- Bxx +11 = 1q_41B$$

$$|- Bxx +11 = q_411B$$

$$|- Bxx +11q_4 =11B$$

$$|- Bxx +1q_41 = 11B$$

$$|- Bxx +q_411 = 11B$$

$$|- Bxxq_4+11 = 11B$$

$$|- Bxq_4x+11 = 11B$$

$$|- Bxxq_0 +11 = 11B$$

$$|- Bxx +q_511 =11B$$

$$|- Bxx +yq_01 = 11B$$

$$|- Bxx +y1q_6 = 11B$$

$$|- Bxx +y1 =q_711B$$

$$|- Bxx +y1 = 1q_71B$$

$$|- Bxx +y1 = 11q_7B$$

$$|- Bxx +y1 =11q_81$$

$$|- Bxx +y1 = 1q_811B$$

$$|- Bxx +y1 =q_8111B$$

$$|- Bxx +y1q_8 =111B$$

$$|- Bxx +yq_51 =111B$$

$$|- Bxx +yyq_6 = 111B$$

$$|- Bxx +yy = q_7111B$$

$$|- Bxx +yy = 1q_711B$$

$$|- Bxx +yy = 11q_71B$$

$$|- Bxx +yy = 111q_7B$$

$$|- Bxx +yy = 111q_81$$

$$\vdash Bxx +yy = 111q_8 1B$$
$$\vdash Bxx +yy = q_8 1111B$$
$$\vdash Bxx +yyq_8 = 1111B$$
$$\vdash Bxx +yq_8 y = 1111B$$
$$\vdash Bxx +yyq_5 = 1111B$$
$$\vdash Bxx +yy = q_9 1111B$$
$$\vdash Bxx +yy = 1q_9 111B$$
$$\vdash Bxx +yy = 11q_9 11B$$
$$\vdash Bxx +yy = 111q_9 1B$$
$$\vdash Bxx +yy = 1111q_9 B$$

$$\therefore Bq_0 11+11=B \vdash Bxxyy=1111q_9 B$$

REVIEW QUESTIONS

1. Explain the ID and move of a Turing Machine indicates the major differences between Turing Machine and Pushdown Automaton. Design a Turing Machine to accept the language $L = \{o^n \ 1^m \ 2^{m+n} / n \geq 0, m \geq 1\}$.

2. Design Turing Machine to recognize the following.
   $L = \{w \ c \ w^R / w \in (a+b)^* \text{ and 'c' is a or b}\}$

3. Construct Turing Machine to accept following language and give state transition table and diagram. Check the machine by taking a suitable instance.
   $l = \{a^n b^n a^n b^n / n \geq 1\}$

4. Giving the basic steps involved in designing a Turing Machine, design a TM which accepts the strings derived from $\{0, 1\}$ and have even number of ones.

5. Design a Turing Machine to recognize the language $L = \{a^n b^{2n}/n \geq 1\}$.

6. Design a Turing Machine to recognize the language $L = \{a^n b^{n+1}/n \geq 1\}$.

7. Design a Turing Machine to find square of an integer.

8. Design a Turing Machine for the following.

   a. To compliment a given binary number.

   b. To compute $f(x, y) = x+y$ for x and y positive integers represented in unary.

9. Design a Turing Machine to compute the function – factorial n i.e. n! or $\angle n$.

Chapter 8

Computability Theory

8.1 INTRODUCTION

Any mathematical problem is computable and it can be solved by a computing device, since an algorithm exists for every computable problem. Such problems are termed to be "solvable" or "decidable" problems. Turing machine is the one which performs computations using some algorithms. The classes of problems which cannot be computable are called as "unsolvable" or "undecidable" problems.

Chomsky's hierarchy of languages discusses about different class of languages which are recognized by different machines. This chapter also discusses about Linear Bounded Automata (LBA) and Context Sensitive Languages (CSL). The significance of LR(0) grammar is also discussed. The decidability and undecidability of a problem is discussed. A general purpose machine called Universal Turing Machine (UTM) is introduced. Finally, the classes of P and NP problems are discussed.

8.2 CHOMSKY'S HIERARCHY OF LANGUAGES

The four classes of languages that we have studied are **regular, context-free, and context-sensitive** and **recursively enumerable** and are often referred to as **Chomsky hierarchy.**

Chomsky classified the grammar into four types in terms of productions as type-3, type-2, type-1 and type-0.

Each level of hierarchy can be characterized by a class of grammars and by a certain type of abstract machine or model of computation. The following picture shows the hierarchy of languages and relationship with the machines.

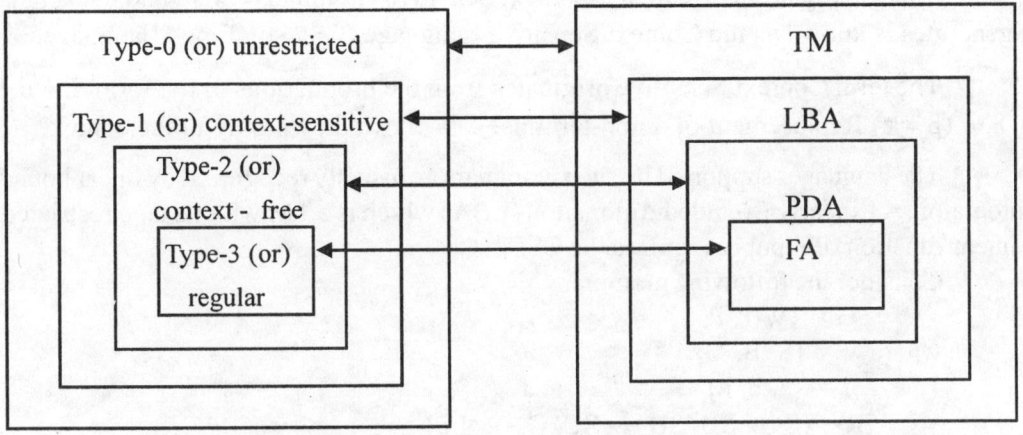

Type-0 (or) Unrestricted Grammars:

There are no restrictions on the production rules of this class. This class generates the largest family of grammars permitting productions of the form.

$$\alpha \rightarrow \beta$$

where α and β are arbitrary strings of symbols with $\alpha \neq \varepsilon$.

Such grammars are known as unrestricted grammars. Every type-0 grammar generates a recursively enumerable set. It is called recursively enumerable language.

These grammars include all the formal grammars. Such grammars remain capable of generating languages which can be recognized by Turing machines.

A Turing machine is constructed to recognize the sentences generated by this grammar Consider the following grammar

$$G = \{V, T, P, S\}$$
$$\text{where} \quad V = \{A, B, C\}$$
$$T = \{a, b, c\}$$

P is the set of productions i.e.,

$P = \{ A \rightarrow AB, AB \rightarrow BC, B \rightarrow a \}$

S is the start Symbol.

This grammar type-0 or unrestricted grammar

The set of productions 'P' consists of

$A \rightarrow AB, AB \rightarrow BC, B \rightarrow a$

Type-1 (or) Context-Sensitive Grammar

In this class, the restrictions on each production rule $\alpha \rightarrow \beta$ is that the length of the consequent β is atleast as much as the antecedent α, except for $S \rightarrow \varepsilon$ ($|\alpha| \leq |\beta|$)

Also the start symbol 'S' does not appear on the right hand side of any production.

This grammar is also called a Context Sensitive Grammar (CSG), and the language it generates is known as the Context Sensitive Language (CSL) or Type-1 language.

The term Context Sensitive originates from the productions of the form $\alpha_1 A \alpha_2 \rightarrow \alpha_1 \beta \alpha_2$ ($\beta \neq \varepsilon$) Replacement of a non-terminal A is allowed by β only in the context.

The languages supported by such grammars are usually recognized by linear bounded automaton. A Linearly Bounded Automaton (LBA) which is a TM with its tape restricted as a linear function of input recognizes the CSL.

Consider the following grammar

$$G = \{V, T, P, S\}$$

where $V = \{S, B, C\}$

$T = \{a, b, c\}$

$P = \{S \rightarrow SB, SB \rightarrow SC, C \rightarrow abS\}$

'S' is start symbol

This grammar is a type-1 or Context-Sensitive Grammar.

Type-2 (or) Context-Free Grammar (CFG)

In this type, the left hand side of each production rule is non-terminal symbol i.e. productions are of the form $A \rightarrow \alpha$, where A is a non-terminal and $\alpha \in (V \cup T)^*$.

The productions of the form $A \rightarrow \varepsilon$ (besides A = S) are also permitted. In addition, S is allowed to appear on right-hand side of a production.

This grammar is referred to as Context-Free Grammar (CFG). The grammars of most PL's approximate to this grammar.

Example: ALGOL

NPDM generates and accepts the context- free languages
consider the following grammar

$$G = \{V, T, P, S\}$$

where $V = \{S, A\}$

$T = \{a, b, c\}$

$P = \{ S \rightarrow A, A \rightarrow abS\}$.

S is start symbol.

The grammar G is a Type-2 or CFG. The languages generated by such grammar are usually recognized by non-deterministic Push down automaton.

Type-3 (or) Regular Grammars

For this type, in each production rule, the left hand side is a non-terminal symbol and the right hand side contains atmost one non-terminal symbol which is the right-most (or) left most symbol.

Here the rule $R \rightarrow \varepsilon$ holds well, provided 'R' does not appear on the right hand side of the given rule. This family of formal languages can be obtained through regular expressions.

This type of grammar is called regular grammar. This grammar is too primitive for PL's. The regular language can be generated and recognized by Finite State Machine (FSM). Consider the following grammar

$$G = (V, T, P, S)$$
$$V = \{S, A, B\}$$
$$T = \{a, b, c\}$$
$$P = \{S \rightarrow aA, S \rightarrow B, A \rightarrow aS\}$$

S is the start symbol
This grammar is a Type-3 or regular grammar

Grammar	Language	Automaton	Production rules	Example
Type-0	Recursively enumerable	Turing machine	No restrictions	n!
Type-1	Context-Sensitive	Linear Bounded Automata	$\alpha\, A\, \beta \rightarrow \alpha\, \gamma\, \beta$ such that $\gamma \neq \varepsilon$	$a^n b^n c^n$
Type-2	Context-Free	Non-deterministic Push Down automata	$A \rightarrow \gamma$	$a^n b^n$
Type-3	Regular language →Right linear → Left linear	Finite state	$A \rightarrow wB, A \rightarrow w$ (or) $A \rightarrow Bw, A \rightarrow w$	a*b*

8.3 LINEAR BOUNDED AUTOMATA AND CONTEXT SENSITIVE LANGUAGE

A Linear Bounded Automaton (LBA) is a non-deterministic Turing machine satisfying the following two conditions:

1) Its input alphabet includes two special symbols ¢ and \$, the left and right end markes respectively.

2) The LBA has no moves left from ¢ or right from \$, nor it prints another symbol over ¢ or \$.

The linear bounded automaton is simply a Turing machine which, instead of having potentially infinite tape on which to compute is restricted to the portion of the tape containing the input 'x' plus two tape squares holding the end markers (¢ and \$).

We shall see that restricting the Turing machine to an amount of tape that, on each input is bounded by some linear function of the length of the input would result in the identical computational ability as restricting the Turing machine to the portion of the tape containing the input – hence the name "Linear Bounded Automaton".

An LBA will be denoted by M = (Q, Σ, Γ, δ, q_0, $\not{c}$, \$, F) where Q, Σ, Γ, δ, q_0, F are as for a non-deterministic TM, $\not{c}$ and \$ are symbols in Σ, the left and right end markers. L(M), the language accepted by M is {w | w is in (Σ - {$\not{c}$, \$})* and $q_0 \not{c}$ w \$ $\vdash^{\text{--}M}$ α q β for some q in F}

Note: The end markers are on the input tape initially but are not considered part of the word to be accepted or rejected.

Since an LBA cannot move off the input, there is no need to suppose that there is a blank tape to the right of \$.

The Model of Linear Bounded Automaton

This model is important because:
  a)   The set of Context-Sensitive Languages is accepted by the model.
  b)   The infinite storage is restricted in size but not in accessibility to the storage in comparison with the Turing machine model

It is called linear bounded automaton (LBA) because a linear function is used to restrict (to bound) the length of the tape.

Here we define the linear bounded automaton and develop the relation between linear bounded automata and Context Sensitive Languages.

It should be noted that the study of Context Sensitive Languages is important from practical point of view because many compiler languages lie between Context-Sensitive and Context-Free Languages.

A linear bounded automaton is a non-deterministic Turing machine which has a single tape whose length is not infinite but bounded by a linear function ℓ of the length of the input string.

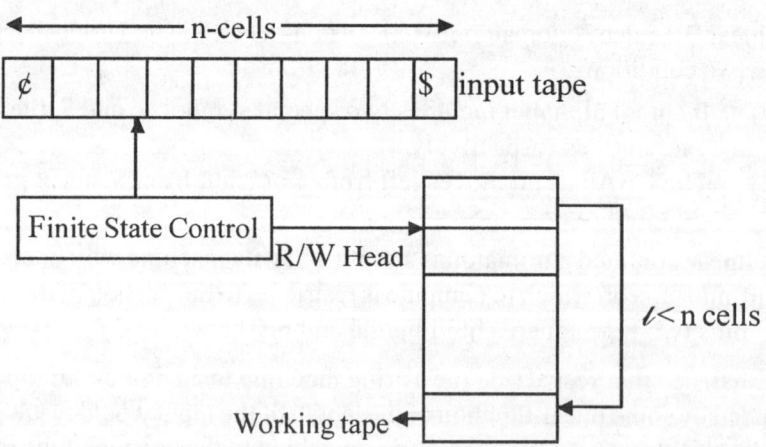

Language Acceptance by Linear Bounded Automaton

The set of strings accepted by non-deterministic LBA is the set of strings generated by the context sensitive grammars, excluding null strings. If L is a context sensitive language then it is accepted by LBA. The converse is also true.

Context-Sensitive Grammar

A grammar $G = (V, T, P, S)$ is said to be Context-Sensitive if all productions are of the form

$$\alpha \to \beta, \text{ where } \alpha, \beta \in (V \cup T)^+ \text{ and } |\alpha| \le |\beta|.$$

Context Sensitive Grammars are associated with a language family with the same name is defined as "A language L is said to be context-sensitive if there exists a context sensitive grammar G", such that

$$L = L(G) \text{ or } L = L(G) \cup \{\varepsilon\}$$

A context sensitive grammar can never generate a language containing the empty string i.e. $A \to \varepsilon$ is not allowed.

Every context free language without ε can be generated by special case of context sensitive grammar, say by one in Chomsky or Greibach Normal form, both of which satisfy the conditions of context sensitive grammar.

The context-free language is a subset of the family of context-sensitive languages, if empty string is included in the definition of context-sensitive language (but not in grammar). A context sensitive grammar is one whose productions are all of the form

$$\alpha A \beta \to \alpha B \beta, \text{ where } A \in V \text{ and } \alpha, \beta, B \in (V \cup T)^* \text{ and } B \ne \varepsilon.$$

The context sensitive grammars are more general than context-free grammars (type-2) and less than unrestricted grammar (type-0) i.e. we apply some restrictions on unrestricted grammar (type-0) and the resultant grammar is known as context-sensitive grammar (type-1).

Similarly, we apply some restrictions on context sensitive grammars and the resultant grammar is known as context-free grammar (type-2).

The language generated by a context sensitive grammar is called as context sensitive language (CSL).

Example 8.1

Show that $L = \{a^n b^n c^n \mid n \ge 1\}$ is a context sensitive language. The context sensitive grammar for the language is shown below.

$$S \to abc \mid aAbc$$
$$Ab \to bA$$
$$Ac \to Bbcc$$
$$bB \to Bb$$
$$aB \to aa \mid aaA$$

Solution:

Now consider the string for n=3, $a^3 b^3 c^3 \in L$ the derivation of $a^3 b^3 c^3$ is as follows

$$S \quad aAbc$$

$\Rightarrow abAc$	(since Ab → bA)
$\Rightarrow abBbcc$	(since Ac → Bbcc)
$\Rightarrow aBbbcc$	(since bB → Bb)
$\Rightarrow aaAbbcc$	(since aB → aaA)
$\Rightarrow aabAbcc$	(since Ab → bA)
$\Rightarrow aabbAcc$	(since Ab → bA)
$\Rightarrow aabbBbccc$	(since Ac → Bbcc)
$\Rightarrow aabBbbccc$	(since bB → Bb)
$\Rightarrow aaBbbbccc$	(since aB → aa)
$\Rightarrow aaabbbccc$	

$$\therefore S \overset{*}{\Rightarrow} aaabbbccc$$
$$S \overset{*}{\Rightarrow} a^3 b^3 c^3$$

L is a Context-Sensitive language.

∴ The grammar is in the form of x → y where x, y ∈ $(V \cup T)^*$ and $|x| \le |y|$, which is a context sensitive grammar.

∴ The language L = { $a^n b^n c^n | n \ge 1$ } generated by the grammar is called context sensitive language.

8.4 LR(0) GRAMMAR

Programming languages are context free languages and parsing (or syntax analysis) is an essential part of any compiler. Various parsing algorithms for CFG are developed to parse the programming constructs.

There are **two** types of parsing techniques: **top-down parsing** and **bottom-up parsing**. In top down parsing, we start from the start symbol S of CFG and reach to the desired string w. On the other hand in bottom up parsing we start with a string and reduce it to start symbol. The **bottom-up** parse finds rightmost derivation of string.

An LR(K) grammar is a subclass of context-free grammar which allows a parser to perform left to right scan of input to produce a rightmost derivation using K symbols of lookahead to select next rule. In LR(K) grammar 'L' stands for left to right scan and 'R' stands for rightmost derivation and 'K' is number of symbols from lookahead.

Concept of Handle and Viable Prefix

Before discussing the definition of LR(K) grammar we will discuss two important concepts i.e. handle and viable prefix.

Let, the rightmost derivation in zero or more number of steps can be denoted as $\overset{*}{\underset{rm}{\Rightarrow}}$

Consider a CFG G for which the rightmost derivation of some string can be given as:

$$S \overset{*}{\underset{rm}{\Rightarrow}} \alpha A w \overset{*}{\underset{rm}{\Rightarrow}} \alpha \beta w$$

Here β is called **handle** of **right sentential form** $\alpha\beta w$. Each prefix of $\alpha\beta$ is called viable prefix of G.

Note: Hence viable prefix is a prefix in right sentential form that does not continue past the right end of right most handle obtained during derivation.

Defnition LR(0) grammar

Example 8.2
Is the following grammar LR(0)

$$S^l \rightarrow S$$
$$S \rightarrow a\,S\,a \mid b\,S\,b \mid c$$

Solution:
Given grammar is

$$S^l \rightarrow S$$
$$S \rightarrow a\,S\,a$$
$$S \rightarrow b\,S\,b$$
$$S \rightarrow c$$

The conical collection of sets of LR(0) items for the above grammar is shown below

I_0: $S^l \rightarrow .\,S$
$\quad S \rightarrow .\,a\,S\,a$
$\quad S \rightarrow .\,b\,S\,b$
$\quad S \rightarrow .\,c$

I_1: (I_0, S)
$\quad S^l \rightarrow S.$ — completely reduced

I_2: (I_0, a)
$\quad S \rightarrow a.\,S\,a$
$\quad S \rightarrow .\,a\,S\,a$
$\quad S \rightarrow .\,b\,S\,b$
$\quad S \rightarrow .\,c$

I_3: (I_0, b)
$\quad S \rightarrow b.\,S\,b$
$\quad S \rightarrow .\,a\,S\,a$
$\quad S \rightarrow .\,b\,S\,b$
$\quad S \rightarrow .\,c$

I_4: (I_0, c)
$\quad S \rightarrow c.$ – completely reduced

I_5: (I_2, S)
       $S \rightarrow a\ S.\ a$
       $(I_2, a) - I_2$
       $(I_2, b) - I_3$
       $(I_2, c) - I_4$
I_6: (I_3, S)
       $S \rightarrow b\ S.\ b$
       $(I_3, a) - I_2$
       $(I_3, b) - I_3$
       $(I_3, c) - I_4$
I_7: (I_5, a)
       $S \rightarrow a\ S\ a$ (completely reduced)
I_8: (I_6, b) (completely reduced)
       $S \rightarrow b\ S\ b.$

$\therefore$ The collection of items = $\{I_0, I_1, I_2, I_3, I_4, I_5, I_6, I_7, I_8\}$

LR(0) items and viable prefixes can be generated. Therefore, it is LR(0) grammar.

The DFA corresponding to the states i.e. items is as follows.

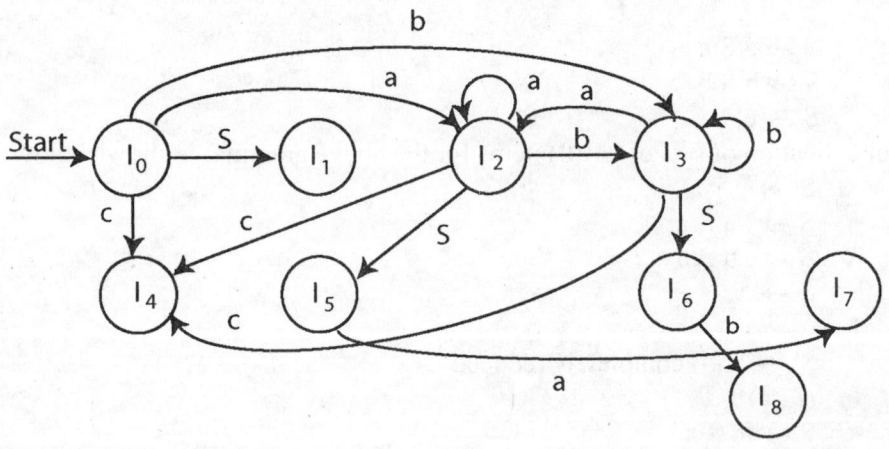

Example 8.3

Acceptance of the input string using parsing.

Derivation of input string from the grammar

$S^1 \Rightarrow S$
   $\Rightarrow a\ S\ a$ (since $S \rightarrow a\ S\ a$)
   $\Rightarrow a\ b\ S\ b\ a$ (since $S \rightarrow b\ S\ b$)
   $\Rightarrow a\ b\ a\ S\ a\ b\ a$ (since $S \rightarrow a\ S\ a$)
   $\Rightarrow a\ b\ a\ c\ a\ b\ a$ (since $S \rightarrow c$)
   $\therefore S^1 \Rightarrow a\ b\ a\ c\ a\ b\ a$

Stack	Input string	Remarks
0	a b a c a b a	-
0 a 2	b a c a b a	Shift
0 a 2 b 3	a c a b a	Shift
0 a 2 b 3 a 2	c a b a	Shift
0 a 2 b 3 a 2 c 4	a b a	Shift
0 a 2 b 3 a 2 S	a b a	Reduce by S → c
0 a 2 b 3 a 2 S 5	a b a	In state 2, on seeing 'S' changes to state 5 shift
0 a 2 b 3 a 2 S 5 a 7	b a	7 on b there is no transition
0 a 2 b 3 a 2 S 5 a 7	b a	Reduce by S → a S a
0 a 2 b 3 S	b a	3 on S ⇒ 6
0 a 2 b 3 S 6	b a	Shift
0 a 2 b 3 S 6 b 8	a	Reduce by S → b S b
0 a 2 S	a	2 on S ⇒ 5
0 a 2 S 5	a #	Shift
0 a 2 S 5 a 7	#	Reduce by S → a S a
0 S 1	#	Reduce by S¹ → S
0 S¹	#	

0 S¹ with # means, the string is accepted by the parser.

Note: If transition is there, uses shift the input symbol and state on to the stack.
If transition is not there, uses reduce by some production.

8.5 DECIDABILITY AND UNDECIDABILITY OF PROBLEMS

A problem is said to be decidable if there exists some algorithm which solves the problem. If no such algorithm exists then the problem is said to be undecidable. Generally, if the problem is recognized by a Turing machine then it is called decidable problem otherwise undecidable problem.x

Note: If a language is recursive then it is called decidable language, otherwise it is called undecidable.

Reduction:

In general, if we have an algorithm to convert instances of a problem p_1 to instances of a problem p_2, that have the same answer, then it is said that p_1 reduces to p_2. This proof is used to show that p_2 is atleast as hard as p_1. Thus if p_1 is not recursive, then p_2 cannot be recursive. If p_2 is non-recursive, then p_1 cannot be recursive. Care to be taken to reduce a known hard problem to prove to be atleast as hard, never the opposite.

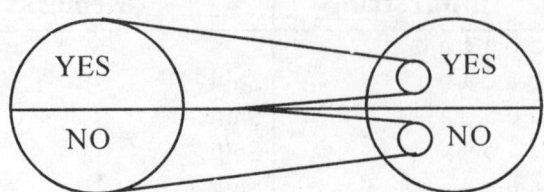

In the above Figure Reductions turn positive instances into positive, and negative to negative

As suggested in the above figure a reduction must turn any instance of 'p_1' that has a 'Yes' answer into an instance of 'p_2' with a 'Yes' answer, and every instance of 'p_1' with a 'No' answer must be turned into an instance of 'p_2' with a 'No' answer.

Note that it is not essential that every instance of p_2 be the target of one or more instances of p_1 and infact it is quite common that only a small fraction of p_2 is a target of the reduction.

Formally a reduction from p_1 to p_2 is a TM, which takes an instance of p_1 written on its tape and halt with an instance of p_2 on its tape.

Generally we describe reductions as if they were computer programs that take an instance of p_1 as input and produce an instance of p_2 as output.
If there is a reduction from p_1 to p_2 then
   i)   If p_1 is undecidable then so is p_2
   ii)   If p_1 is non-recursive then so is p_2

Undecidable problems:

"Decidable" is a synonym for "recursive" i.e. tend to refer languages as recursive and problems as decidable. If a language is not recursive, then the problem expressed by that language is called "undecidable".

All non-trivial problems about the language accepted by a TM are undecidable. Problems which cannot be solved using any algorithm approach are referred as undecidable or problems for which no algorithm exists are undecidable.

A class of problems with two outputs (Yes/No) is said to be decidable there exist some definite algorithm which always terminates (halts) with two outputs (Yes/No), otherwise the class of problems is said to be undecidable i.e. a given problem is undecidable if
   i)   Its corresponding language is non-recursive and
   ii)   It has no solution or answer or algorithm.

Example-1: It is easy to write a program that will search through all positive integers x, y, z and number $n > 2$ for a solution to the equation $x^n + y^n = z^n$. The problem is to determine the program will halt or not. The program will halt iff when a solution is found.

Example-2: The halting problem of Turing Machine is undecidable.

8.6 UNIVERSAL TURING MACHINE (UTM)

A universal TM, a UTM, is a TM that can be fed as input a string composed of two parts: the first is the encoded program of any TM, T followed by a marker, the second part is a string that will be called data. The operation of the UTM is that, no matter what machine T is, and no matter what the data string is, the UTM will operate on the data as if it were T. If T would have crashed on this input, it will crash; If T could loop forever, it will loop forever and if T would accept the input, the UTM does so too. Not only that but the UTM will leave on its tape the encoded T, the marker, and the contents of what T would leave on its tape when it accepts this very input string. It is a programmable TM instead of having to build a different electronic device for each algorithm.

It will be carrier to describe the working of a UTM employing a different encoding algorithm one that is slightly less universal as it makes restrictions on the number of states the TM to be simulated can have and on the size of that TM's tape character set.

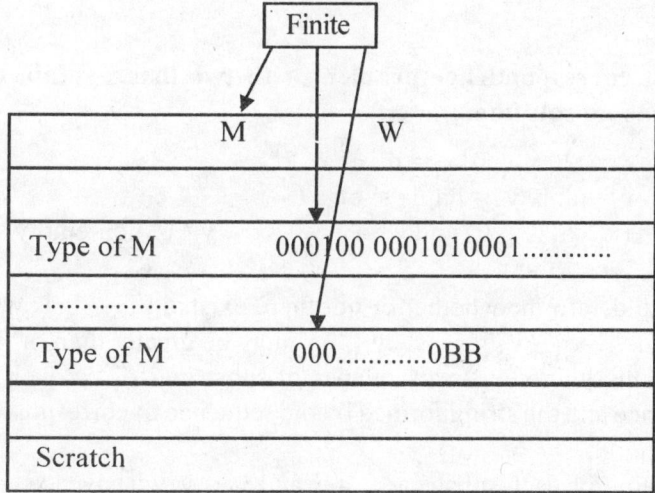

The universal Turing Machine (UTM) should have the capability of limiting any TM. T had given the following information in its tape.

i) The description of T in terms of it's FM cooperation or program area of the tape 1
ii) The initial configuration of T i.e., starting or current state and the symbol scanned. (state area of the tape)
iii) The processing data to be fed to T (data area of tape).

8.7 UNDECIDABILITY OF POST'S CORRESPONDENCE PROBLEM

Post's correspondence problem:

An instance of Post's correspondence problem (PCP) consists of two lists of strings over some alphabet Σ, the two lists must be of equal length, generally refer them as A and B lists, and write list A $= w_1, w_2 .... w_n$ for some integer n and write list B $= x_1, x_2 .... x_k$ for some integer k. For each i, the pair (w_i, x_i) is said to be a correspondence pair.

This instance of PCP has a solution, if there is a sequence of one or more integers $i_1, i_2, ... i_m$ that, when interpreted as indexes for strings in the A and B lists, yield the same string i.e.,

$$w_{i1}, w_{i2}, ... w_{im} = x_{i1}, x_{i2}, ... x_{im}$$

The sequence $i_1, i_2, ...i_m$ is a solution to this instance of a PCP, so PCP is decidable.

Example 8.4

Prove that the post correspondence problem with two lists x = (ab, b, b) and y = (abb, ba, bb) has no solution.

Solution: Given that $x_1 = ab$, $x_2 = b$, $x_3 = b$

$y_1 = abb$, $y_2 = ba$, $y_3 = bb$

Consider two lists $x = (x_1, x_2, ...x_n)$ and $y = (y_1, y_2, ... y_n)$ of non-empty strings over an alphabet $\Sigma = \{0, 1\}$

The PCP is to determine whether or not there exists $i_1, i_2, ...i_m$ where $1 \le i_j \le n$ such that $x_{i1} x_{i2} ....x_{im} = y_{i1} y_{i2} ....y_{im}$, 'm' may be greater than 'n' i.e. we have to determine whether or not there exists a sequence of substrings of 'x' such that the string formed by this sequence and the string formed by the sequence of corresponding substrings of y are identical.

In this problem, for each substring $x_i \in x$ and $y_i \in y$, we have $|x_i| < |y_i|, \forall$ i.

Hence the string generated by a sequence of substrings of 'x' is shorter than the string generated by the sequence of corresponding substrings of 'y' (or) the length of elements of 'y' are greater than the lengths of the corresponding elements of 'x'.

$\therefore$ The given PCP has no solution.

Example 8.5

Show that the following post correspondence problem has a solution and given the solution

	List A	**List B**
i	w$_i$	x$_i$
1	11	111
2	100	001
3	111	11

Solution:

Post Correspondence Problem: Consider two lists A = (w$_1$, w$_2$,....w$_n$) and B = (x$_1$, x$_2$,...x$_n$) of non-empty strings over an alphabet Σ = {0, 1}

The PCP is to determine whether or not there exists i$_1$, i$_2$, ...i$_m$ where $1 \leq i_j \leq n$ such that   w$_{i1}$ w$_{i2}$......w$_{im}$ = x$_{i1}$ x$_{i2}$....x$_{im\text{-}}$, 'm' may be greater than 'n';.

The required sequence is given by i$_1$ = 1, i$_2$ = 2, i$_3$ = 3 i.e. (1, 2, 3) and n=3. The corresponding strings are as follows

1	2	3		1	2	3
$\boxed{1\ 1}$	$\boxed{1\ 0\ 0}$	$\boxed{1\ 1\ 1}$		$\boxed{1\ 1\ 1}$	$\boxed{0\ 0}$	$\boxed{1\ 1}$
w$_1$	w$_2$	w$_3$		x$_1$	x$_2$	x$_3$

$\therefore$ The sequence is

$$w_1\ w_2\ w_3 = 1\ 1\ 1\ 0\ 0\ 1\ 1\ 1$$
$$x_1\ x_2\ x_3\ \ = 1\ 1\ 1\ 0\ 0\ 1\ 1\ 1$$
$$\therefore\ w_1\ w_2\ w_3 = x_1\ x_2\ x_3 = 1\ 1\ 1\ 0\ 0\ 1\ 1\ 1$$

$\therefore$ The given PCP has a solution.

8.7.1 MODIFIED VERSION OF PCP

We show that PCP is undecidable by showing that if it were decidable, we would have an algorithm for Lu.

First, we show that, if PCP were decidable, a modified version of PCP would also be decidable. The modified post correspondence problem (MPCP) is the following Given lists A and B, of k strings each from Σ^*, say

$$A = w_1, w_2, w_3 \ldots \ldots w_k$$

and   B = x$_1$, x$_2$, x$_3$............x$_k$ does there exists a sequence of integers i$_1$, i$_2$........i$_r$ such that   w$_1$ w$_{i1}$ w$_{i2}$............w$_{ir}$ = x$_1$ x$_{i1}$ x$_{i2}$............x$_{ir}$

The difference between the MPCP and PCP is that in the MPCP, a solution is required to start with the first string on each list.

Theorem : If PCP were decidable, then MPCP would be decidable that is, MPCP reduces to PCP

Proof: Let $A = w_1, w_2 \ldots \ldots w_k$

and $B = x_1, x_2, \ldots \ldots x_k$ be an instance of the MPCP

We convert this instance of MPCP to an instance of PCP that has a solution if our MPCP instance has a solution. If PCP were decidable, we would then be able to solve the MPCP, proving the lemma.

Let Σ be the smallest alphabet containing all the symbols in lists A and B, and let ¢ and $ not be in Σ. Let y_i be obtained from w_i by inserting the symbol ¢ after each character of w_i and let z_i be obtained from x_i by inserting the symbol ¢ ahead of each character of x_i. Create new words $y_0 = \text{¢ } y_1$, $z_0 = z_1$, $y_{k+1} = \$$, $z_{k+1} = \text{¢ }\$$

Let $C = y_0, y_1, y_2 \ldots \ldots \ldots y_{k+1}$ and $D = z_0, z_1, z_2 \ldots \ldots \ldots z_{k+1}$

For example, the lists C and D are constructed from the lists A and B are shown in the following figure

I	List A	List B
	w_i	x_i
1	1	111
2	10111	10
3	10	0

MPCP

I	List C	List D
	y_i	z_i
0	¢ 1 ¢	¢ 1 ¢1¢1
1	1 ¢	¢ 1 ¢1¢1
2	1 ¢ 0 ¢1¢1¢1¢	¢ 1 ¢ 0
3	1 ¢ 0 ¢	¢ 0
4	$	¢ $

In general, the lists C and D represent an instance of PCP. We claim that this instance of PCP has a solution iff the instance of MPCP represented by lists A and B has a solution.

To see this, note that if $1, i_1, i_2, \ldots \ldots i_r$ is a solution to MPCP with lists A and B, then $0, i_1, i_2, \ldots \ldots i_r, k+1$ is a solution to PCP with lists C and D.

Like wise, if $i_1, i_2 \ldots \ldots i_r$ is a solution to PCP with lists C and D then $i_1 = 0$ and $i_r = k+1$, since y_0 and z_0 are the only words with the same index that begin with the same symbol and y_{k+1} and z_{k+1} are the only words with the same index end with the same symbol.

Let j be the smallest integer such that $i_j = k+1$. Then $i_1, i_2 \ldots i_j$ is also a solution, since the symbol $ occurs only as the last symbol of y_{k+1} and z_{k+1} and for no l, where $1 \leq l < j$, is $i_l = k+1$. Clearly $1, i_2, i_3 \ldots i_{j-1}$ is a solution of MPCP for lists A and B.

If there is an algorithm to decide PCP, we can construct an algorithm to decide MPCP by converting any instance of MPCP to PCP as above.

8.8 TURING REDUCIBILITY

Definition:

Let, A and B be the two sets such that A, B $\subseteq$ N of natural numbers. Then A is Turing reducible to B and denoted as A $\leq_T$ B.

If there is an oracle machine that computes the characteristic function of A when it is executed with oracle machine for B.

This is also called as A is B-recursive and B-computable. The oracle machine is an abstract machine used to study decision problem. It is also called as **Turing machine** with **black box**. We say that A is Turing equivalent to B and write A $\equiv_T$ B if A $\leq_T$ B and B $\leq_T$ A.

Some Properties

1. Every set is Turing equivalent to its complement.
2. Every computable set is Turing equivalent to every other computable set.
3. If A $\leq_T$ B and B $\leq_T$ C then A $\leq_T$ C.

8.9 THE CLASSES P AND NP PROBLEMS

8.9.1 THE CLASS OF LANGUAGES P

If the Turing Machine M has some polynomial P(n) and the machine M never makes more than P(n) moves when an input of length n is given to M then such a Turing machine is said to be polynomial time TM.

The class P is set of languages accepted by **polynomial-time** Turing machine. However, the class P is also a set of problems that can be solved by real computer in polynomial time. There are many such problems.

For example – Transitive closure, matrix multiplication, Kruskal's algorithm for minimum spanning tree.

Kruskal's Algorithm:

In Kruskal's algorithm the minimum weight is obtained. In this algorithm also the circuit should not be formed. Each time the edge of minimum weight has to be selected from the graph. It is not necessary in this algorithm to have edges of minimum weights to be adjacent. Let us solve one example by Kruskal's algorithm.

Example 1:

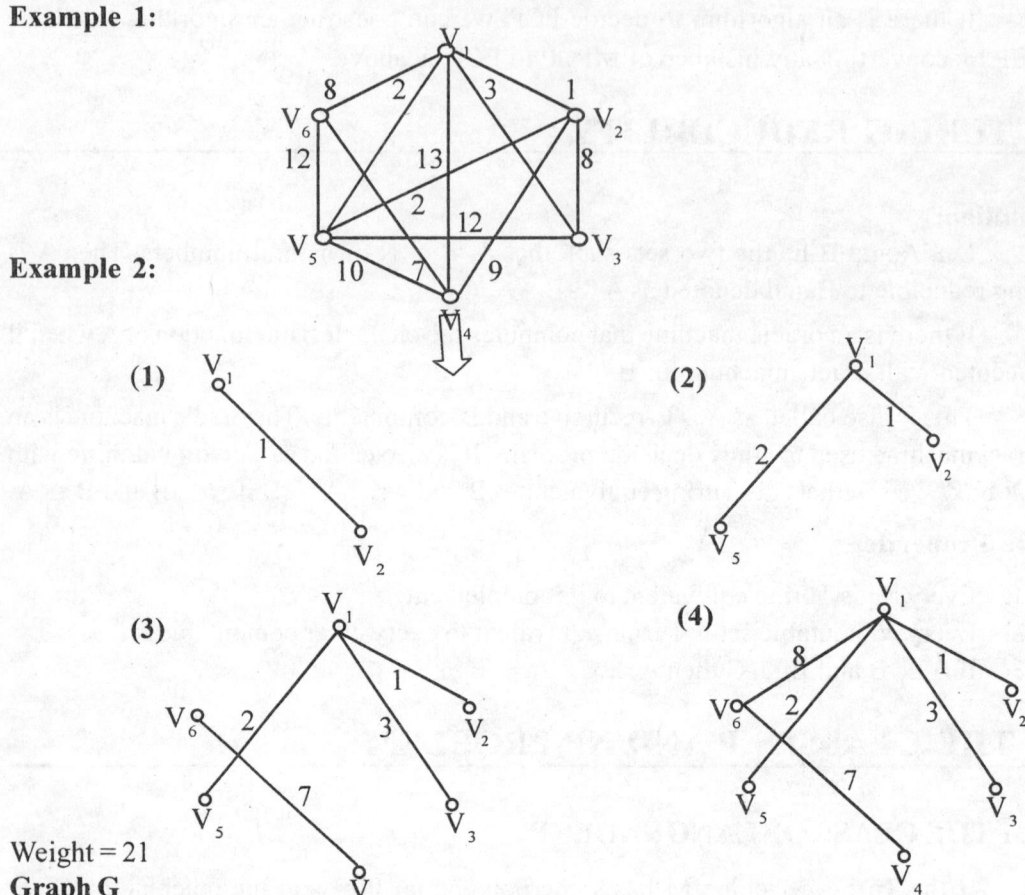

Example 2:

(1)

(2)

(3)

(4)

Weight = 21
Graph G

8.9.2 THE CLASS OF LANGUAGES NP

If the non- deterministic Turing machine M has some polynomial P(n) and the machine M never makes more than P(n) moves when an input of length n is given to M then such a Turing machine is said to be polynomial time (non deterministic TM) NTM. The class NP is set of languages accepted by polynomial-time non deterministic Turing machine. There are many problems in NP which are not in P if we can't resolve the **P =NP** question, we can atleast demonstrate that certain problems in NP are hardest, in the sense that if any one of them were in P, then P = NP. Such problems are also called as **NP complete problems.**
Travelling Salesman's Problem

Given a set of cities, and a cost to travel between each pair of cities, determine whether there is a path that visits every city once and returns to the first city, such that the cost travelled is less.

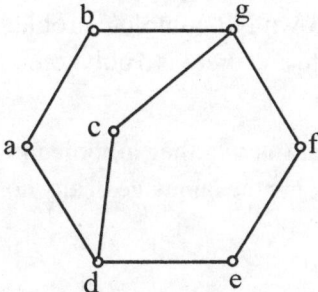

This problem is a NP problem as there may exist some path with shortest distance between the cities. If you get the solution by applying certain algorithm then Travelling salesman problem is **NP complete problem.** If we get no solution at all by applying an algorithm then that travelling salesman problem belongs to **NP hard** class.

The **P** is a class of problems which can be solved in polynomial time whereas **NP** is a class of problems for which is it is not possible to determine whether they are solved in polynomial time or not. Hence NP class problems can be modeled by Deterministic Turing machines. The NP class problems are of two kinds **NP complete** and **NP hard problems.** If for solving some problems there may be some algorithm a problem is assigned to the NP class if it is solvable in polynomial time by a non deterministic Turing machine. One can also say P ⊆ NP. A problem is NP-hard make it possible to solve all problems in class NP in polynomial time would called NP complete if it may or may not be solved in polynomial time. Therefore NP problems are undecidable problems. In this chapter we have seen various proofs regarding undecidability. We have also seen the concept of posts correspondence problem and obtained solution to some of these problems. Finally we have got introduced with the class of languages called P and NP. Undecidability is mainly associated with the NP class problems.

8.9.3 NP COMPLETE AND NP HARD PROBLEMS

NP complete Problem:

Let L be a language (problem) in NP. We say L is NP complete if the following statements are true about L.

i) L is in NP

ii) For every language L in NP, there is a polynomial time reduction of L[1] to L generally view a proof of NP completeness for a problem as a proof that the problem is not in P.

To prove first problems, called SAT (for Boolean satisfiability) to be NP-complete by showing that the language of every polynomial time NTM has a polynomial time reduction to SAT.

However, once if there are some NP complete problems, to prove a new problem to be NP-complete by reducing some known NP-complete problem to it, using a polynomial time reduction i.e. "If P_1 is NP complete and there is a polynomial time reduction of P_1 to P_2, then P_2 is NP complete".

The first NP complete problem is whether a boolean expression is satisfiable is proved NP-complete by explicitly reducing the language of any non-deterministic polynomial time TM to the satisfiability problem.

NP-Hard Problem:

Any decision problem, whether or not in NP, to which we can transform as NP complete problem is not solvable in polynomial times unless NP = P, then such a problem is called "NP hard problem".

NP hard problems are atleast as hard as the NP complete problems, whereas NP complete problem can be said to be solvable in polynomial time iff P = NP. All it is said with certainty about an NP hard problem is that it cannot be solved in polynomial time unless P = NP.

Some problems in L are so hard that although condition (2) can be proved of the definition of NP completeness but not condition (1).

That L is in NP. If so, call L as NP-hard is sufficient to show that L is very likely to require exponential time or worse. However if L is not in NP, then if apparent difficulty does not support the argument that all NP-complete problems are difficult i.e. it could turn out that P = NP and yet L still require exponential time.

Example: Integer linear programming is NP hard

Proof: We reduce 3-CNF satisfiability to integer linear programming.

Let $E = F_1 \wedge F_2 \ldots \ldots \ldots \wedge F_r$ be an expression in 3-CNF, and let $x_1, x_2 \ldots \ldots x_n$ be the variables of E. The matrix A will have a column for each literal x_i or x_i, $1 \leq i \leq n$.

Thus the inequality $Ax \geq b$ is viewed as a set of linear inequalities among the literals.

For each i, $1 \leq i \leq n$, we have the inequalities

$$x_i + x_i \geq 1, \quad x_i \geq 0$$
$$-x_i - x_i \geq -1, \, x_i \geq 0$$

Which has the effect of saying one of x_i and x_i is '0', the other is '1'.

For each clause $\alpha_1 + \alpha_2 + \alpha_3 \geq 1$

It says that atleast one literal in each clause has value 1. The inequalities are all satisfied if E is satisfiable. Thus linear integer programming is NP-hard.

SOLVED PROBLEMS

Problem 1: Construct LR(0) items for the grammar given, find its equivalent DFA. Check the parsing by taking a suitable derived string.

$$S^l \to S$$
$$S \to AS \mid \lambda \ or \ \varepsilon$$
$$A \to aA \mid b, \quad \varepsilon/\lambda \ is \ null$$

Solution:

Given grammar is

$$S^l \to S$$
$$S \to AS \mid \varepsilon$$
$$A \to aA \mid b$$

The items for the above grammar can be found by placing a dot on the left side of the right part of the production, we have

$$S^l \to .S$$
$$S \to .AS$$
$$S \to .$$
$$A \to .aA$$
$$A \to .b$$

Constructing the set of LR(0) items:

The closure function is applied to the starting production to get the collection of items.

Initially to get the fisrt item I_0, apply closure $(\{S^l \to . S\})$

Closure $(\{S^l \to . S\})$ =

$$\left.\begin{array}{l} S \to .AS \\ S \to . \\ A \to .aA \\ A \to .b \end{array}\right\} I_0$$

To get next items, we apply goto function for I_0 on the grammar symbols preceded by a dot and apply closure on the resulting set.

$$I_0: \quad S^l \to .S$$
$$S \to .AS$$

$$S \rightarrow .$$
$$A \rightarrow .aA$$
$$A \rightarrow .b$$

$I_1:$ (I_0, S)

    $S^1 \rightarrow S.$—completely reduced

$I_2:$ (I_0, A)

    $S \rightarrow A.S$

    $S \rightarrow .AS$

    $S \rightarrow .$

    $A \rightarrow .aA$

    $A \rightarrow .b$

$I_3:$ (I_0, a)

    $A \rightarrow a.A$

    $A \rightarrow .aA$

    $A \rightarrow .b$

$I_4:$ (I_0, b)

    $A \rightarrow b.$ – completely reduced

$I_5:$ (I_2, S)

    $S \rightarrow AS.$ – completely reduced

    $(I_2, A) - I_2$

    $(I_2, a)\ \ - I_3$

    $(I_2, b) - I_4$

$I_6:$ (I_3, A)

    $A \rightarrow aA.$ – completely reduced

    $(I_3, a) - I_3$

    $(I_3, b) - I_4$

$\therefore$ The collection of items is as follows

$$C = \{I_0, I_1, I_2, I_3, I_4, I_5, I_6\}$$

Considering I_0 to I_6 as different states, the equivalent DFA can be constructed as follows.

$(I_0, S) - I_1$	$(I_2, A) - I_2$
$(I_0, A) - I_2$	$(I_2, a) - I_3$

$(I_0, a) - I_3$ $(I_2, b) - I_4$

$(I_0, b) - I_4$ $(I_3, A) - I_6$

$(I_2, S) - I_5$ $(I_3, a) - I_3$

$(I_3, b) - I_4$

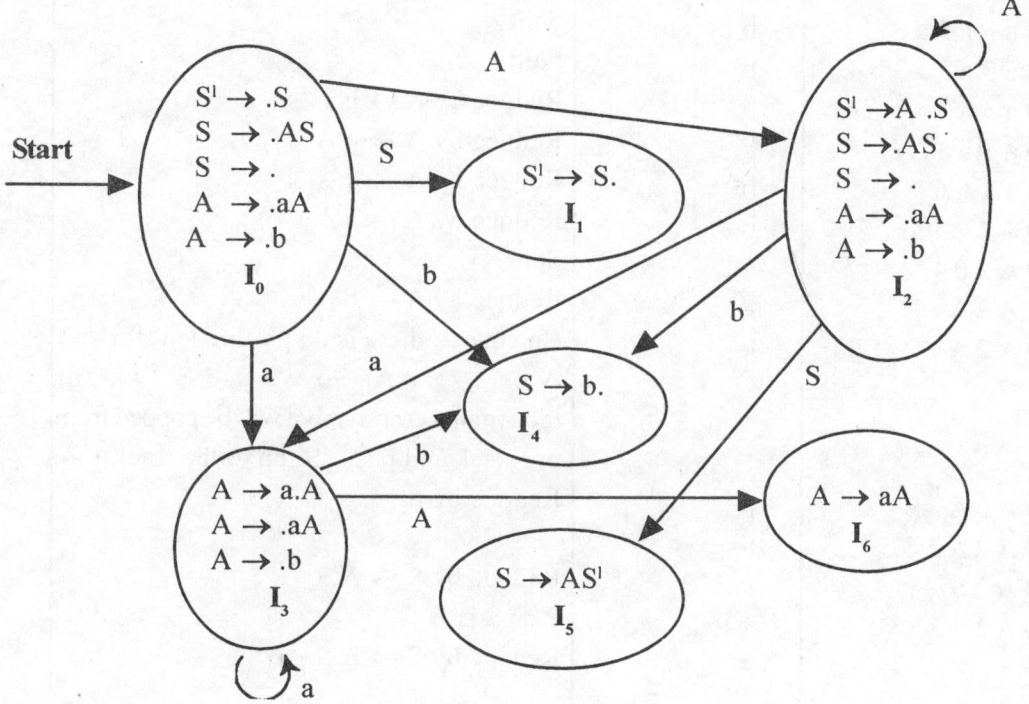

Acceptance of a string using parsing

Derivation of a string from the grammar

$$S^1 \Rightarrow S$$
$$\Rightarrow AS$$
$$\Rightarrow aAS$$
$$\Rightarrow aaAS$$
$$\Rightarrow aaaAS$$
$$\Rightarrow aaabAS$$
$$\Rightarrow aaabbS \qquad (\text{since } S \rightarrow \varepsilon)$$
$$\Rightarrow aaabb$$
$$\therefore \quad S^1 \Rightarrow aaabb$$

Stack	Input string	Remarks
0	a a a b b #	-
0 a 3	a a b b	Shift
0 a 3 a 3	a b b	Shift
0 a 3 a 3 a 3	b b	Shift
0 a 3 a 3 a 3 b 4	b	Shift
0 a 3 a 3 a 3 A 6	b	Reduce by A → b
0 a 3 a 3 A 6	b	Reduce by A → a A
0 a 3 A 6	b	Reduce by A → a A
0 A 2	b #	Reduce by A → a A
0 A 2 b 4	#	Shift
0 A 2 A	#	Reduce by A → b
0 A 2 A 2	#	(In state 2, there is no production for - → A or - → A A. So we use S → . for reduction, i.e no symbol will be popped from the stack and push 'S' on to the stack)
0 A 2 A 2 S	#	Reduce by S → .
0 A 2 A 2 S 5	#	-
0 A 2 S	#	Reduce by S → A S
0 A 2 S 5	#	2 on S is 5
0 S 1	#	Reduce by S → A S
0 S¹	#	--

∴   0 S¹ with # means acceptance of the input string.

Problem 2: *Construct LR(0) items for the grammar given find its equivalent DFA. Check the parsing by taking a suitable derived string.*

$$E \rightarrow T \, E^l$$
$$E^l \rightarrow + T \, E^l \mid \varepsilon$$
$$T \rightarrow F \, T^l$$
$$T^l \rightarrow * F \, T^l \mid \varepsilon$$
$$F \rightarrow id, \quad \varepsilon \text{ is null.}$$

Solution: The given grammar
$$E \rightarrow T \, E^l$$
$$E^l \rightarrow + T \, E^l \mid \varepsilon$$
$$T \rightarrow F \, T^l$$

$T^1 \rightarrow *\ F\ T^1\ |\ \varepsilon$

$F\ \rightarrow id$ is obtained from the following grammar

E $\rightarrow$ E + T | T

T $\rightarrow$ T * F | F

F $\rightarrow$ id

If we eliminate left-recursion from the production $A \rightarrow A\ \alpha\ |\ \beta$, then its equivalent productions without left recursion is $A \rightarrow \beta A^1$

$A^1 \rightarrow \alpha\ A^1\ |\ \varepsilon$

Hence, we will use the following grammar

$E \rightarrow E + T\ |\ T$

$T \rightarrow T * F\ |\ F$

$F \rightarrow id$

Without loss of generality, augmenting the above grammar, we get

$E^1 \rightarrow E$

$E \rightarrow E + T$

$E \rightarrow T$

$T \rightarrow T * F$

$T \rightarrow F$

$F \rightarrow id$

The canonical collection of sets of LR(0) items for the above grammar is shown below.

I_0: $E^1 \rightarrow .E$

       $E \rightarrow .E + T$

       $E \rightarrow .T$

       $T \rightarrow .T * F$

       $T \rightarrow .F$

       $F \rightarrow .id$

I_1: (I_0, E)

       $E^1 \rightarrow E.$

       $E \rightarrow E. + T$

I_2: (I_0, T)

       $E \rightarrow T.$

       $T \rightarrow T. * F$

I_3: (I_0, F)

       $T \rightarrow F.$ – completely reduced

I_4: (I_0, id)

       $F \rightarrow id.$ – completely reduced

I_5: $(I_1, +)$

       $E \rightarrow E +.T$

       $T \rightarrow .T *F$

$$T \rightarrow .F$$
$$F \rightarrow .id$$

I_6: $(I_2, *)$
$$T \rightarrow T *.F$$
$$F \rightarrow .id$$

I_7: (I_5, T)
$$E \rightarrow E + T.$$
$$T \rightarrow T.*F$$
$$(I_5, F) - I_3$$
$$(I_5, id) - I_4$$

I_8: (I_6, F)
$$T \rightarrow T*F. - \text{completely reduced}$$
$$(I_6, id) - I_4$$
$$(I_7, *) - I_6$$

∴ The collection of items, C = {$I_0, I_1, I_2, I_3, I_4, I_5, I_6, I_7, I_8$}

Consider I_0 to I_8 as different states, the equivalent DFA is shown below

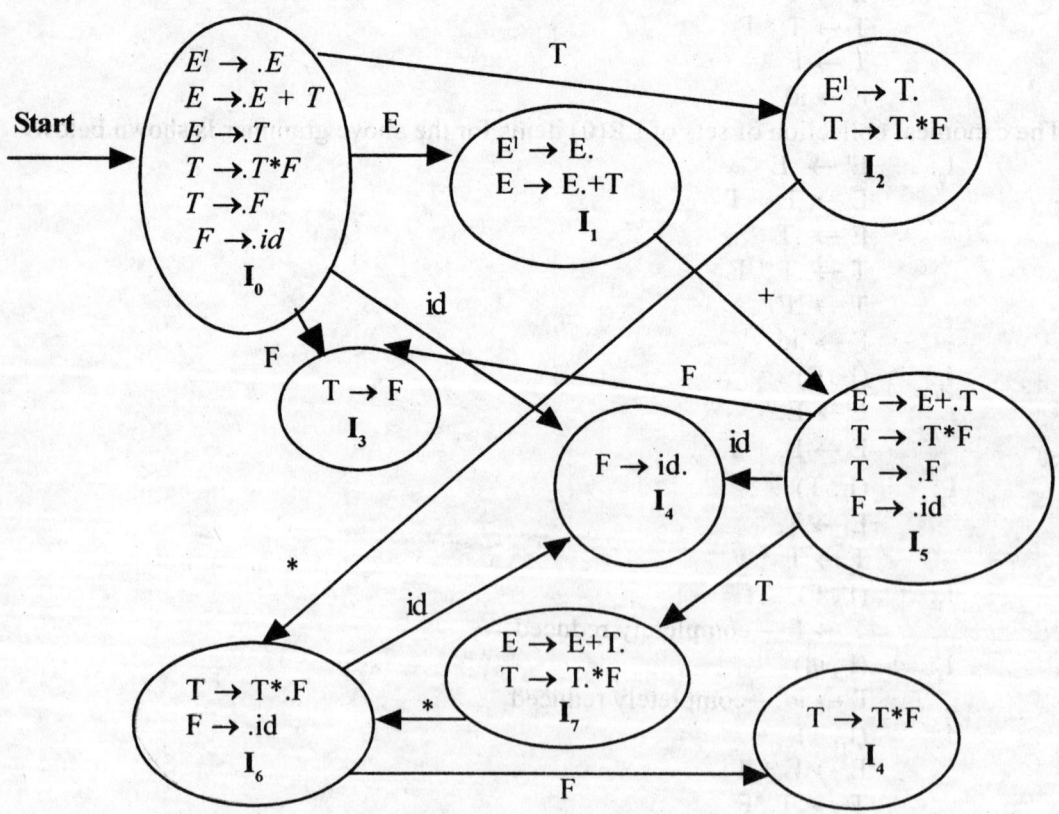

Acceptance of a string using parsing:

Derive a string from the grammar

$$E \Rightarrow E + T$$
$$\Rightarrow E + T * F$$
$$\Rightarrow E + T * id$$
$$\Rightarrow E + F * id$$
$$\Rightarrow E + id * id$$
$$\Rightarrow T + id * id$$
$$\Rightarrow F + id * id$$
$$E \Rightarrow id + id * id$$

Stack	Input string	Remarks
0	id + id * id #	-
0 id 4	+ id * id #	Shift
0 F 3	+ id * id #	Reduce by F → id
0 T 2	+ id * id #	Reduce by T → F
0 E 1	+ id * id #	Reduce by E → T
0 E 1 + 5	id * id #	Shift
0 E 1 + 5 id 4	* id #	Shift
0 E 1 + 5 F 3	* id #	Reduce by F → id
0 E 1 + 5 T 7	* id #	Reduce by T → F
0 E 1 + 5 T 7 * 6	id #	Shift
0 E 1 +5 T 7 * 6 id 4	#	Shift
0 E 1 + 5 T 7 * 6 F 8	#	Reduce by F → id
0 E 1 + 5 T 7	#	Reduce by T → T * F
0 E	#	Reduce by E → E + T
0 E 1	#	
0 E¹	#	Reduce by E¹ → E

OE^1 with # means, the input string is accepted

Problem 3: Construct LR(0) items for the grammar given. Find its equivalent DFA. Check the parsing by taking a suitable derived string.

$S \rightarrow E$ *(Augmentation given)*

$E \rightarrow E + T \mid T$

$T \rightarrow a \mid (E)$

Solution: Given grammar is

$$S \rightarrow E$$
$$E \rightarrow E + T$$
$$E \rightarrow T$$
$$T \rightarrow a$$
$$T \rightarrow (E)$$

The items for the above grammar can be found by placing a dot on the left side of the right part of the production, we have

$$S \rightarrow .E$$
$$E \rightarrow .E + T$$
$$E \rightarrow .T$$
$$T \rightarrow .a$$
$$T \rightarrow .(E)$$

The closure function is applied to the starting production to get the collection of items. Initially to get the first item I_0, apply closure $(S \rightarrow .E)$.

I_0: $S \rightarrow .E$

$E \rightarrow .E + T$

$E \rightarrow .T$

$T \rightarrow .a$

$T \rightarrow .(E)$

I_1: (I_0, E)

$S \rightarrow E.$

$E \rightarrow E. + T$

I_2: (I_0, T)

$E \rightarrow T.$ – completely reduced

I_3: (I_0, a)

$T \rightarrow a.$ – completely reduced

I_4: $(I_0, ( )$

   $T \rightarrow (.E)$

   $E \rightarrow .E + T$

   $E \rightarrow .T$

   $T \rightarrow .a$

   $T \rightarrow .(E)$

I_5: $(I_1, +)$

   $E \rightarrow E +.T$

   $T \rightarrow .a$

   $T \rightarrow .(E)$

I_6: (I_4, E)

   $T \rightarrow (E.)$

   $E \rightarrow E. + T$

   $(I_4, T) - I_2$

   $(I_4, a) - I_3$

   $(I_4, ( ) - I_4$

I_7: (I_5, T)

   $E \rightarrow E + T.$ – completely reduced

   $(I_5, a) - I_3$

   $(I_5, ( ) - I_4$

I_8: $(I_6,))$

   $T \rightarrow (E).$ – completely reduced

   $(I_6, +) - I_5$

$\therefore$ The collection of LR(0) items is as follows $C = \{I_0, I_1, I_2, I_3, I_4, I_5, I_6, I_7\text{-}, I_8\}$

Consider I_0 to I_8 as different states, the equivalent DFA is shown below

$(I_0, E) - I_1$	$(I_4, T) - I_2$
$(I_0, T) - I_2$	$(I_4, a) - I_3$
$(I_0, a) - I_3$	$(I_4, ( ) - I_4$
$(I_0, ( ) - I_4$	$(I_5, T) - I_7$
$(I_1, +) - I_5$	$(I_5, a) - I_3$
$(I_4, E) - I_6$	$(I_5, ( ) - I_4$
$(I_6,)) - I_8$	$(I_6, +) - I_5$

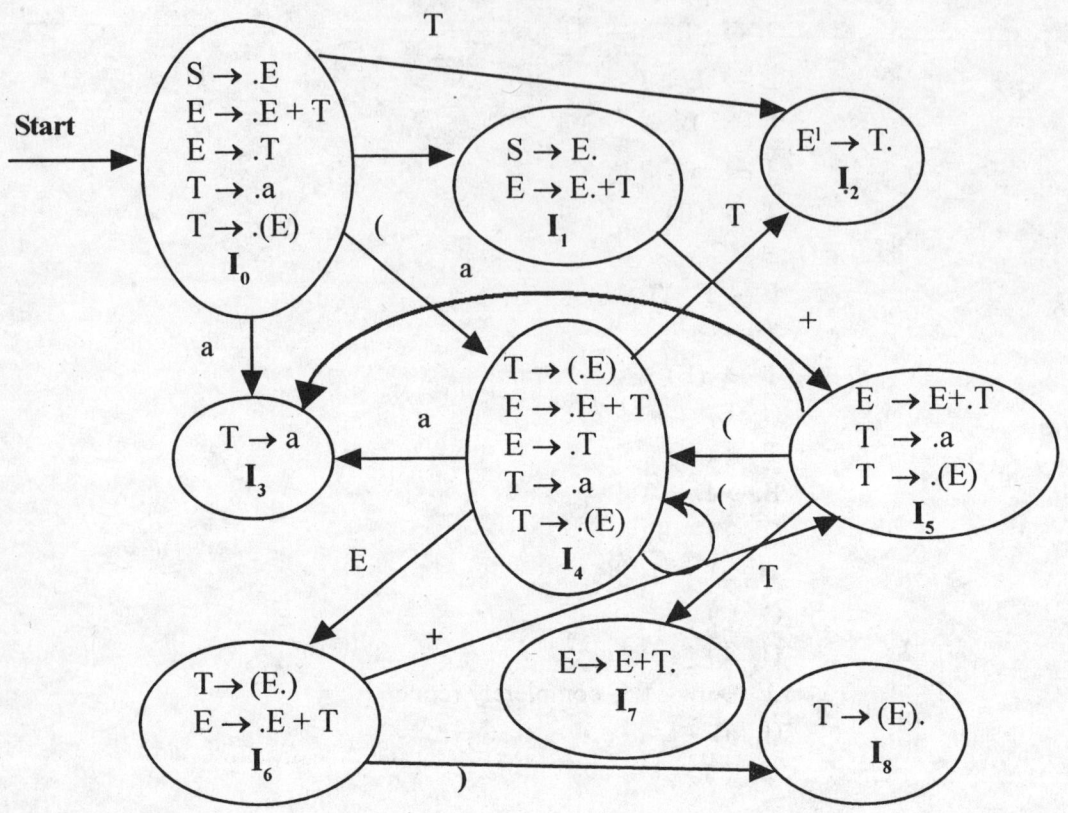

Acceptance of a string using parsing:

Derive a string from the grammar

$$S \Rightarrow E$$
$$\Rightarrow E + T$$
$$\Rightarrow T + T$$
$$\Rightarrow (E) + T$$
$$\Rightarrow (E + T) + T$$
$$\Rightarrow (T + T) + T$$
$$\Rightarrow (a + a) + a$$
$$\therefore S \Rightarrow (a + a) + a$$

Consider the string (a + a) + a, and apply parsing technique

Stack	Input string	Remarks
0	(a + a) + a #	-
0 (4	a + a) + a	Shift
0 (4 a 3	+ a) + a	Shift
0 (4 T 2	+ a) + a	Reduce by T → a
0 (4 E 6	+ a) + a	Reduce by E → T
0 (4 E 6 + 5	a) + a	Shift
0 (4 E 6 + 5 a 3	) + a	Shift
0 (4 E 6 + 5 T 7	) + a	Reduce by T → a
0 (4 E	) + a	Reduce by E → E + T
0 (4 E 6	) + a	Shift
0 (4 E 6) 8	+ a	Shift
0 T 2	+ a	Reduce by T → (E)
0 E 1	+ a #	Reduce by E → T
0 E 1 + 5	a #	Shift
0 E 1 + 5 a 3	#	Shift
0 E 1 + 5 T 7	#	Reduce by T → a
0 E	#	Reduce by E → T + E
0 E 1	#	
0 S	#	Reduce by S → E

∴ 0S with # means, the input string (a+a)+a is accepted.

Problem 4. *Let $\Sigma = \{0, 1\}$ and A and B be lists of three strings each as given below. Does post correspondence problem has a solution if so give the solution.*

I	List A w_i	List B x_i
1	1	111
2	10111	10
3	10	0

Solution: Let

$w_1 = 1$	$x_1 = 111$
$w_2 = 10111$	$x_2 = 10$
$w_3 = 10$	$x_3 = 0$

Post Correspondence Problem:

Consider two lists $A = (w_1, w_2, w_3 \ldots\ldots w_n)$, $B = (x_1, x_2, \ldots\ldots\ldots x_n)$ of non-empty strings over an alphabet $\Sigma = \{0, 1\}$.

The PCP is to determine whether or not there exists $i_1, i_2, \ldots\ldots i_m$ where $1 \le i_j \le n$ such that $x_{i1} x_{i2} \ldots\ldots x_{im} = w_{i1} w_{i2} \ldots\ldots\ldots w_{im}$. The indices i_j's need not be distinct and 'm' may be greater than 'n'.

Also, if there exists a solution to PCP, there exists infinitely many solutions. The above PCP has a solution, since $w_2 w_1 w_1 w_3 = x_2 x_1 x_1 x_3 = 101111110$, $<2, 1, 1, 3> = <i_1,$ $i_2, i_3, i_4>$ and $m = 4$.

Here $m = 4$, $n = 3$.

$\therefore m > n$

Problem 5. What is post correspondence problem? Is there any solution for the following PCP problem? If so give the solution, if not discuss why?

I	List A w_i	List B x_i
1	00	0
2	011	11
3	1000	011

Solution:

Post Correspondence Problem (PCP):

Consider two lists $A = (w_1, w_2 \ldots\ldots w_n)$ and $B = (x_1, x_2 \ldots\ldots x_n)$ of non-empty strings over an alphabet $\Sigma = \{0, 1\}$.

The PCP is to determine whether or not there exists $i_1, i_2 \ldots\ldots i_m$ where $1 \le i_j \le n$ such that $w_{i1} w_{i2} \ldots.. w_{im} = x_{i1} x_{i2} \ldots\ldots x_{im}$ $(m > n)$ 'm' may be greater than 'n'.

Here we have to determine whether or not there exists a sequence of substrings of 'A' such that the strings formed by the sequence of corresponding substrings of 'B' are identical.

The given problem has no solution, because any string composed of elements of 'A' will be longer than the corresponding string from 'B', i.e. to get same string for same sequence of $w_1, w_2 \ldots..$ and $x_1, x_2 \ldots.$ are difficult. We cannot get solution sequence.

$\therefore$ The given PCP has no solution.

Problem 6. Find the solution for PCP problem given below

I	List A w_i	List B x_i
1	a	aaa
2	abaaa	ab
3	ab	b

Solution:

Post Correspondence Problem (PCP):

Consider two lists $A = (w_1, w_2........w_n)$ and $B = (x_1, x_2......x_n)$ of non-empty strings over an alphabet $\Sigma = \{0, 1\}$.

The PCP is to determine whether or not there exists $i_1, i_2.....i_n$ where $1 \le i_j \le n$ such that x_{i1} $x_{i2}...x_{im} = w_{i1} w_{i2}........w_{im}$, 'm' may be greater than 'n'.

In the given lists A and B, each list consists of three strings w_1, w_2, w_3 and x_1, x_2, x_3.

$\therefore$ $w_1 = a$ $x_1 = aaa$

  $w_2 = abaaa$ $x_2 = ab$

  $w_3 = ab$ $x_3 = b$.

The given PCP has a solution, because $w_2 w_1 w_1 w_3 = x_2 x_1 x_1 x_3$

                          abaaa a a ab = ab aaa aaa b

$\therefore$ The strings in the sequence $<2, 1, 1, 3>$ are equal. Here m = 4, n = 3.

$\therefore$ m > n.

Problem 7. Show that the PCP with two lists $x = (b, bab^3, ba)$ and $y = (b^3, ba, a)$ has a solution. Give the solution sequence.

Solution:

Post Correspondence Problem (PCP):

Consider two lists $A = (x_1, x_2........x_n)$ and $B = (y_1, y_2......y_n)$ of non-empty strings over an alphabet $\Sigma = \{0, 1\}$.

The PCP is to determine whether or not there exists $i_1, i_2.....i_m$ where $1 \le i_j \le n$ such that x_{i1} $x_{i2}...x_{im} = y_{i1} y_{i2}........y_{im}$, 'm' may be greater than 'n' i.e. we have to determine whether or not there exists a sequence of substrings of 'x' such that the string formed by this sequence and the string formed by the sequence of corresponding substrings of 'y' are identical.

Given that $x = (b, bab^3, ba)$

                $y = (b^3, ba, a)$

The required sequence is given by $i_1 = 2$, $i_2 = 1$, $i_3 = 1$, $i_4 = 3$ i.e. (2, 1, 1, 3) and m = 4. The correspondence strings are

bab^3	b	b	ba		ba	b^3	b^3	a
x_2	x_1	x_1	x_3		y_2	y_1	y_1	y_3

$x_2 \, x_1 \, x_1 \, x_3 = y_2 \, y_1 \, y_1 \, y_3 = $ babbbbbba

$\therefore$ The given PCP has a solution.

Problem 8. Find whether the post correspondence problem $P = \{(10, 101), (011, 11), (101, 011)\}$ has a match. Give the solution.

Solution:

Given   x = (10, 011, 101)

   y = (101, 11, 011) where x and y are two lists i.e.

   i.e, $x_1 = 10$, $y_1 = 101$

      $x_2 = 011$, $y_2 = 11$

      $x_3 = 101$, $y_3 = 011$

Consider x_1 x_3 x_3 x_3 y_1 y_3 y_3 y_3

         10 101 101 101   101 011 011 011

      10101101101

      101011011011

The problem is unsolvable, since the string from list 'y; exceeds the string from the list 'x' by single symbol '1'.

For 'x_1' it goes wrong and for 'x_3' the string repeats on.

$\therefore$ The problem is unsolvable.

Suppose that this instance of PCP has a solution i_1, i_2......i_m. Clearly $i_1 = 1$, since no string beginning with $x_2 = 011$ can equal to a string beginning with $y_2 = 11$ no string beginning with $x_3 = 101$ can equal to a string beginning $y_3 = 011$.

we write the string from list A above the corresponding string from list B.

so far, we have   10

         101

The next selection from A must begin with a '1'.

Thus $i_2 = 1$ or $i_2 = 3$.

But $i_2 = 1$ will not do, since no string beginning with $x_1\, x_1 = 1010$ can equal to a string beginning

with $\quad y_1\, y_1 = 101101$

with $\quad i_2 = 3$, we have

$$10101$$
$$101011$$

Since the string from list 'B' again exceeds the string from the list 'A' by the single symbol '1', a similar argument shows that $i_3 = i_4 = \ldots = 3$.

Thus there is only one sequence of choices that generates compatible string and for this sequence string B is always one character longer.

Thus this instance of PCP has no solution.

Problem 9. Explain why PCP with two lists $x = (01, 1, 1)$ and $y = (0101, 10, 11)$ has no solution. What is modified PCP problem?

Solution:

Post Correspondence Problem:

Consider two lists $x = (x_1, x_2 \ldots \ldots x_n)$ and $B = (y_1, y_2 \ldots \ldots y_n)$ of non-empty strings over an alphabet $\Sigma = \{0, 1\}$.

The PCP is to determine whether or not there exists $i_1, i_2 \ldots \ldots i_n$ where $1 \le i_j \le n$ such that $x_{i1}\, x_{i2} \ldots x_{im} = y_{i1}\, y_{i2} \ldots \ldots y_{im}$, 'm' may be greater than 'n' i.e. we have to determine whether or not there exists a sequence of substrings of 'x' such that the string formed by this sequence and the string formed by the sequence of corresponding substrings of 'y' are identical.

Here given that $\quad x = (01, 1, 1)$

$$y = (0101, 10, 11)$$

For each substring $x_i \in x$ and $y_i \in y$, we have $|x_i| < |y_i|$ for all i. Hence the string generated by a sequence of substrings of 'x' is shorter than the string generated by the sequence of corresponding substrings of 'y'. Therefore, the PCP has no solution.

Modified PCP: If the first substring used in PCP is always x_1 and y_1, then the PCP is known as Modified Post Correspondence Problem

REVIEW QUESTIONS

1. Construct LR(0) items for the grammar given. Find its equivalent DFA. Check the parsing by taking a suitable derived string.

    $S \rightarrow a\,A\,B$

    $A \rightarrow a\,A\,b \mid a\,b$

    $B \rightarrow !a\,B \mid a.$

2. For the grammar shown below, construct the set of LR(0) items

    $S^1 \rightarrow S\,\$$

    $S \rightarrow a\,S\,b \mid a\,b$

3. What is decidability . Explain any two undecidable problems.

4. Define the 'handler' for the derivation 'aabb' find the viable prefixes of the grammar shown below

    $S^1 \rightarrow S$

    $S^1 \rightarrow asb/ab$

5. Explain the classes of P and NP and state two NP-Complete problems. Given some examples for these problems.

6. What are type 0,1,2,3 grammars . Compare them in different aspects.

7. Show that the grammar G is given by G = ({S} , {a,b} , P,S) where

    P = {(S $\rightarrow$ aSb) , (S $\rightarrow$ a)}   is LR (1)

8. Explain Context-Sensitive languages.

9. Describe various formal languages with examples.

10. Write short notes on LR grammars.

11. Discuss the hierarchy theorem.

12. What is decidability. Explain any two undecidable problems.

13. When an item is said to be complete, find the sets of LR(0) items.

14. Describe the linear bounded automaton and the language it accepts.

15. Write the design procedure of shift-reduce parser by taking a suitable example.